American Society of Civil Engineers

Minimum Design Loads for Buildings and Other Structures

Revision of ANSI/ASCE 7-95

This document uses both Système International (SI) units and customary units.

Structural Engineering Institute

Published by the American Society of Civil Engineers
1801 Alexander Bell Drive
Reston, Virginia 20191-4400

ABSTRACT

ASCE standard, *Minimum Design Loads for Buildings and Other Structures* (ASCE 7-98 a revision of ANSI/ASCE 7-95), gives requirements for dead, live, soil, flood, wind, snow, rain, ice, and earthquake loads, and their combinations, that are suitable for inclusion in building codes and other documents. The major revision of this standard involves the section on wind loads. This section has been greatly expanded to include the latest information in the field of wind load engineering. Requirements have been added for flood loads and ice loads. An appendix on serviceability requirements has also been added. The structural load requirements provided by this standard are intended for use by architects, structural engineers, and those engaged in preparing and administering local building codes.

Library of Congress Cataloging-in-Publication Data

Minimum design loads for buildings and other structures / American Society of Civil Engineers
    p. cm.
   "ASCE 7-98"
   "Revision of ANSI/ASCE 7-95."
   "Approved October 1999."
   "Published January 2000."
   Includes bibliographical references and index.
   ISBN 0-7844-0445-3
  1. Structural engineering—United States. 2. Standards, Engineering—United States. I. American Society of Civil Engineers.
  TH851.M56 1996
  624.1′72—dc21

94-3854
CIP

STANDARDS

In April 1980, the Board of Direction approved ASCE Rules for Standards Committees to govern the writing and maintenance of standards developed by the Society. All such standards are developed by a consensus standards process managed by the Management Group F (MGF), Codes and Standards. The consensus process includes balloting by the balanced standards committee made up of Society members and nonmembers, balloting by the membership of ASCE as a whole, and balloting by the public. All standards are updated or reaffirmed by the same process at intervals not exceeding 5 years.

The following Standards have been issued.

ANSI/ASCE 1-82 N-725 Guideline for Design and Analysis of Nuclear Safety Related Earth Structures

ANSI/ASCE 2-91 Measurement of Oxygen Transfer in Clean Water

ANSI/ASCE 3-91 Standard for the Structural Design of Composite Slabs and ANSI/ASCE 9-91 Standard Practice for the Construction and Inspection of Composite Slabs

ANSE 4-86 Seismic Analysis of Safety-Related Nuclear Structures

Building Code Requirements for Masonry Structures (ACI530-99/ASCE5-99/TMS402-99) and Specifications for Masonry Structures (ACI530.1-99/ASCE6-99/TMS602-99)

ANSI/ASCE 7-98 Minimum Design Loads for Buildings and Other Structures

ANSI/ASCE 8-90 Standard Specification for the Design of Cold-Formed Stainless Steel Structural Members

ANSI/ASCE 9-91 listed with ASCE 3-91

ANSI/ASCE 10-97 Design of Latticed Steel Transmission Structures

ANSI/ASCE 11-90 Guideline for Structural Condition Assessment of Existing Buildings

ANSI/ASCE 12-91 Guideline for the Design of Urban Subsurface Drainage

ASCE 13-93 Standard Guidelines for Installation of Urban Subsurface Drainage

ASCE 14-93 Standard Guidelines for Operation and Maintenance of Urban Subsurface Drainage

ANSI/ASCE 15-93 Standard Practice for Direct Design of Buried Precast Concrete Pipe Using Standard Installations (SIDD)

ASCE 16-95 Standard for Load and Resistance Factor Design (LRFD) of Engineered Wood Construction

ASCE 17-96 Air-Supported Structures

ASCE 18-96 Standard Guidelines for In-Process Oxygen Transfer Testing

ASCE 19-96 Structural Applications of Steel Cables for Buildings

ASCE 20-96 Standard Guidelines for the Design and Installation of Pile Foundations

ASCE 21-96 Automated People Mover Standards—Part 1

ASCE 21-98 Automated People Mover Standards—Part 2

ASCE 22-97 Independent Project Peer Review

ASCE 23-97 Specification for Structural Steel Beams with Web Openings

ASCE 24-98 Flood Resistant Design and Construction

ASCE 25-97 Earthquake-Actuated Automatic Gas Shut-Off Devices

FOREWORD

The material presented in this publication has been prepared in accordance with recognized engineering principles. This Standard and Commentary should not be used without first securing competent advice with respect to their suitability for any given application. The publication of the material contained herein is not intended as a representation or warranty on the part of the American Society of Civil Engineers, or of any other person named herein, that this information is suitable for any general or particular use or promises freedom from infringement of any patent or patents. Anyone making use of this information assumes all liability from such use.

ACKNOWLEDGEMENTS

The American Society of Civil Engineers (ASCE) acknowledges the work of the Minimum Design Loads on Buildings and Other Structures Standards Committee of the Codes and Standards Activities Division of the Structural Engineering Institute. This group comprises individuals from many backgrounds including: consulting engineering, research, construction industry, education, government, design and private practice.

This revision of the standard began in 1995 and incorporates information as described in the commentary.

This Standard was prepared through the consensus standards process by balloting in compliance with procedures of ASCE's Codes and Standards Activities Committee. Those individuals who serve on the Standards Committee are:

Kharaiti L. Abrol
Shirin D. Ader
Demirtas C. Bayar
John E. Breen
David G. Brinker
Ray A. Bucklin
James R. Cagley
Jack E. Cermak
Kevin C.K. Cheung
Edward Cohen
James S. Cohen
Jay H. Crandell
Stanley W. Crawley
Majed Dabdoub
Amitabha Datta
Charles A. De Angelis
James M. Delahay
Bradford K. Douglas
John F. Duntemann
Donald Dusenberry
Bruce R. Ellingwood, Vice-Chair
Edward R. Estes
David A. Fanella
Theodore V. Galambos
Satyendra K. Ghosh
Lorenzo Gonzalez
Dennis W. Graber
Lawrence G. Griffis
David S. Gromala
Robert D. Hanson
James R. Harris, Chair
Joseph P. Hartman
Steven R. Hemler
Nicholas Isyumov
Christopher P. Jones
John H. Kapmann, Jr.
A. Harry Karabinis
D.J. Laurie Kennedy
Jon P. Kiland
Randy Kissell
Uno Kula

Edward M. Laatsch
John V. Loscheider
Ian Mackinlay
Harry W. Martin
Rusk Masih
George M. Matsumura
Robert R. McCluer
Richard McConnell
Kishor C. Mehta
Rick Mendlen
Joseph J. Messersmith, Jr.
Joe N. Nunnery
Michael O'Rourke
Clifford Oliver
Frederick J. Palmer
Alan B. Peabody
David B. Peraza
Dale C. Perry
Clarkson W. Pinkham
Robert D. Prince
Robert T. Ratay
Massandra K. Ravindra
Lawrence D. Reaveley
Abraham J. Rokach
William D. Rome
James A. Rossberg
Julie A. Ruth
Herbert S. Saffir
Phillip J Samblanet
Suresh C. Satsangi
Andrew Scanlon
William L. Shoemaker
Emil Simiu
Thomas D. Skaggs
Thomas L. Smith
James G. Soules
Theodore Stathopoulos
Frank W. Stockwell, Jr.
Donald R. Strand
Edgar L. Sutton, Jr.
Harry B. Thomas

Wayne N. Tobiasson
Brian E. Trimble
David P. Tyree
Thomas R. Tyson
Joseph W. Vellozzi
Richard A. Vognild
Francis J. Walter, Jr.
Marius B. Wechsler
Yi Kwei Wen
Peter J.G. Willse
Lyle L. Wilson
Joseph A. Wintz, III

Task Committee on General Structural Requirements
James S. Cohen
John F. Duntemann
Donald Dusenberry, Chair
John L. Gross
Mark B. Hogan
Anatol Longinow
Robert T. Ratay
James G. Soules

Task Committee on Strength
Bruce R. Ellingwood, Chair
Theodore V. Galambos
David S. Gromala
James R. Harris
D.J. Laurie Kennedy
Clarkson Pinkham
Andrew Scanlon
James G. Soules
Massandra K. Ravindra
Yi Kwei Wen

Task Committee on Live Loads
James R. Cagley
Raymond A. Cook
Ross B. Corotis
Charles A. De Angelis

CONTENTS

Appendix A

Appendix B

Minimum Design Loads for Buildings and Other Structures

1.0 GENERAL

1.1 SCOPE

This standard provides minimum load requirements for the design of buildings and other structures that are subject to building code requirements. Loads and appropriate load combinations, which have been developed to be used together, are set forth for strength design and allowable stress design. For design strengths and allowable stress limits, design specifications for conventional structural materials used in buildings and modifications contained in this standard shall be followed.

1.2 DEFINITIONS

The following definitions apply to the provisions of the entire standard.

Allowable stress design: A method of proportioning structural members such that elastically computed stresses produced in the members by nominal loads do not exceed specified allowable stresses (also called working stress design).

Authority having jurisdiction: The organization, political subdivision, office or individual charged with the responsibility of administering and enforcing the provisions of this standard.

Buildings: Structures, usually enclosed by walls and a roof, constructed to provide support or shelter for an intended occupancy.

Design strength: The product of the nominal strength and a resistance factor.

Essential facilities: Buildings and other structures that are intended to remain operational in the event of extreme environmental loading from wind, snow or earthquakes.

Factored load: The product of the nominal load and a load factor.

Limit state: A condition beyond which a structure or member becomes unfit for service and is judged either to be no longer useful for its intended function (serviceability limit state) or to be unsafe (strength limit state).

Load effects: Forces and deformations produced in structural members by the applied loads.

Load factor: A factor that accounts for deviations of the actual load from the nominal load, for uncertainties in the analysis that transforms the load into a load effect, and for the probability that more than one extreme load will occur simultaneously.

Loads: Forces or other actions that result from the weight of all building materials, occupants and their possessions, environmental effects, differential movement, and restrained dimensional changes. Permanent loads are those loads in which variations over time are rare or of small magnitude. All other loads are variable loads. (See also nominal loads.)

Nominal loads: The magnitudes of the loads specified in Sections 3 through 9 (dead, live, soil, wind, snow, rain, flood and earthquake) of this standard.

Nominal strength: The capacity of a structure or member to resist the effects of loads, as determined by computations using specified material strengths and dimensions and formulas derived from accepted principles of structural mechanics or by field tests or laboratory tests of scaled models, allowing for modeling effects and differences between laboratory and field conditions.

Occupancy: The purpose for which a building or other structure, or part thereof, is used or intended to be used.

Other structures: Structures, other than buildings, for which loads are specified in this standard.

P-delta effect: The second order effect on shears and moments of frame members induced by axial loads on a laterally displaced building frame.

Resistance factor: A factor that accounts for deviations of the actual strength from the nominal strength and the manner and consequences of failure (also called strength reduction factor).

Strength design: A method of proportioning structural members such that the computed forces produced in the members by the factored loads do not exceed the member design strength (also called load and resistance factor design).

Temporary facilities: Buildings or other structures that are to be in service for a limited time and have a limited exposure period for environmental loadings.

1.3 BASIC REQUIREMENTS

1.3.1 Strength

Buildings and other structures, and all parts thereof, shall be designed and constructed to support safely the factored loads in load combinations defined in this document without exceeding the appropriate strength limit states for the materials of construction. Alternatively, buildings and other structures, and all parts thereof, shall be designed and constructed to support safely the nominal loads in load combinations defined in this document without exceeding the appropriate specified allowable stresses for the materials of construction.

1.3.2 Serviceability

Structural systems and members thereof shall be designed to have adequate stiffness to limit deflections, lateral drift, vibration, or any other deformations that adversely affect the intended use and performance of buildings and other structures.

1.3.3 Self-Straining Forces

Provision shall be made for anticipated self-straining forces arising from differential settlements of foundations and from restrained dimensional changes due to temperature, moisture, shrinkage, creep, and similar effects.

1.3.4 Analysis

Load effects on individual structural members shall be determined by methods of structural analysis that take into account equilibrium, general stability, geometric compatibility, and both short- and long-term material properties. Members that tend to accumulate residual deformations under repeated service loads shall have included in their analysis the added eccentricities expected to occur during their service life.

1.3.5 Counteracting Structural Actions

All structural members and systems, and all components and cladding in a building or other structure, shall be designed to resist forces due to earthquake and wind, with consideration of overturning, sliding, and uplift, and continuous load paths shall be provided for transmitting these forces to the foundation. Where sliding is used to isolate the elements, the effects of friction between sliding elements shall be included as a force. Where all or a portion of the resistance to these forces is provided by dead load, the dead load shall be taken as the minimum dead load likely to be in place during the event causing the considered forces. Consideration shall be given to the effects of vertical and horizontal deflections resulting from such forces.

1.4 GENERAL STRUCTURAL INTEGRITY

Buildings and other structures shall be designed to sustain local damage with the structural system as a whole remaining stable and not being damaged to an extent disproportionate to the original local damage. This shall be achieved through an arrangement of the structural elements that provides stability to the entire structural system by transferring loads from any locally damaged region to adjacent regions capable of resisting those loads without collapse. This shall be accomplished by providing sufficient continuity, redundancy, or energy-dissipating capacity (ductility), or a combination thereof, in the members of the structure.

1.5 CLASSIFICATION OF BUILDINGS AND OTHER STRUCTURES

Buildings and other structures shall be classified, based on the nature of occupancy, according to Table 1-1 for the purposes of applying flood, wind, snow, and earthquake provisions. The categories range from I to IV, where Category I represents buildings and other structures with a low hazard to human life in the event of failure and Category IV represents essential facilities. Each building or other structure shall be assigned to the highest applicable category or categories. Assignment of the same structure to multiple categories, based on use and the type of load condition being evaluated (e.g. wind, seismic, etc.), shall be permissible.

When buildings or other structures have multiple uses (occupancies), the relationship between the uses of various parts of the building or other structure and the independence of the structural systems for those various parts shall be examined. The classification for each independent structural system of a multiple use building or other structure shall be that of the highest usage group in any part of the building or other structure which is dependent on that basic structural system.

1.6 ADDITIONS AND ALTERATIONS TO EXISTING STRUCTURES

When an existing building or other structure is enlarged or otherwise altered, structural members affected shall be strengthened if necessary so that the factored loads defined in this document will be supported without exceeding the specified design strength for the materials of construction. When using allowable stress design, strengthening is required when the stresses due to nominal loads exceed the specified allowable stresses for the materials of construction.

1.7 LOAD TESTS

A load test of any construction shall be conducted when required by the authority having jurisdiction whenever there is reason to question its safety for the intended occupancy or use.

TABLE 1-1. Classification of Buildings and Other Structures for Flood, Wind, Snow, and Earthquake Loads

Nature of Occupancy	Category
Buildings and other structures that represent a low hazard to human life in the event of failure including, but not limited to: • Agricultural facilities • Certain temporary facilities • Minor storage facilities	I
All buildings and other structures except those listed in Categories I, III and IV	II
Buildings and other structures that represent a substantial hazard to human life in the event of failure including, but not limited to: • Buildings and other structures where more than 300 people congregate in one area • Buildings and other structures with day-care facilities with capacity greater than 150 • Buildings and other structures with elementary or secondary school facilities with capacity greater than 150 • Buildings and other structures with a capacity greater than 500 for colleges or adult education facilities • Health care facilities with a capacity of 50 or more resident patients but not having surgery or emergency treatment facilities • Jails and detention facilities • Power generating stations and other public utility facilities not included in Category IV Buildings and other structures containing sufficient quantities of toxic, explosive or other hazardous substances to be dangerous to the public if released including, but not limited to: • Petrochemical facilities • Fuel storage facilities • Manufacturing or storage facilities for hazardous chemicals • Manufacturing or storage facilities for explosives	III
Buildings and other structures that are equipped with secondary containment of toxic, explosive or other hazardous substances (including, but not limited to double wall tank, dike of sufficient size to contain a spill, or other means to contain a spill or a blast within the property boundary of the facility and prevent release of harmful quantities of contaminants to the air, soil, ground water, or surface water) or atmosphere (where appropriate) shall be eligible for classification as a Category II structure. In hurricane prone regions, buildings and other structures that contain toxic, explosive, or other hazardous substances and do not qualify as Category IV structures shall be eligible for classification as Category II structures for wind loads if these structures are operated in accordance with mandatory procedures that are acceptable to the authority having jurisdiction and which effectively diminish the effects of wind on critical structural elements or which alternatively protect against harmful releases during and after hurricanes. Buildings and other structures designated as essential facilities including, but not limited to: • Hospitals and other health care facilities having surgery or emergency treatment facilities • Fire, rescue and police stations and emergency vehicle garages • Designated earthquake, hurricane, or other emergency shelters • Communications centers and other facilities required for emergency response • Power generating stations and other public utility facilities required in an emergency • Ancillary structures (including, but not limited to communication towers, fuel storage tanks, cooling towers, electrical substation structures, fire water storage tanks or other structures housing or supporting water or other fire-suppression material or equipment) required for operation of Category IV structures during an emergency • Aviation control towers, air traffic control centers and emergency aircraft hangars • Water storage facilities and pump structures required to maintain water pressure for fire suppression • Buildings and other structures having critical national defense functions	IV

2.0 COMBINATIONS OF LOADS

2.1 GENERAL

Buildings and other structures shall be designed using the provisions of either Section 2.3 or 2.4. Either Section 2.3 or 2.4 shall be used exclusively for proportioning elements of a particular construction material throughout the structure.

2.2 SYMBOLS AND NOTATION

D = dead load;
E = earthquake load;
F = load due to fluids with well-defined pressures and maximum heights;
F_a = flood load;
H = load due to lateral earth pressure, ground water pressure, or pressure of bulk materials;
L = live load;
L_r = roof live load;
R = rain load;
S = snow load;
T = self-straining force;
W = wind load.

2.3 COMBINING FACTORED LOADS USING STRENGTH DESIGN

2.3.1 Applicability

The load combinations and load factors given in Section 2.3.2 shall be used only in those cases in which they are specifically authorized by the applicable material design standard.

2.3.2 Basic Combinations

Structures, components, and foundations shall be designed so that their design strength equals or exceeds the effects of the factored loads in the following combinations:

1. $1.4(D + F)$
2. $1.2(D + F + T) + 1.6(L + H) + 0.5(L_r \text{ or } S \text{ or } R)$
3. $1.2D + 1.6(L_r \text{ or } S \text{ or } R) + (0.5L \text{ or } 0.8W)$
4. $1.2D + 1.6W + 0.5L + 0.5(L_r \text{ or } S \text{ or } R)$
5. $1.2D + 1.0E + 0.5L + 0.2S$
6. $0.9D + 1.6W + 1.6H$
7. $0.9D + 1.0E + 1.6H$

Exceptions:

1. The load factor on L in combinations (3), (4), and (5) shall equal 1.0 for garages, areas occupied as places of public assembly, and all areas where the live load is greater than 100 lb/ft² (pounds-force per square foot) (4.79 kN/m²).
2. The load factor on H shall be set equal to zero in combinations (6) and (7) if the structural action due to H counteracts that due to W or E. Where lateral earth pressure provides resistance to structural actions from other forces, it shall not be included in H but shall be included in the design resistance.

Each relevant strength limit state shall be investigated. Effects of one or more loads not acting shall be investigated. The most unfavorable effects from both wind and earthquake loads shall be investigated, where appropriate, but they need not be considered to act simultaneously. Refer to Section 9.2.2 for specific definition of the earthquake load effect E.[1]

2.3.3 Load Combinations Including Flood Load

When a structure is located in a flood zone (Section 5.3.1), the following load combinations shall be considered:

1. In V Zones or coastal A Zones, $1.6W$ in combinations (4) and (6) shall be replaced by $1.6W + 2.0F_a$.
2. In non-coastal A Zones, $1.6W$ in combinations (4) and (6) shall be replaced by $0.8W + 1.0F_a$.

2.4 COMBINING NOMINAL LOADS USING ALLOWABLE STRESS DESIGN

2.4.1 Basic Combinations

Loads listed herein shall be considered to act in the following combinations, whichever produces the most unfavorable effect in the building, foundation, or structural member being considered. Effects of one or more loads not acting shall be considered.

1. D
2. $D + L + F + H + T + (L_r \text{ or } S \text{ or } R)$
3. $D + (W \text{ or } 0.7E) + L + (L_r \text{ or } S \text{ or } R)$
4. $0.6D + W + H$
5. $0.6D + 0.7E + H$

[1]The same E from Section 9 is used for Sections 2.3.2 and 2.4.1. Refer to the Commentary for Section 9.

The most unfavorable effects from both wind and earthquake loads shall be considered, where appropriate, but they need not be assumed to act simultaneously. Refer to Section 9.2.2 for the specific definition of the earthquake load effect E.[2]

2.4.2 Load Combinations Including Flood Load

When a structure is located in a flood zone, the following load combinations shall be considered:

1. In V Zones or coastal A Zones (Section 5.3.1), $1.5F_a$ shall be added to other loads in combinations (3) and (4), and E shall be set equal to zero in (3).
2. In non-coastal A Zones, $0.75F_a$ shall be added to combinations (3) and (4), and E shall be set equal to zero in (3).

2.4.3 Load Reduction

When structural effects due to two or more loads in combination with dead load, but excluding earth-

---

[2]The same E from Section 9 is used for Sections 2.3.2 and 2.4.1. Refer to the Commentary for Section 9.

quake load, are investigated in load combinations of Sections 2.4.1 and 2.4.2, the combined effects due to the two or more loads multiplied by 0.75 plus effects due to dead loads shall not be less than the effects from the load combination of the dead load plus the load producing the largest effects.

Increases in allowable stress shall not be used with these loads or load combinations unless it can be demonstrated that such an increase is justified by structural behavior caused by rate or duration of load.

The load combinations including earthquake loads shall follow the requirements in Section 9.

2.5 LOAD COMBINATIONS FOR EXTRAORDINARY EVENTS

Where required by the applicable code, standard, or the authority having jurisdiction, strength and stability shall be checked to ensure that structures are capable of withstanding the effects of extraordinary (i.e. low-probability) events such as fires, explosions and vehicular impact.

3.0 DEAD LOADS

3.1 DEFINITION

Dead loads consist of the weight of all materials of construction incorporated into the building including but not limited to walls, floors, roofs, ceilings, stairways, built-in partitions, finishes, cladding and other similarly incorporated architectural and structural items, and fixed service equipment including the weight of cranes.

3.2 WEIGHTS OF MATERIALS AND CONSTRUCTIONS

In determining dead loads for purposes of design, the actual weights of materials and constructions shall be used, provided that in the absence of definite information, values approved by the authority having jurisdiction shall be used.

3.3 WEIGHT OF FIXED SERVICE EQUIPMENT

In determining dead loads for purposes of design, the weight of fixed service equipment, such as plumbing stacks and risers, electrical feeders, and heating, ventilating, and air conditioning systems, shall be included.

4.0 LIVE LOADS

4.1 DEFINITION

Live loads are those loads produced by the use and occupancy of the building or other structure and do not include construction or environmental loads such as wind load, snow load, rain load, earthquake load, flood load, or dead load. Live loads on a roof are those produced (1) during maintenance by workers, equipment, and materials; and (2) during the life of the structure by movable objects such as planters and by people.

4.2 UNIFORMLY DISTRIBUTED LOADS

4.2.1 Required Live Loads

The live loads used in the design of buildings and other structures shall be the maximum loads expected by the intended use or occupancy but shall in no case be less than the minimum uniformly distributed unit loads required by Table 4-1.

4.2.2 Provision for Partitions

In office buildings or other buildings, where partitions will be erected or rearranged, provision for partition weight shall be made, whether or not partitions are shown on the plans, unless the specified live load exceeds 80 lb/ft^2 (3.83 kN/m^2).

4.3 CONCENTRATED LOADS

Floors and other similar surfaces shall be designed to support safely the uniformly distributed live loads prescribed in Section 4.2 or the concentrated load, in pounds (kilonewtons), given in Table 4-1, whichever produces the greater load effects. Unless otherwise specified, the indicated concentration shall be assumed to be uniformly distributed over an area 2.5 ft^2 (762 mm) [6.25 ft^2 (0.58 m^2)] and shall be located so as to produce the maximum load effects in the structural members.

Any single panel point of the lower chord of exposed roof trusses or any point along the primary structural members supporting roofs over manufacturing, commercial storage and warehousing, and commercial garage floors shall be capable of carrying safely a suspended concentrated load of not less than 2,000 lb (poundforce) (8.90 kN) in addition to dead load. For all other occupancies, a load of 200 lb (0.89 kN) shall be used instead of 2,000 lb (8.90 kN).

4.4 LOADS ON HANDRAILS, GUARDRAIL SYSTEMS, GRAB BAR SYSTEMS, VEHICLE BARRIER SYSTEMS, AND FIXED LADDERS

4.4.1 Definitions

Handrail: A rail grasped by hand for guidance and support. A handrail assembly includes the handrail, supporting attachments and structures.

Fixed Ladder: A ladder that is permanently attached to a structure, building, or equipment.

Guardrail system: A system of building components near open sides of an elevated surface for the purpose of minimizing the possibility of a fall from the elevated surface by people, equipment or material.

Grab bar system: A bar provided to support body weight in locations such as toilets, showers and tub enclosures.

Vehicle barrier system: A system of building components near open sides of a garage floor or ramp, or building walls which act as restraints for vehicles.

4.4.2 Loads

(a) Handrail assemblies and guardrail systems shall be designed to resist a load of 50 lb/ft (poundforce per linear foot) (0.73 kN/m) applied in any direction at the top and to transfer this load through the supports to the structure. For one- and two-family dwellings, the minimum load shall be 20 lb/ft (0.29 kN/m).

Further, all handrail assemblies and guardrail systems shall be able to resist a single concentrated load of 200 lb (0.89 kN), applied in any direction at any point along the top, and have attachment devices and supporting structure to transfer this loading to appropriate structural elements of the building. This load need not be assumed to act concurrently with the loads specified in the preceding paragraph.

Intermediate rails (all those except the handrail), balusters, and panel fillers shall be designed to withstand a horizontally applied normal load of 50 lb (0.22 kN) on an area not to exceed 1 ft^2 (305 mm^2) including openings and space between rails. Reactions due to this loading are not required to be superimposed with those of either preceding paragraph.

(b) Grab bar systems shall be designed to resist a single concentrated load of 250 lb (1.11 kN) applied in any direction at any point.

(c) Vehicle barrier systems for passenger cars shall be designed to resist a single load of 6,000 lb (26.70 kN) applied horizontally in any direction to the barrier system, and shall have anchorages or attachments capable of transferring this load to the structure. For design of the system the load shall be assumed to act at a minimum height of 1 ft 6 in. (460 mm) above the floor or ramp surface on an area not to exceed 1 ft^2 (305 mm^2), and is not required to be assumed to act concurrently with any handrail or guardrail loadings specified in the preceding paragraphs of Section 4.4.2. Garages accommodating trucks and buses shall be designed in accordance with an approved method which contains provision for traffic railings.

(d) The minimum design live load on fixed ladders with rungs shall be a single concentrated load of 300 lb, and shall be applied at any point to produce the maximum load effect on the element being considered. The number and position of additional concentrated live load units shall be a minimum of one unit of 300 lb for every 10 ft of ladder height. Ship ladders, with treads instead of rungs, shall have minimum design loads as stairs, defined in Table 4-1.

(e) Where rails of fixed ladders extend above a floor or platform at the top of the ladder, each side rail extension shall be designed to resist a concentrated live load of 100 lb in any direction at any height up to the top of the side rail extension.

4.5 LOADS NOT SPECIFIED

For occupancies or uses not designated in Section 4.2. or 4.3, the live load shall be determined in accordance with a method approved by the authority having jurisdiction.

4.6 PARTIAL LOADING

The full intensity of the appropriately reduced live load applied only to a portion of a structure or member shall be accounted for if it produces a more unfavorable effect than the same intensity applied over the full structure or member.

4.7 IMPACT LOADS

The live loads specified in Sections 4.2.1 and 4.4.2 shall be assumed to include adequate allowance for ordinary impact conditions. Provision shall be made in the structural design for uses and loads that involve unusual vibration and impact forces.

4.7.1 Elevators
All elevator loads shall be increased by 100% for impact and the structural supports shall be designed within the limits of deflection prescribed by ANSI/ASME A17.1 [2] and ANSI/ASME A17.2 [1].

4.7.2 Machinery
For the purpose of design, the weight of machinery and moving loads shall be increased as follows to allow for impact: (1) elevator machinery, 100%; (2) light machinery, shaft- or motor-driven, 20%; (3) reciprocating machinery or power-driven units, 50%; and (4) hangers for floors or balconies, 33%. All percentages shall be increased where specified by the manufacturer.

4.8 REDUCTION IN LIVE LOADS

The minimum uniformly distributed live loads, L_o in Table 4-1, may be reduced according to the following provisions.

4.8.1 General
Subject to the limitations of Sections 4.8.2 through 4.8.5, members for which a value of $K_{LL}A_T$ is 400 ft^2 (37.16 m^2) or more are permitted to be designed for a reduced live load in accordance with the following formula:

$$L = L_o \left(0.25 + \frac{4.57}{\sqrt{K_{LL}A_T}} \right) \qquad \text{(Eq. 4-1)}$$

in SI:

$$L = L_o \left(0.25 + \frac{15}{\sqrt{K_{LL}A_T}} \right)$$

where

L = reduced design live load per square foot (meter) of area supported by the member;

L_o = unreduced design live load per square foot (meter) of area supported by the member (see Table 4-1);

K_{LL} = live load element factor (see Table 4-2); and

A_T = tributary area, in square feet (square meters).

L shall not be less than 0.50 L_o for members supporting one floor and L shall not be less than 0.40 L_o for members supporting two or more floors.

4.8.2 Heavy Live Loads

Live loads that exceed 100 lb/ft² (4.79 kN/m²) shall not be reduced except the live loads for members supporting two or more floors may be reduced by 20%.

4.8.3 Passenger Car Garages

The live loads shall not be reduced in passenger car garages except the live loads for members supporting two or more floors may be reduced by 20%.

4.8.4 Special Occupancies

Live loads of 100 lb/ft² (4.79 kN/m²) or less shall not be reduced in public assembly occupancies.

4.8.5 Special Structural Elements

Live loads shall not be reduced for one-way slabs except as permitted in Section 4.8.2. Live loads of 100 lb/ft² (4.79 kN/m²) or less shall not be reduced for roof members except as specified in Section 4.9.

4.9 MINIMUM ROOF LIVE LOADS

4.9.1 Flat, Pitched, and Curved Roofs

Ordinary flat, pitched, and curved roofs shall be designed for the live loads specified in Eq. 4-2 or other controlling combinations of loads as discussed in Section 2, whichever produces the greater load. In structures such as greenhouses, where special scaffolding is used as a work surface for workmen and materials during maintenance and repair operations, a lower roof load than specified in Eq. 4-2 shall not be used unless approved by the authority having jurisdiction.

$$L_r = 20R_1R_2 \quad \text{where} \quad 12 \leq L_r \leq 20 \quad \text{(Eq. 4-2)}$$

in SI:

$$L_r = 0.96R_1R_2 \quad \text{where} \quad 0.58 \leq L_r \leq 0.96$$

where L_r = roof live load per square foot of horizontal projection in pounds per square foot (kN/m²).

The reduction factors R_1 and R_2 shall be determined as follows:

$$R_1 = \begin{cases} 1 & \text{for } A_t \leq 200 \text{ ft}^2 \ (18.58 \text{ m}^2) \\ 1.2 - 0.001A_t & \text{for } 200 \text{ ft}^2 < A_t < 600 \text{ ft}^2 \\ 0.6 & \text{for } A_t \geq 600 \text{ ft}^2 \ (55.74 \text{ m}^2) \end{cases}$$

in SI:

$$R_1 = \begin{cases} 1 & \text{for } A_t \leq 18.58 \text{ m}^2 \\ 1.2 - 0.01076A_t & \text{for } 18.58 \text{ m}^2 < A_t < 55.74 \text{ m}^2 \\ 0.6 & \text{for } A_t \geq 55.74 \text{ m}^2 \end{cases}$$

where A_t = tributary area in square feet (square meters) supported by any structural member and

$$R_2 = \begin{cases} 1 & \text{for } F \leq 4 \\ 1.2 - 0.05F & \text{for } 4 < F < 12 \\ 0.6 & \text{for } F \geq 12 \end{cases}$$

where, for a pitched roof, F = number of inches of rise per foot (in SI: $F = 0.12 \times$ slope, with slope expressed in percentage points) and, for an arch or dome, F = rise-to-span ratio multiplied by 32.

4.9.2 Special-Purpose Roofs

Roofs used for promenade purposes shall be designed for a minimum live load of 60 lb/ft² (2.87 kN/m²). Roofs used for roof gardens or assembly purposes shall be designed for a minimum live load of 100 lb/ft² (4.79 kN/m²). Roofs used for other special purposes shall be designed for appropriate loads, as approved by the authority having jurisdiction.

4.10 CRANE LOADS

The crane live load shall be the rated capacity of the crane. Design loads for the runway beams, including connections and support brackets, of moving bridge cranes and monorail cranes shall include the maximum wheel loads of the crane and the vertical impact, lateral, and longitudinal forces induced by the moving crane.

4.10.1 Maximum Wheel Load

The maximum wheel loads shall be the wheel loads produced by the weight of the bridge, as applicable, plus the sum of the rated capacity and the weight of the trolley with the trolley positioned on its runway at the location where the resulting load effect is maximum.

4.10.2 Vertical Impact Force

The maximum wheel loads of the crane shall be increased by the percentages shown below to determine the induced vertical impact or vibration force:

Monorail cranes (powered) 25
Cab-operated or remotely operated bridge cranes
(powered) 25
Pendant-operated bridge cranes (powered) 10
Bridge cranes or monorail cranes with hand-geared
bridge, trolley and hoist 0

4.10.3 Lateral Force

The lateral force on crane runway beams with electrically powered trolleys shall be calculated as 20% of the sum of the rated capacity of the crane and the weight of the hoist and trolley. The lateral force shall be assumed to act horizontally at the traction surface of a runway beam, in either direction perpendicular to the beam, and shall be distributed with due regard to the lateral stiffness of the runway beam and supporting structure.

4.10.4 Longitudinal Force

The longitudinal force on crane runway beams, except for bridge cranes with hand-geared bridges, shall be calculated as 10% of the maximum wheel loads of the crane. The longitudinal force shall be assumed to act horizontally at the traction surface of a runway beam, in either direction parallel to the beam.

4.11 REFERENCES

[1] American National Standard Practice for the Inspection of Elevators, Escalators, and Moving Walks (Inspectors' Manual), ANSI A17.2, 1988.
[2] American National Standard Safety Code for Elevators and Escalators, ANSI/ASME A17.1, 1993.
[3] American National Standard for Assembly Seating, Tents, and Air-Supported Structures, ANSI/ NFPA 102, 1992.

TABLE 4-1. Minimum Uniformly Distributed Live Loads, L_o and Minimum Concentrated Live Loads

Occupancy or Use	Uniform psf (kN/m²)	Concentration lb (kN)
Apartments (see residential)		
Access floor systems		
Office use	50 (2.4)	2,000 (8.9)
Computer use	100 (4.79)	2,000 (8.9)
Armories and drill rooms	150 (7.18)	
Assembly areas and theaters		
Fixed seats (fastened to floor)	60 (2.87)	
Lobbies	100 (4.79)	
Movable seats	100 (4.79)	
Platforms (assembly)	100 (4.79)	
Stage floors	150 (7.18)	
Balconies (exterior)	100 (4.79)	
On one- and two-family residences only, and not exceeding 100 ft² (9.3 m²)	60 (2.87)	
Bowling alleys, poolrooms and similar recreational areas	75 (3.59)	
Catwalks for maintenance access	40 (1.92)	300 (1.33)
Corridors		
First floor	100 (4.79)	
Other floors, same as occupancy served except as indicated		
Dance halls and ballrooms	100 (4.79)	
Decks (patio and roof)		
Same as area served, or for the type of occupancy accommodated		
Dining rooms and restaurants	100 (4.79)	
Dwellings (see residential)		
Elevator machine room grating [on area of 4 in.² (2,580 mm²)]		300 (1.33)
Finish light floor plate construction [on area of 1 in.² (645 mm²)]		200 (0.89)
Fire escapes	100 (4.79)	
On single-family dwellings only	40 (1.92)	
Fixed Ladders		See Section 4.4
Garages (passenger cars only)	50 (2.40)	—[1]
Trucks and buses		—[2]
Grandstands (see stadium and arena bleachers)		
Gymnasiums, main floors and balconies	100 (4.79)[4]	
Handrails, guardrails and grab bars	See Section 4.4	
Hospitals		
Operating rooms, laboratories	60 (2.87)	1,000 (4.45)
Private rooms	40 (1.92)	1,000 (4.45)
Wards	40 (1.92)	1,000 (4.45)
Corridors above first floor	80 (3.83)	1,000 (4.45)
Hotels (see residential)		
Libraries		
Reading rooms	60 (2.87)	1,000 (4.45)
Stack rooms	150 (7.18)[3]	1,000 (4.45)
Corridors above first floor	80 (3.83)	1,000 (4.45)
Manufacturing		
Light	125 (6.00)	2,000 (8.90)
Heavy	250 (11.97)	3,000 (13.40)
Marquees and Canopies	75 (3.59)	
Office Buildings		
File and computer rooms shall be designed for heavier loads based on anticipated occupancy		
Lobbies and first floor corridors	100 (4.79)	2,000 (8.90)
Offices	50 (2.40)	2,000 (8.90)
Corridors above first floor	80 (3.83)	2,000 (8.90)

TABLE 4-1. Minimum Uniformly Distributed Live Loads, L_o and Minimum Concentrated Live Loads (*Continued*)

Occupancy or Use	Uniform psf (kN/m²)	Concentration lb (kN)
Penal Institutions		
Cell blocks	40 (1.92)	
Corridors	100 (4.79)	
Residential		
Dwellings (one- and two-family)		
Uninhabitable attics without storage	10 (0.48)	
Uninhabitable attics with storage	20 (0.96)	
Habitable attics and sleeping areas	30 (1.44)	
All other areas except stairs and balconies	40 (1.92)	
Hotels and multifamily houses		
Private rooms and corridors serving them	40 (1.92)	
Public rooms and corridors serving them	100 (4.79)	
Reviewing stands, grandstands and bleachers	100 (4.79)[4]	
Roofs	See Sections 4.3 and 4.9	
Schools		
Classrooms	40 (1.92)	1,000 (4.45)
Corridors above first floor	80 (3.83)	1,000 (4.45)
First floor corridors	100 (4.79)	1,000 (4.45)
Scuttles, skylight ribs, and accessible ceilings		200 (9.58)
Sidewalks, vehicular driveways, and yards, subject to trucking	250 (11.97)[5]	8,000 (35.60)[6]
Stadiums and Arenas		
Bleachers	100 (4.79)[4]	
Fixed Seats (fastened to floor)	60 (2.87)[4]	
Stairs and exitways	100 (4.79)	—[7]
One- and two-family residences only	40 (1.92)	
Storage areas above ceilings	20 (0.96)	
Storage warehouses (shall be designed for heavier loads if required for anticipated storage)		
Light	125 (6.00)	
Heavy	250 (11.97)	
Stores		
Retail		
First floor	100 (4.79)	1,000 (4.45)
Upper floors	73 (3.59)	1,000 (4.45)
Wholesale, all floors	125 (6.00)	1,000 (4.45)
Vehicle barriers	See Section 4.4	
Walkways and elevated platforms (other than exitways)	60 (2.87)	
Yards and terraces, pedestrians	100 (4.79)	

[1]Floors in garages or portions of building used for the storage of motor vehicles shall be designed for the uniformly distributed live loads of Table 4-1 or the following concentrated load: (1) for passenger cars accommodating not more than nine passengers, 2,000 lb (8.90 kN) acting on an area of 20 in.² (12,900 mm²); (2) mechanical parking structures without slab or deck, passenger car only, 1,500 lb (6.70 kN) per wheel.

[2]Garages accommodating trucks and buses shall be designed in accordance with an approved method which contains provisions for truck and bus loadings.

[3]The weight of books and shelving shall be computed using an assumed density of 65 lb/ft³ (pounds per cubic foot, sometimes abbreviated pcf) (10.21 kN/m³) and converted to a uniformly distributed load; this load shall be used if it exceeds 150 lb/ft² (7.18 kN/m²).

[4]In addition to the vertical live loads, horizontal swaying forces parallel and normal to the length of seats shall be included in the design according to the requirements of ANSI/NFPA 102 [3].

[5]Other uniform loads in accordance with an approved method which contains provisions for truck loadings shall also be considered where appropriate.

[6]The concentrated wheel load shall be applied on an area of 20 in.² (12,900 mm²).

[7]Minimum concentrated load on stair treads [on area of 4 in.² (2,580 mm²)] is 300 lb (1.33 kN).

TABLE 4-2. Live Load Element Factor, K_{LL}

Element	K_{LL}[1]
Interior columns	4
Exterior columns without cantilever slabs	4
Edge columns with cantilever slabs	3
Corner columns with cantilever slabs	2
Edge beams without cantilever slabs	2
Interior beams	2
All other members not identified above including:	1
Edge beams with cantilever slabs	
Cantilever beams	
Two-way slabs	
Members without provisions for continuous shear transfer normal to their span	

[1]In lieu of the values above, K_{LL} is permitted to be calculated.

5.0 SOIL AND HYDROSTATIC PRESSURE AND FLOOD LOADS

5.1 PRESSURE ON BASEMENT WALLS

In the design of basement walls and similar approximately vertical structures below grade, provision shall be made for the lateral pressure of adjacent soil. Due allowance shall be made for possible surcharge from fixed or moving loads. When a portion or the whole of the adjacent soil is below a free-water surface, computations shall be based on the weight of the soil diminished by buoyancy, plus full hydrostatic pressure.

Basement walls shall be designed to resist lateral soil loads. Soil loads specified in Table 5-1 shall be used as the minimum design lateral soil loads unless specified otherwise in a soil investigation report approved by the authority having jurisdiction. The lateral pressure from surcharge loads shall be added to the lateral earth pressure load. The lateral pressure shall be increased if soils with expansion potential are present at the site as determined by a geotechnical investigation.

5.2 UPLIFT ON FLOORS AND FOUNDATIONS

In the design of basement floors and similar approximately horizontal elements below grade, the upward pressure of water, where applicable, shall be taken as the full hydrostatic pressure applied over the entire area. The hydrostatic head shall be measured from the underside of the construction. Any other upward loads shall be included in the design.

Where expansive soils are present under foundations or slabs-on-ground, the foundations, slabs, and other components shall be designed to tolerate the movement or resist the upward pressures caused by the expansive soils, or the expansive soil shall be removed or stabilized around and beneath the structure.

5.3 FLOOD LOADS

The provisions of this section apply to buildings and other structures located in areas prone to flooding as defined on a flood hazard map.

5.3.1 Definitions

The following definitions apply to the provisions of Section 5.3.

Approved: Acceptable to the authority having jurisdiction.

Base Flood: The flood having a 1% chance of being equalled or exceeded in any given year.

Base Flood Elevation (BFE): The elevation of flooding, including wave height, having a 1% chance of being equaled or exceeded in any given year.

Breakaway Wall: Any type of wall using approved materials and construction techniques, which does not provide structural support to a structure, and which is designed and constructed to fail under specified circumstances without damage to the structure or to the supporting foundation system.

Coastal A Zone: An area within a Special Flood Hazard Area, landward of a V Zone or landward of an open coast without mapped V Zones. To be classified as a coastal A Zone, the principal source of flooding must be astronomical tides, storm surges, seiches or tsunamis, not riverine flooding.

Coastal High Hazard Area (V Zone): An area within a Special Flood Hazard Area, extending from offshore to the inland limit of a primary frontal dune along an open coast, and any other area which is subject to high velocity wave action from storms or seismic sources. This area is designated on FIRMs as V, VE, VO, or V1-30.

Design Flood: The greater of the following two flood events: 1) the Base Flood, affecting those areas identified as Special Flood Hazard Areas on the community's FIRM; or 2) the flood corresponding to the area designated as a Flood Hazard Area on a community's Flood Hazard Map or otherwise legally designated.

Design Flood Elevation (DFE): The elevation of the Design Flood, including wave height, relative to the datum specified on a community's Flood Hazard Map.

Flood Hazard Area: The area subject to flooding during the Design Flood.

Flood Hazard Map: The map delineating Flood Hazard Areas adopted by the authority having jurisdiction.

Flood Insurance Rate Map (FIRM): An official map of a community, on which the Federal Insurance Administration has delineated both Special Flood Hazard Areas and the risk premium zones applicable to the community.

Special Flood Hazard Area (Area of Special Flood Hazard): The land in the floodplain subject to a 1% or greater chance of flooding in any given year. These areas are delineated on a community's Flood Insurance Rate Map (FIRM) as A Zones (A, AE, A1-30, A99, AR, AO, or AH) or V Zones (V, VE, VO, or V1-30).

5.3.2 Design Requirements

5.3.2.1 Design Loads

Structural systems of buildings or other structures shall be designed, constructed, connected and anchored to resist flotation, collapse and permanent lateral movement due to action of wind loads and loads from flooding associated with the design flood including hydrostatic, hydrodynamic and impact loads (see Section 2).

5.3.2.2 Breakaway Walls

When walls and partitions located below the design flood elevation in a coastal high hazard area are required to break away, such walls and their connections to the structure shall be designed for not less than 10 psf (0.48 kN/m²) nor more than 20 psf (0.96 kN/m²), except if the design wind load is greater, on the vertical projected area. Breakaway walls which exceed a design loading resistance of 20 psf (0.96 kN/m²) shall not be used unless the design meets the following conditions:

1. Breakaway wall collapse shall result from a water load less than that which occurs during the base flood; and
2. The elevated portion of the building and supporting foundation system shall resist collapse, displacement, and other structural damage due to the effects of wind and water loads acting simultaneously on all building components (structural and non-structural).

5.3.3 Loads During Flooding

5.3.3.1 Load Basis

In flood hazard areas the structural design shall be based on the design flood.

5.3.3.2 Hydrostatic Loads

Hydrostatic loads caused by a depth of water to the level of the design flood elevation shall be applied over all surfaces involved, both above and below ground level, except that for surfaces exposed to free water, the design depth shall be increased by 1 ft (0.30 m).

Reduced uplift and lateral loads on surfaces of enclosed spaces below the design flood elevation shall apply only if provision is made for entry and exit of floodwater (see Section 5.3.4.4).

5.3.3.3 Hydrodynamic Loads

Where water velocities do not exceed 10 ft/s (3.05 m/s), dynamic effects of the moving water shall be converted into equivalent hydrostatic loads by increasing the design flood elevation for design purposes by an equivalent surcharge depth, d_h, on the headwater side and above the ground level only, equal to:

$$d_h = \frac{aV^2}{2g} \qquad \text{(Eq. 5-1)}$$

where

V = average velocity of water in feet per second (meters per second);
g = acceleration due to gravity, 32.2 ft/s² (9.81 m/s²); and
a = coefficient of drag or shape factor (not less than 1.25).

The equivalent surcharge depth shall be added to the design flood elevation design depth and the resultant hydrostatic pressures applied to, and uniformly distributed across, the vertical projected area of the building or structure which is perpendicular to the flow. Surfaces parallel to the flow or surfaces wetted by the tailwater shall be subject to the hydrostatic pressures for depths to the design flood elevation only.

Where water velocities exceed 10 ft/s (3.05 m/s), dynamic effects of the moving water shall be determined by a detailed analysis utilizing basic concepts of fluid mechanics.

5.3.3.4 Wave Loads

Wave loads shall be determined by one of the following three methods: (1) using the analytical procedures outlined in this section; (2) by more advanced numerical modeling procedures; or (3) by laboratory test procedures (physical modeling).

Wave loads are those loads which result from water waves propagating over the water surface and striking a building or other structure. Design and construction of buildings and other structures subject to wave loads shall account for the following loads:

waves breaking on any portion of the building or structure; uplift forces caused by shoaling waves beneath a building or structure, or portion thereof; wave rump striking any portion of the building or structure; wave-induced drag and inertia forces; wave-induced scour at the base of a building or structure, or its foundation. Wave loads shall be included for both V Zones and A Zones. In V Zones, waves are 3 ft (0.91 m) high, or higher; in coastal floodplains landward of the V Zone, waves are less than 3 ft high (0.91 m).

Non-breaking and broken wave loads shall be calculated using the procedures described in Sections 5.3.3.2 and 5.3.3.3 to calculate hydrostatic and hydrodynamic loads.

Breaking wave loads shall be calculated using the procedures described in Sections 5.3.3.4.1 through 5.3.3.4.4. Breaking wave heights used in the procedures described in Sections 5.3.3.4.1 through 5.3.3.4.4 shall be calculated for V Zones and coastal A Zones using Eqs. 5-2 and 5-3.

$$H_b = 0.78 d_s \qquad \text{(Eq. 5-2)}$$

where

H_b = breaking wave height in feet (m); and
d_s = local stillwater depth in feet (m).

The local stillwater depth shall be calculated using Eq. 5-3, unless more advanced procedures or laboratory tests permitted by this section are used.

$$d_s = 0.65(\text{BFE} - \text{G}) \qquad \text{(Eq. 5-3)}$$

where

BFE = Base Flood Elevation in feet (m); and
G = Ground elevation in feet (m).

5.3.3.4.1 Breaking wave loads on vertical pilings and columns

The net force resulting from a breaking wave acting on a rigid vertical pile or column shall be assumed to act at the stillwater elevation and shall be calculated by the following:

$$F_D = 0.5 \gamma_w C_D D H_b^2 \qquad \text{(Eq. 5-4)}$$

where

F_D = net wave force, in pounds (kN);
γ_w = unit weight of water, in pounds per cubic foot (kN/m^3), = 62.4 pcf (9.80 kN/m^3) for fresh water and 64.0 pcf (10.05 kN/m^3) for salt water;
C_D = coefficient of drag for breaking waves, = 1.75 for round piles or columns, and = 2.25 for square piles or columns;
D = pile or column diameter, in feet (m) for circular sections, or for a square pile or column, 1.4 times the width of the pile or column in feet (m); and
H_b = breaking wave height, in feet (m).

5.3.3.4.2 Breaking wave loads on vertical walls

Maximum pressures and net forces resulting from a normally incident breaking wave acting on a rigid vertical wall shall be calculated by the following:

$$P_{\max} = C_p \gamma_w d_s + 1.2 \gamma_w d_s \qquad \text{(Eq. 5-5)}$$

and

$$F_t = 1.1 C_p \gamma_w d_s^2 + 2.4 \gamma_w d_s^2 \qquad \text{(Eq. 5-6)}$$

where

$P_{\max}$ = maximum combined dynamic ($C_p \gamma_w d_s$) and static ($1.2 \gamma_w d_s$) wave pressures, also referred to as shock pressures in pounds per square foot (kN/m^2);
F_t = total breaking wave force per unit length of structure, also referred to as shock, impulse or wave impact force in pounds/foot (kN/m), acting near the stillwater elevation;
C_p = dynamic pressure coefficient [$1.6 < C_p < 3.5$ (see Table 5-2)];
γ_w = unit weight of water, in pounds per cubic foot (kN/m^3), = 62.4 pcf (9.80 kN/m^3) for fresh water and 64.0 pcf (10.05 kN/m^3) for salt water; and
d_s = stillwater depth in feet (m) at base of building or other structure where the wave breaks.

This procedure assumes the vertical wall causes a reflected or standing wave against the waterward side of the wall, with the crest of the wave at a height of $1.2d_s$ above the stillwater level. Thus, the dynamic, static and total pressure distributions against the wall, and the resulting force, are as shown in Fig. 5-1.

This procedure also assumes the space behind the vertical wall is dry, with no fluid balancing the static component of the wave force on the outside of the wall. If free water exists behind the wall, the hydrostatic component of the wave pressure and force disappears and the dynamic wave pressure and net force shall be computed by:

$$P_{max} = C_p \gamma_w d_s \qquad \text{(Eq. 5-7)}$$

and

$$F_t = 1.1 C_p \gamma_w d_s^2 \qquad \text{(Eq. 5-8)}$$

where

P_{max} = maximum dynamic wave pressure in pounds per square foot (kN/m^2);

F_t = total breaking wave force per unit length of structure, also referred to as shock, impulse or wave impact force in pounds/foot (kN/m), acting near the stillwater elevation;

C_p = dynamic pressure coefficient [$1.6 < C_p < 3.5$ (see Table 5-2)];

γ_w = unit weight of water, in pounds per cubic foot (kN/m^3), = 62.4 pcf (9.80 kN/m^3) for fresh water and 64.0 pcf (10.05 kN/m^3) for salt water; and

d_s = stillwater depth in feet (m) at base of building or other structure where the wave breaks.

5.3.3.4.3 Breaking wave loads on non-vertical walls

Breaking wave forces given by Eqs. 5-6 and 5-8 shall be modified in instances where the walls or surfaces upon which the breaking waves act are non-vertical. The horizontal component of breaking wave force shall be given by:

$$F_{nv} = F_t \sin^2\theta \qquad \text{(Eq. 5-9)}$$

where

F_{nv} = horizontal component of breaking wave force in pounds/foot (kN/m);

F_t = total breaking wave force acting on a vertical surface in pounds/foot (kN/m); and

θ = vertical angle between non-vertical surface and the horizontal.

5.3.3.4.4 Breaking wave loads from obliquely incident waves

Breaking wave forces given by Eqs. 5-6 and 5-8 shall be modified in instances where waves are obliquely incident. Breaking wave forces from non-normally incident waves shall be given by:

$$F_{oi} = F_t \sin^2\alpha \qquad \text{(Eq. 5-10)}$$

where

F_{oi} = horizontal component of obliquely incident breaking wave force in pounds/foot (kN/m);

F_t = total breaking wave force (normally incident waves) acting on a vertical surface in pounds/foot (kN/m); and

α = horizontal angle between the direction of wave approach and the vertical surface.

5.3.3.5 Impact Loads

Impact loads are those which result from debris, ice and any object transported by floodwaters striking against buildings and structures or parts thereof.

Minimum impact load is a concentrated load acting horizontally at the most critical location at or below the base flood elevation produced by a 1,000 lb (4.5 kN) object travelling at the velocity of the floodwater and acting on a 1 ft^2 (0.09 m^2) surface of the structure. It shall be assumed that the velocity of the object is reduced to 0 in 1 s.

5.3.4 Special Flood Hazard Areas—A Zones

The following aspects of A Zone design and construction shall be accounted for in the design: elevation above the design flood elevation, anchorage, floodproofing, enclosures below the design flood elevation, and scour.

5.3.4.1 Elevation

Buildings or structures within an A Zone shall be elevated so that the lowest habitable floor, including basement, is located at or above the elevation shown in Table 5-3.

Exception: Non-residential buildings floodproofed in accordance with Section 5.3.4.3 or 5.3.4.4.

All structural components subject to hydrostatic and hydrodynamic loads and impact loads from water-borne objects during the occurrence of flooding to the design flood elevation shall be capable of resisting such forces, including the effect of buoyancy.

5.3.4.2 Anchorage

The structural systems of buildings or other structures shall be designed, connected and anchored to prevent flotation, collapse and permanent lateral movement resulting from wind loads, impact loads, hydrodynamic loads and hydrostatic loads, including the effects of buoyancy, from flooding equal to the design flood elevation.

5.3.4.3 Non-Residential Flood-Resistant Construction

As an alternative to meeting the elevation provision of Section 5.3.4.1, non-residential buildings or other structures located in non-coastal high hazard areas shall be floodproofed so that the structure is watertight. Walls and floors below an elevation 1 ft (0.30 m) above the elevation prescribed in Table 5-3 shall be substantially impermeable to the passage of water. Openings in floodproofed walls and floors shall be provided with watertight closures and shall have adequate structural capacity to resist all applicable loads.

5.3.4.4 Enclosures Below Design Flood Elevation

Enclosed spaces below the design flood elevation shall not be used for any purpose other than parking of vehicles, building access or storage. Enclosed spaces which do not meet the requirements of Section 5.3.4.3 shall be provided with vents, valves or other openings which will automatically equalize the hydrostatic forces on exterior and interior walls by allowing for the entry and exit of floodwaters.

To provide for equalization of hydrostatic forces a minimum of two openings having a total net area of not less than 1 in.^{2} for every square foot (0.007 m^2 for every square meter) of enclosed area subject to flooding shall be provided. The bottom of all openings shall not be higher than 12 in. (0.30 m) above grade. Openings shall not be equipped with screens, louvers, valves or other coverings or devices unless they permit the automatic entry and exit of floodwaters.

5.3.4.5 Scour

The effects of scour shall be included in the design of the foundations of buildings or other structures in special flood hazard areas—A Zones.

Foundation embedment shall be below the depth of potential scour.

5.3.5 Coastal High Hazard Areas—V Zones

Loadings in V Zones are more severe than loadings in A Zones, and the design shall take into account the following: elevation above the design flood elevation, foundation type, obstructions below the design flood elevation, and the effects of erosion and scour.

5.3.5.1 Elevation

Buildings or structures erected within a coastal high hazard area shall be elevated so that the lowest portion of the lowest horizontal structural members supporting the lowest floor with the exception of footings, mat or raft foundation, piles, pile caps, columns, grade beams and bracing shall be located at or above the elevation shown in Table 5-4.

Buildings or structures erected in coastal high hazard areas shall be supported on piles or columns. The piles or columns and their foundation and structure attached thereto shall be anchored to resist flotation, collapse and permanent lateral movement due to the effects of wind, water and impact loads acting simultaneously on all building components.

All structural components subject to wind loads, hydrostatic and hydrodynamic loads and impact loads from waterborne objects during the occurrence of flooding to the design flood elevation shall be capable of resisting such forces, including the effects of buoyancy.

5.3.5.2 Space Below Design Flood Elevation

Spaces below the design flood elevation shall be free of obstruction.

Exceptions:

1. Footings, mat or raft foundations, piles, pile caps, columns, grade beams and bracing that provide structural support for the building.
2. Structural systems of entrances and required exits.
3. Incidental storage of portable or mobile items that are readily moveable in the event of a storm.
4. Walls or partitions shall not be used to enclose all or part of the space, unless they are not part of the structural support of the building and are designed to breakaway or collapse without causing collapse, displacement or other damage to the structural system of the building in accordance with Section 5.3.2.2. Insect screening, open wood lattice, and similar screening which allow the passage of water shall not be used unless these systems comply with Section 5.3.2.2.

5.3.5.3 Erosion and Scour

The effects of long-term erosion, storm-induced erosion and local scour shall be included in the design of foundations of buildings or other structures in coastal high hazard areas. Foundation embedment shall be below the depth of potential scour.

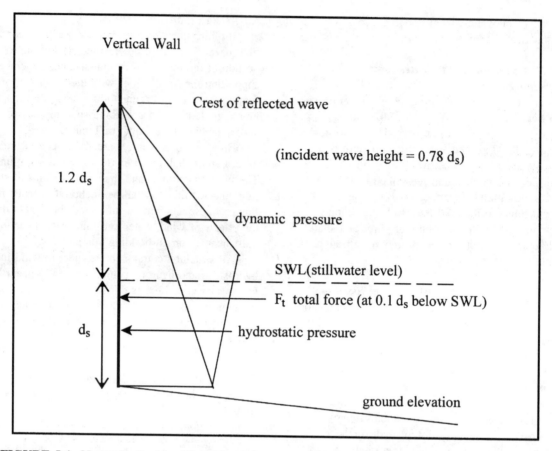

FIGURE 5-1. Normally Incident Breaking Wave Pressures and Net Force Against a Vertical Wall

TABLE 5-1. Design Lateral Soil Load

Description of Backfill Material	Unified Soil Classification	Design Lateral Soil Load[1] [psf per foot of depth (kN/m² per meter of depth)]
Well-graded, clean gravels; gravel-sand mixes	GW	35 (5.50)[2]
Poorly graded clean gravels; gravel-sand mixes	GP	35 (5.50)[2]
Silty gravels, poorly graded gravel-sand mixes	GM	35 (5.50)[2]
Clayey gravels, poorly graded gravel-and-clay mixes	GC	45 (7.07)[2]
Well-graded, clean sands; gravel-sand mixes	SW	35 (5.50)[2]
Poorly graded clean sands; sand-gravel mixes	SP	35 (5.50)[2]
Silty sands, poorly graded sand-silt mixes	SM	45 (7.07)[2]
Sand-silt clay mix with plastic fines	SM-SC	85 (13.35)[3]
Clayey sands, poorly graded sand-clay mixes	SC	85 (13.35)[3]
Inorganic silts and clayey silts	ML	85 (13.35)[3]
Mixture of inorganic silt and clay	ML-CL	85 (13.35)[3]
Inorganic clays of low to medium plasticity	CL	100 (15.71)
Organic silts and silt-clays, low plasticity	OL	—[4]
Inorganic clayey silts, elastic silts	MH	—[4]
Inorganic clays of high plasticity	CH	—[4]
Organic clays and silty clays	OH	—[4]

[1]Design lateral soil loads are given for moist conditions for the specified soils at their optimum densities. Actual field conditions shall govern. Submerged or saturated soil pressures shall include the weight of the buoyant soil plus the hydrostatic loads.

[2]For relatively rigid walls, as when braced by floors, the design lateral soil load shall be increased for sand and gravel type soils to 60 psf (9.43 kN/m²) per foot (meter) of depth. Basement walls extending not more than 8 ft (2.44 m) below grade and supporting light floor systems are not considered as being relatively rigid walls.

[3]For relatively rigid walls, as when braced by floors, the design lateral load shall be increased for silt and clay type soils to 100 psf (15.71 kN/m²) per foot (meter) of depth. Basement walls extending not more than 8 ft (2.44 m) below grade and supporting light floor systems are not considered as being relatively rigid walls.

[4]Unsuitable as backfill material.

TABLE 5-2. Value of Dynamic Pressure Coefficient, C_p

Building Category	C_p
I	1.6
II	2.8
III	3.2
IV	3.5

TABLE 5-3. Minimum Elevation of Lowest Floor, A Zones

Building Category	Minimum Elevation of Lowest Habitable Floor
I	DFE
II	DFE
III	DFE
IV	BFE + 1.0 ft (0.30 m), or DFE, whichever is higher

TABLE 5-4. Minimum Elevation of Bottom of Lowest Supporting Horizontal Structural Member of Lowest Floor, V Zones

Building Category	Minimum Elevation of Bottom of Lowest Supporting Horizontal Structural Member of Lowest Floor	
	Member Orientation Relative to Direction of Wave Approach	
	Parallel[1]	Perpendicular[1]
I	DFE	DFE
II	DFE	BFE + 1.0 ft (0.30 m) or DFE, whichever is higher
III	BFE + 1.0 ft (0.30 m) or DFE, whichever is higher	BFE + 2.0 ft (0.60 m) or DFE, whichever is higher
IV	BFE + 1.0 ft (0.30 m) or DFE, whichever is higher	BFE + 2.0 ft (0.60 m) or DFE, whichever is higher

[1]Orientation of lowest horizontal structural member relative to the general direction of wave approach; parallel shall mean less than or equal to 20 degrees from the direction of wave approach; perpendicular shall mean greater than 20 degrees from the direction of approach.

6.0 WIND LOADS

6.1 GENERAL

6.1.1 Scope

Buildings and other structures, including the main wind force resisting system and all components and cladding thereof, shall be designed and constructed to resist wind loads as specified herein.

6.1.2 Allowed Procedures

The design wind loads for buildings and other structures, including the main wind force resisting system and component and cladding elements thereof, shall be determined using one of the following procedures: (1) Method 1—Simplified Procedure as specified in Section 6.4 for buildings meeting the requirements specified therein; (2) Method 2—Analytical Procedure as specified in Section 6.5 for buildings meeting the requirements specified therein; and (3) Method 3—Wind Tunnel Procedure as specified in Section 6.6.

6.1.3 Wind Pressures Acting on Opposite Faces of Each Building Surface

In the calculation of design wind loads for the main wind force resisting system and for components and cladding for buildings, the algebraic sum of the pressures acting on opposite faces of each building surface shall be taken into account.

6.1.4 Minimum Design Wind Loading

The design wind load, determined by any one of the procedures specified in Section 6.1.2, shall be not less than specified in this section.

6.1.4.1 Main Wind Force Resisting System

The wind load to be used in the design of the main wind force resisting system for an enclosed or partially enclosed building or other structure shall not be less than 10 lb/ft^2 (0.48 kN/m^2) multiplied by the area of the building or structure projected onto a vertical plane normal to the assumed wind direction. The design wind force for open buildings and other structures shall be not less than 10 lb/ft^2 (0.48 kN/m^2) multiplied by the area A_f.

6.1.4.2 Components and Cladding

The design wind pressure for components and cladding of buildings shall be not less than a net

pressure of 10 lb/ft^2 (0.48 kN/m^2) acting in either direction normal to the surface.

6.2 DEFINITIONS

The following definitions apply only to the provisions of Section 6:

Approved: Acceptable to the authority having jurisdiction.

Basic wind speed, V: 3-second gust speed at 33 ft (10 m) above the ground in Exposure C (see Section 6.5.6.1) as determined in accordance with Section 6.5.4.

Building, enclosed: A building that does not comply with the requirements for open or partially enclosed buildings.

Building envelope: Cladding, roofing, exterior walls, glazing, door assemblies, window assemblies, skylight assemblies, and other components enclosing the building.

Building and other structure, flexible: Slender buildings and other structures that have a fundamental natural frequency less than 1 Hz.

Building, low-rise: Enclosed or partially enclosed buildings which comply with both of the following conditions:

1. mean roof height h less than or equal to 60 ft (18 m); and
2. mean roof height h does not exceed least horizontal dimension.

Building, open: A building having each wall at least 80% open. This condition is expressed for each wall by the equation $A_o \geq 0.8 A_g$ where:

A_o = total area of openings in a wall that receives positive external pressure, in ft^2 (m^2)

A_g = the gross area of that wall in which A_o is identified, in ft^2 (m^2)

Building, partially enclosed: A building which complies with both of the following conditions:

1. the total area of openings in a wall that receives positive external pressure exceeds the sum of the areas of openings in the balance of the building envelope (walls and roof) by more than 10%, and
2. the total area of openings in a wall that receives positive external pressure exceeds 4 ft^2 (0.37 m^2)

or 1% of the area of that wall, whichever is smaller, and the percentage of openings in the balance of the building envelope does not exceed 20%.

These conditions are expressed by the following equations:

1. $A_o > 1.10 A_{oi}$
2. $A_o > 4$ ft^2 (0.37 m^2) or $>0.01 A_g$, whichever is smaller, and $A_{oi}/A_{gi} \leq 0.20$ where:

A_o, A_g are as defined for Open Building

A_{oi} = the sum of the areas of openings in the building envelope (walls and roof) not including A_o, in ft^2 (m^2)

A_{gi} = the sum of the gross surface areas of the building envelope (walls and roof) not including A_g, in ft^2 (m^2)

Building or other structure, regular shaped: A building or other structure having no unusual geometrical irregularity in spatial form.

Building or other structures, rigid: A building or other structure whose fundamental frequency is greater than or equal to 1 Hz.

Building, simple diaphragm: An enclosed or partially enclosed building in which wind loads are transmitted through floor and roof diaphragms to the vertical main wind force resisting system.

Components and cladding: Elements of the building envelope that do not qualify as part of the main wind-force resisting system.

Design force, F: Equivalent static force to be used in the determination of wind loads for open buildings and other structures.

Design pressure, p: Equivalent static pressure to be used in the determination of wind loads for buildings.

Effective wind area: The area used to determine GC_p. For component and cladding elements, the effective wind area in Figs. 6-5 through 6-8 is the span length multiplied by an effective width that need not be less than one-third the span length. For cladding fasteners, the effective wind area shall not be greater than the area that is tributary to an individual fastener.

Escarpment: Also known as scarp, with respect to topographic effects in Section 6.5.7, a cliff or steep slope generally separating two levels or gently sloping areas.

Glazing: Glass or transparent or translucent plastic sheet used in windows, doors, or skylights.

Glazing, impact resistant: Glazing which has been shown by an approved test method to withstand the impact of wind borne missiles likely to be generated in wind borne debris regions during design winds.

Hill: With respect to topographic effects in Section 6.5.7, a land surface characterized by strong relief in any horizontal direction (see Fig. 6.2).

Hurricane prone regions: Areas vulnerable to hurricanes; in the United States and its territories defined as:

1. the U.S. Atlantic Ocean and Gulf of Mexico coasts where the basic wind speed is greater than 90 mph, and
2. Hawaii, Puerto Rico, Guam, Virgin Islands, and American Samoa.

Impact resistant covering: A covering designed to protect glazing, which has been shown by an approved test method to withstand the impact of wind borne missiles likely to be generated in wind borne debris regions during design winds.

Importance factor, I: A factor that accounts for the degree of hazard to human life and damage to property.

Main wind-force resisting system: An assemblage of structural elements assigned to provide support and stability for the overall structure. The system generally receives wind loading from more than one surface.

Mean roof height, h: The average of the roof eave height and the height to the highest point on the roof surface, except that, for roof angles of less than or equal to 10, the mean roof height shall be the roof heave height.

Openings: Apertures or holes in the building envelope which allow air to flow through the building envelope and which are designed as "open" during design winds as defined by these provisions.

Recognized literature: Published research findings and technical papers that are approved.

Ridge: With respect to topographic effects in Section 6.5.7 an elongated crest of a hill characterized by strong relief in two directions (see Fig. 6.2).

Wind borne debris regions: Areas within hurricane prone regions located.

1. within one mile of the coastal mean high water line where the basic wind speed is equal to or greater than 110 mph and in Hawaii; or
2. in areas where the basic wind speed is equal to or greater than 120 mph.

6.3 SYMBOLS AND NOTATIONS

The following symbols and notation apply only to the provisions of Section 6:

A = effective wind area, in ft^2 (m^2);

A_f = area of open buildings and other structures either normal to the wind direction or projected on a plane normal to the wind direction, in ft^2 (m^2);

A_g = the gross area of that wall in which A_o is identified, in ft^2 (m^2);

A_{gi} = the sum of the gross surface areas of the building envelope (walls and roof) not including A_g, in ft^2 (m^2);

A_o = total area of openings in a wall that receives positive external pressure, in ft^2 (m^2);

A_{oi} = the sum of the areas of openings in the building envelope (walls and roof) not including A_o, in ft^2 (m^2);

A_{og} = total area of openings in the building envelope ft^2 (m^2);

a = width of pressure coefficient zone, in ft (m);

B = horizontal dimension of building measured normal to wind direction, in ft (m);

$\bar{b}$ = mean hourly wind speed factor in Eq. 6-12 from Table 6-4;

$\hat{b}$ = 3-second gust speed factor from Table 6-4;

C_f = force coefficient to be used in determination of wind loads for other structures;

C_p = external pressure coefficient to be used in determination of wind loads for buildings;

c = turbulence intensity factor in Eq. 6-3 from Table 6-4;

D = diameter of a circular structure or member, in ft (m);

D' = depth of protruding elements such as ribs and spoilers, in ft (m);

G = gust effect factor;

G_f = gust effect factor for main wind-force resisting systems of flexible buildings and other structures;

GC_p = product of external pressure coefficient and gust effect factor to be used in determination of wind loads for buildings;

GC_{pf} = product of the equivalent external pressure coefficient and gust effect factor to be used in determination of wind loads for main wind-force resisting system of low-rise buildings;

GC_{pi} = product of internal pressure coefficient and gust effect factor to be used in determination of wind loads for buildings;

g_Q = peak factor for background response in Eqs. 6-2 and 6-6;

g_R = peak factor for resonant response in Eq. 6-6;

g_v = peak factor for wind response in Eqs. 6-2 and 6-6;

H = height of hill or escarpment in Fig. 6-2, in ft (m);

h = mean roof height of a building or height of other structure, except that eave height shall be used for roof angle θ of less than or equal to 10°, in ft (m);

I = importance factor;

I_z = intensity of turbulence from Eq. 6-3;

K_1, K_2, K_3 = multipliers in Fig. 6-2 to obtain K_{zt};

K_d = wind directionality factor in Table 6-6;

K_h = velocity pressure exposure coefficient evaluated at height $z = h$;

K_z = velocity pressure exposure coefficient evaluated at height z;

K_{zt} = topographic factor;

L = horizontal dimension of a building measured parallel to the wind direction, in ft (m);

L_h = distance upwind of crest of hill or escarpment in Fig. 6-2 to where the difference in ground elevation is half the height of hill or escarpment, in ft (m);

$L_{\bar{z}}$ = integral length scale of turbulence, in ft (m);

ℓ = integral length scale factor from Table 6-4, in ft (m);

M = larger dimension of sign, in ft (m);

N = smaller dimension of sign, in ft (m);

N_1 = reduced frequency from Eq. 6-10;

n_1 = building natural frequency, Hz;

p = design pressure to be used in determination of wind loads for buildings, in lb/ft^2 (N/m^2);

p_L = wind pressure acting on leeward face in Fig. 6-9;

p_W = wind pressure acting on windward face in Fig. 6-9;

Q = background response from Eq. 6-4;

q = velocity pressure, in lb/ft^2 (N/m^2);

q_h = velocity pressure evaluated at height $z = h$, in lb/ft^2 (N/m^2);

q_i = velocity pressure for internal pressure determination;

q_z = velocity pressure evaluated at height z above ground, in lb/ft^2 (N/m^2);

R = resonant response factor from Eq. 6-8;

R_B, R_h, R_L = values from Eq. 6-11;

R_i = reduction factor from Eq. 6-14;

R_n = value from Eq. 6-9;

r = rise-to-span ratio for arched roofs;

V = basic wind speed obtained from Fig. 6-1, in miles per hour (meters per second). The basic wind speed corresponds to a 3-second gust speed at 33 ft (10 m) above ground in exposure Category C;

V_i = unpartitioned internal volume in ft^3 (m^3);

$\bar{V}_{\bar{z}}$ = mean hourly wind speed at height $\bar{z}$, ft/s (m/s);

W = width of building in Figs. 6-5C and 6-7A, and width of span in Figs. 6-6 and 6-7B, in ft (m);

X = distance to center of pressure from windward edge in Table 6-6, in ft (m);

x = distance upwind or downwind of crest in Fig. 6-2, in ft (m);

z = height above ground level, in ft (m);

$\bar{z}$ = equivalent height of structure, in ft (m);

z_g = nominal height of the atmospheric boundary layer used in this standard. Values appear in Table 6-4;

z_{min} = exposure constant from Table 6-4;

α = 3-second gust speed power law exponent from Table 6-4;

$\hat{\alpha}$ = reciprocal of α from Table 6-4;

$\bar{\alpha}$ = mean hourly wind speed power law exponent in Eq. 6-12 from Table 6-4;

β = damping ratio, percent critical for buildings or other structures;

ϵ = ratio of solid area to gross area for open sign, face of a trussed tower, or lattice structure;

$\bar{\epsilon}$ = integral length scale power law exponent in Eq. 6-5 from Table 6-4;

η = value used in Eq. 6-11 (see Section 6.5.8.2);

θ = angle of plane of roof from horizontal, in degrees;

ν = height-to-width ratio for solid sign.

6.4 METHOD 1—SIMPLIFIED PROCEDURE

6.4.1 Scope

An enclosed or partially enclosed building whose design wind loads are determined in accordance with this section shall meet all the following conditions:

1. the building is a simple diaphragm building as defined in Section 6.2,
2. the building has roof slopes less than 10°,
3. the mean roof height of the building is less than or equal to 30 ft (9 m),
4. the building is a regular shaped building or structure as defined in Section 6.2,
5. the building is not classified as a flexible building as defined in Section 6.2,
6. the building structure has no expansion joints or separations, and
7. the building is not subject to the topographic effects of Section 6.5.7 (i.e., K_{zt} = 1.0).

6.4.2 Design Procedure

1. The *basic wind speed V* shall be determined in accordance with Section 6.5.4. The wind shall be assumed to come from any horizontal direction.
2. An *importance factor I* shall be determined in accordance with Section 6.5.5.
3. An *exposure category or categories* shall be determined in accordance with Section 6.5.6.
4. An *enclosure classification* shall be determined in accordance with Section 6.5.9.
5. The design *wind loads for the main wind force resisting system* shall be determined from Table 6-2. The design wind loads shall be applied normal to the surface. The design wind loads shall be applied simultaneously, with the net combined wall pressure applied on all windward wall surfaces and the net roof pressure applied on all roof surfaces.
6. *The design wind load for component and cladding elements* shall be determined from Table 6-3. These net design pressures shall be applied to each exterior surface.

6.4.3 Air Permeable Cladding

Design wind loads determined from Table 6-3 shall be used for all air permeable cladding unless approved test data or recognized literature demonstrate lower loads for the type of air permeable cladding being considered.

6.5 METHOD 2—ANALYTICAL PROCEDURE

6.5.1 Scope

A building or other structure whose design wind loads are determined in accordance with this section shall meet all of the following conditions:

1. The building or other structure is a regular shaped building or structure as defined in Section 6.2, and
2. The building or other structure does not have response characteristics making it subject to across wind loading, vortex shedding, instability due to galloping or flutter; or does not have a site location for which channeling effects or buffeting in the wake of upwind obstructions warrant special consideration.

6.5.2 Limitations

The provisions of Section 6.5 take into consideration the load magnification effect caused by gusts in resonance with along-wind vibrations of flexible buildings or other structures. Buildings or other structures not meeting the requirements of Section 6.5.1, or having unusual shapes or response characteristics shall be designed using recognized literature documenting such wind load effects or shall use the wind tunnel procedure specified in Section 6.6.

6.5.2.1 Shielding

There shall be no reductions in velocity pressure due to apparent shielding afforded by buildings and other structures or terrain features.

6.5.2.2 Air Permeable Cladding

Design wind loads determined from Section 6.5 shall be used for air permeable cladding unless approved test data or recognized literature demonstrate lower loads for the type of air permeable cladding being considered.

6.5.3 Design Procedure

1. *The basic wind speed V* and *wind directionality factor K_d* shall be determined in accordance with Section 6.5.4.
2. An *importance factor I* shall be determined in accordance with Section 6.5.5.
3. An *exposure category or exposure categories* and *velocity pressure exposure coefficient K_z or K_h*, as applicable, shall be determined for each wind direction in accordance with Section 6.5.6.
4. A *topographic factor K_{zt}* shall be determined in accordance with Section 6.5.7.
5. A *gust effect factor G or G_f*, as applicable, shall be determined in accordance with Section 6.5.8.
6. An *enclosure classification* shall be determined in accordance with Section 6.5.9.
7. *Internal pressure coefficient GC_{pi}* shall be determined in accordance with Section 6.5.11.1.

8. *External pressure coefficients C_p or GC_{pf}, or force coefficients C_f*, as applicable, shall be determined in accordance with Section 6.5.11.2 or 6.5.11.3, respectively.
9. *Velocity pressure q_z or q_h*, as applicable, shall be determined in accordance with Section 6.5.10.
10. *Design wind load P or F* shall be determined in accordance with Sections 6.5.12 and 6.5.13, as applicable.

6.5.4 Basic Wind Speed

The basic wind speed, *V*, used in the determination of design wind loads on buildings and other structures shall be as given in Fig. 6-1 except as provided in Sections 6.5.4.1 and 6.5.4.2. The wind shall be assumed to come from any horizontal direction.

6.5.4.1 Special Wind Regions

The basic wind speed shall be increased where records or experience indicate that the wind speeds are higher than those reflected in Fig. 6-1. Mountainous terrain, gorges, and special regions shown in Fig. 6-1 shall be examined for unusual wind conditions. The authority having jurisdiction shall, if necessary, adjust the values given in Fig. 6-1 to account for higher local wind speeds. Such adjustment shall be based on meteorological information and an estimate of the basic wind speed obtained in accordance with the provisions of Section 6.5.4.2.

6.5.4.2 Estimation of Basic Wind Speeds from Regional Climatic Data

Regional climatic data shall only be used in lieu of the basic wind speeds given in Fig. 6-1 when: (1) approved extreme-value statistical-analysis procedures have been employed in reducing the data; and (2) the length of record, sampling error, averaging time, anemometer height, data quality, and terrain exposure of the anemometer have been taken into account.

In hurricane prone regions, wind speeds derived from simulation techniques shall only be used in lieu of the basic wind speeds given in Fig. 6-1 when: (1) approved simulation or extreme value statistical analysis procedures are used (the use of regional wind speed data obtained from anemometers is not permitted to define the hurricane wind speed risk along the Gulf and Atlantic coasts, the Caribbean or Hawaii); and (2) the design wind speeds resulting from the study shall not be less than the resulting 500-year return period wind speed divided by $\sqrt{1.5}$.

6.5.4.3 Limitation

Tornadoes have not been considered in developing the basic wind-speed distributions.

6.5.4.4 Wind Directionality Factor

The wind directionality factor, K_d, shall be determined from Table 6-6. This factor shall only be applied when used in conjunction with load combinations specified in Sections 2.3 and 2.4.

6.5.5 Importance Factor

An importance factor, I, for the building or other structure shall be determined from Table 6-1 based on building and structure categories listed in Table 1-1.

6.5.6 Exposure Categories

6.5.6.1 General

For each wind direction considered, an exposure category that adequately reflects the characteristics of ground surface irregularities shall be determined for the site at which the building or structure is to be constructed. For a site located in the transition zone between categories, the category resulting in the largest wind forces shall apply. Account shall be taken of variations in ground surface roughness that arise from natural topography and vegetation as well as from constructed features. For any given wind direction, the exposure in which a specific building or other structure is sited shall be assessed as being one of the following categories:

1. **Exposure A.** Large city centers with at least 50% of the buildings having a height in excess of 70 ft (21.3 m). Use of this exposure category shall be limited to those areas for which terrain representative of Exposure A prevails in the upwind direction for a distance of at least 1/2 mi (0.8 km) or 10 times the height of the building or other structure, whichever is greater. Possible channeling effects or increased velocity pressures due to the building or structure being located in the wake of adjacent buildings shall be taken into account.
2. **Exposure B.** Urban and suburban areas, wooded areas, or other terrain with numerous closely spaced obstructions having the size of single-family dwellings or larger. Use of this exposure category shall be limited to those areas for which terrain representative of Exposure B prevails in the upwind direction for a distance of at least 1,500 ft (460 m) or 10 times the height of the building or other structure, whichever is greater.

EXP B SHALL BE ASSUMED UNLESS THE SITE MEETS THE DEFINITION OF ANOTHER EXPOSURE R *ADOP IN IBC 2000*

3. **Exposure C.** Open terrain with scattered obstructions having heights generally less than 30 ft (9.1 m). This category includes flat open country, grasslands and shorelines in hurricane prone regions. *EXPANDED — SEE IBC*
4. **Exposure D.** Flat, unobstructed areas exposed to wind flowing over open water (excluding shorelines in hurricane prone regions) for a distance of at least 1 mi (1.61 km). Shorelines in Exposure D include inland waterways, the Great Lakes and coastal areas of California, Oregon, Washington and Alaska. This exposure shall apply only to those buildings and other structures exposed to the wind coming from over the water. Exposure D extends inland from the shoreline a distance of 1,500 ft (460 m) or 10 times the height of the building or structure, whichever is greater.

6.5.6.2 Exposure Category for Main Wind-Force Resisting Systems

6.5.6.2.1 Buildings and other structures

For each wind direction considered, wind loads for the design of the main wind-force resisting system determined from Fig. 6-3 shall be based on the exposure categories defined in Section 6.5.6.1.

6.5.6.2.2 Low-rise buildings

Wind loads for the design of the main wind-force resisting systems for low-rise buildings shall be determined using a velocity pressure q_h based on the exposure resulting in the highest wind loads for any wind direction at the site when external pressure coefficients GC_{pf} given in Fig. 6-4 are used.

6.5.6.3 Exposure Category for Components and Cladding

6.5.6.3.1 Buildings with mean roof height h less than or equal to 60 ft (18 m)

Components and cladding for buildings with a mean roof height h of 60 ft (18 m) or less shall be designed using a velocity pressure q_h based on the exposure resulting in the highest wind loads for any wind direction at the site.

6.5.6.3.2 Buildings with mean roof height h greater than 60 ft (18 m) and other structures

Components and cladding for buildings with a mean roof height h in excess of 60 ft (18 m) and for other structures shall be designed using the exposure

DELETED IN IBC 2000

resulting in the highest wind loads for any wind direction at the site.

6.5.6.4 Velocity Pressure Exposure Coefficient

Based on the exposure category determined in Section 6.5.6.1, a velocity pressure exposure coefficient K_z or K_h, as applicable, shall be determined from Table 6-5.

6.5.7 Topographic Effects

6.5.7.1 Wind Speed-Up over Hills, Ridges, and Escarpments

Wind speed-up effects at isolated hills, ridges, and escarpments constituting abrupt changes in the general topography, located in any exposure category, shall be included in the design when buildings and other site conditions and locations of structures meet all of the following conditions:

1. The hill, ridge, or escarpment is isolated and unobstructed upwind by other similar topographic features of comparable height for 100 times the height of the topographic feature (100 H) or 2 mi (3.22 km), whichever is less. This distance shall be measured horizontally from the point at which the height H of the hill, ridge, or escarpment is determined;
2. The hill, ridge, or escarpment protrudes above the height of upwind terrain features within a 2 mi (3.22 km) radius in any quadrant by a factor of two or more;
3. The structure is located as shown in Fig. 6-2 in the upper one-half of a hill or ridge or near the crest of an escarpment;
4. $H/L_h \geq 0.2$; and
5. H is greater than or equal to 15 ft (4.5 m) for Exposures C and D and 60 ft (18 m) for Exposures A and B.

6.5.7.2 Topographic Factor

The wind speed-up effect shall be included in the calculation of design wind loads by using the factor K_{zt}:

$$K_{zt} = (1 + K_1 K_2 K_3)^2 \qquad \text{(Eq. 6-1)}$$

where K_1, K_2, and K_3 are given in Fig. 6-2.

6.5.8 Gust Effect Factor

6.5.8.1 Rigid Structures

For rigid structures as defined in Section 6.2, the gust effect factor shall be taken as 0.85 or calculated by the formula:

$$G = 0.925 \left(\frac{1 + 1.7 g_Q I_{\bar{z}} Q}{1 + 1.7 g_v I_{\bar{z}}} \right) \qquad \text{(Eq. 6-2)}$$

(handwritten: 3.4)

$$I_{\bar{z}} = c(33/\bar{z})^{1/6} \qquad \text{(Eq. 6-3)}$$

(handwritten: 3.4; 0.2 FOR EXP C)

where $I_{\bar{z}}$ = the intensity of turbulence at height $\bar{z}$ and where $\bar{z}$ = the equivalent height of the structure defined as 0.6 h but not less than z_{min} for all building heights h. z_{min} and c are listed for each exposure in Table 6-4; g_Q and g_v shall be taken as 3.4. The background response Q is given by

(handwritten: $z_{min} = 15'$ FOR EXP C)

$$Q = \sqrt{\frac{1}{1 + 0.63 \left(\dfrac{B + h}{L_{\bar{z}}} \right)^{0.63}}} \qquad \text{(Eq. 6-4)}$$

where B, h are defined in Section 6.3; and $L_{\bar{z}}$ = the integral length scale of turbulence at the equivalent height given by

(handwritten: 500, EXP C; 0.2, EXP C; .6h)

$$L_{\bar{z}} = l(\bar{z}/33)^{\bar{\epsilon}} \qquad \text{(Eq. 6-5)}$$

in which l and $\bar{\epsilon}$ are constants listed in Table 6-4.

6.5.8.2 Flexible or Dynamically Sensitive Structures

For flexible or dynamically sensitive structures as defined in Section 6.2, the gust effect factor shall be calculated by:

$$G_f = 0.925 \left(\frac{1 + 1.7 I_{\bar{z}} \sqrt{g_Q^2 Q^2 + g_R^2 R^2}}{1 + 1.7 g_v I_{\bar{z}}} \right) \qquad \text{(Eq. 6-6)}$$

g_Q and g_v shall be taken as 3.4 and g_R is given by

$$g_R = \sqrt{2 \ln(3,600 n_1)} + \frac{0.577}{\sqrt{2 \ln(3,600 n_1)}} \qquad \text{(Eq. 6-7)}$$

R, the resonant response factor, is given by

$$R = \sqrt{\frac{1}{\beta} R_n R_h R_B (0.53 + 0.47 R_L)} \qquad \text{(Eq. 6-8)}$$

$$R_n = \frac{7.47 N_1}{(1 + 10.3 N_1)^{5/3}} \qquad \text{(Eq. 6-9)}$$

$$N_1 = \frac{n_1 L_{\bar{z}}}{\bar{V}_{\bar{z}}} \qquad \text{(Eq. 6-10)}$$

$$R_h = \frac{1}{4.6 n_1 h / V_{\bar{z}}} - \frac{1}{2(4.6 n_1 h / V_{\bar{z}})^2}\left(1 - e^{-2(4.6 n_1 h / V_{\bar{z}})}\right) \quad \text{for } \eta > 0$$

$$R_\ell = \frac{1}{\eta} - \frac{1}{2\eta^2}(1 - e^{-2\eta}) \quad \text{for } \eta > 0 \quad \text{(Eq. 6-11a)}$$

$$R_\ell = 1 \quad \text{for } \eta = 0 \quad \text{(Eq. 6-11b)}$$

where the subscript ℓ in Eq. 6-11 shall be taken as h, B, and L respectively.

n_1 = building natural frequency;
$R_\ell = R_h$ setting $\eta = 4.6 n_1 h / \bar{V}_{\bar{z}}$;
$R_\ell = R_B$ setting $\eta = 4.6 n_1 B / \bar{V}_{\bar{z}}$;
$R_\ell = R_L$ setting $\eta = 15.4 n_1 L / \bar{V}_{\bar{z}}$;
β = damping ratio, percent of critical h, B, L are defined in Section 6.3; and
$\bar{V}_{\bar{z}}$ = mean hourly wind speed (ft/s) at height $\bar{z}$ determined from Eq. 6-12.

$$\bar{V}_{\bar{z}} = \bar{b}\left(\frac{\bar{z}}{33}\right)^{\bar{\alpha}} V\left(\frac{88}{60}\right) \quad \text{(Eq. 6-12)}$$

where $\bar{b}$ and $\bar{\alpha}$ are constants listed in Table 6-4 and V is the basic wind speed in mph.

6.5.8.3 Rational Analysis

In lieu of the procedure defined in Sections 6.5.8.1 and 6.5.8.2, determination of the gust effect factor by any rational analysis defined in the recognized literature is permitted.

6.5.8.4 Limitations

Where combined gust effect factors and pressure coefficients (GC_p, GC_{pi}, and GC_{pf}) are given in figures and tables, the gust effect factor shall not be determined separately.

6.5.9 Enclosure Classifications

6.5.9.1 General

For the purpose of determining internal pressure coefficients, all buildings shall be classified as enclosed, partially enclosed, or open as defined in Section 6.2.

6.5.9.2 Openings

A determination shall be made of the amount of openings in the building envelope in order to determine the enclosure classification as defined in Section 6.5.9.1.

6.5.9.3 Wind Borne Debris

Glazing in the lower 60 ft (18.3 m) of Category II, III, or IV buildings sited in wind borne debris regions shall be impact resistant glazing or protected with an impact resistant covering, or such glazing that receives positive external pressure shall be assumed to be openings.

6.5.9.4 Multiple Classifications

If a building by definition complies with both the "open" and "partially enclosed" definitions, it shall be classified as an "open" building. A building that does not comply with either the "open" or "partially enclosed" definitions shall be classified as an "enclosed" building.

6.5.10 Velocity Pressure

Velocity pressure, q_z, evaluated at height z shall be calculated by the following equation:

$$q_z = 0.00256 K_z K_{zt} K_d V^2 I \ (\text{lb/ft}^2) \quad \text{(Eq. 6-13)}$$

[in SI: $q_z = 0.613 K_z K_{zt} K_d V^2 I \ (\text{N/m}^2)$]

where K_d is the wind directionality factor defined in Section 6.5.4.4, K_z is the velocity pressure exposure coefficient defined in Section 6.5.6.4, K_{zt} is the topographic factor defined in Section 6.5.7.2, and q_h is the velocity pressure calculated using Eq. 6-13 at mean roof height h.

The numerical coefficient 0.00256 (0.613 in SI) shall be used except where sufficient climatic data are available to justify the selection of a different value of this factor for a design application.

6.5.11 Pressure and Force Coefficients

6.5.11.1 Internal Pressure Coefficients

Internal pressure coefficients, GC_{pi}, shall be determined from Table 6-7 based on building enclosure classifications determined from Section 6.5.9.

6.5.11.1.1 Reduction factor for large volume buildings, R_i

For a partially enclosed building containing a single, unpartitioned large volume, the internal pressure coefficient, GC_{pi}, shall be multiplied by the following reduction factor, R_i:

$$R_i = 1.0 \text{ or } R_i = 0.5\left(1 + \frac{1}{\sqrt{1 + \frac{V_i}{22,800 A_{og}}}}\right) \leq 1.0 \quad \text{(Eq. 6-14)}$$

where

A_{og} = total area of openings in the building envelope (walls and roof, in ft^2);

V_i = unpartitioned internal volume, in ft^3.

6.5.11.2 External Pressure Coefficients

6.5.11.2.1 Main wind force resisting systems

External pressure coefficients for main wind force resisting systems C_p are given in Fig. 6-3 and Table 6-8. Combined gust effect factor and external pressure coefficients, GC_{pf}, are given in Fig. 6-4 for low-rise buildings. The pressure coefficient values and gust effect factor in Fig. 6-4 shall not be separated.

6.5.11.2.2 Components and cladding

Combined gust effect factor and external pressure coefficients for components and cladding GC_p are given in Figs. 6-5 through 6-8. The pressure coefficient values and gust effect factor shall not be separated.

6.5.11.3 Force Coefficients

Force coefficients C_f are given in Tables 6-9 through 6-13.

6.5.11.4 Roof Overhangs

6.5.11.4.1 Main wind-force resisting system

Roof overhangs shall be designed for a positive pressure on the bottom surface of windward roof overhangs corresponding to $C_p = 0.8$ in combination with the pressures determined from using Figs. 6-3 and 6-4.

6.5.11.4.2 Components and cladding

For all buildings, roof overhangs shall be designed for pressures determined from pressure coefficients given in Fig. 6-5B.

6.5.12 Design Wind Loads on Enclosed and Partially Enclosed Buildings

6.5.12.1 General

6.5.12.1.1 Sign convention

Positive pressure acts toward the surface and negative pressure acts away from the surface.

6.5.12.1.2 Critical load condition

Values of external and internal pressures shall be combined algebraically to determine the most critical load.

6.5.12.1.3 Tributary areas greater than 700 ft^2 (65 m^2)

Component and cladding elements with tributary areas greater than 700 ft^2 (65 m^2) shall be permitted to be designed using the provisions for main wind force resisting systems.

6.5.12.2 Main Force Resisting Systems

6.5.12.2.1 Rigid buildings of all heights

Design wind pressures for the main wind force resisting system of buildings of all heights shall be determined by the following equation:

$$p = qGC_p - q_i(GC_{pi}) \text{ (lb/ft}^2) \text{ (N/m}^2) \quad \text{(Eq. 6-15)}$$

where

$q = q_z$ for windward walls evaluated at height z above the ground;

$q = q_h$ for leeward walls, side walls, and roofs, evaluated at height h;

$q_i = q_h$ for windward walls, side walls, leeward walls, and roofs of enclosed buildings and for negative internal pressure evaluation in partially enclosed buildings;

$q_i = q_z$ for positive internal pressure evaluation in partially enclosed buildings where height z is defined as the level of the highest opening in the building that could affect the positive internal pressure. For buildings sited in wind borne debris regions, glazing in the lower 60 ft (18.3 m) that is not impact resistant or protected with an impact resistant covering, the glazing shall be treated as an opening in accordance with Section 6.5.9.3. For positive internal pressure evaluation, q_i may conservatively be evaluated at height h ($q_i = q_h$);

G = gust effect factor from 6.5.8;

C_p = external pressure coefficient from Fig. 6-3 or Table 6-8;

(GC_{pi}) = internal pressure coefficient from Table 6-7.

q and q_i shall be evaluated using exposure defined in Section 6.5.6.2.1.

6.5.12.2.2 Low-rise buildings

Alternatively, design wind pressures for the main wind force resisting system of low-rise buildings shall be determined by the following equation:

$$p = q_h[(GC_{pf}) - (GC_{pi})] \ (\text{lb/ft}^2) \ (\text{N/m}^2) \qquad \text{(Eq. 6-16)}$$

where

q_h = velocity pressure evaluated at mean roof height h using exposure defined in Section 6.5.6.2.2;

(GC_{pf}) = external pressure coefficient from Fig. 6-4; and

(GC_{pi}) = internal pressure coefficient from Table 6-7.

6.5.12.2.3 Flexible buildings

Design wind pressures for the main wind force resisting system of flexible buildings shall be determined from the following equation:

$$p = qG_fC_p - q_i(GC_{pi}) \ (\text{lb/ft}^2) \ (\text{N/m}^2) \qquad \text{(Eq. 6-17)}$$

where q, q_i, C_p and (GC_{pi}) are as defined in Section 6.5.12.2.1, and G_f = gust effect factor defined in Section 6.5.8.2.

6.5.12.3 Full and Partial Loading

The main wind-force resisting system of buildings with mean roof height h greater than 60 ft (18 m) shall be designed for the torsional moments resulting from design wind loads calculated from Section 6.5.12 acting in the combinations indicated in Fig. 6-9.

6.5.12.4 Components and Cladding

6.5.12.4.1 Low-rise buildings and buildings with $h \leq$ 60 ft (18.3 m)

Design wind pressures on component and cladding elements of low-rise buildings and buildings with $h \leq 60$ ft (18.3 m) shall be determined from the following equation:

$$p = q_h[(GC_p) - (GC_{pi})] \ (\text{lb/ft}^2) \ (\text{N/m}^2) \qquad \text{(Eq. 6-18)}$$

where

q_h = velocity pressure evaluated at mean roof height h using exposure defined in Section 6.5.6.3.1;

(GC_p) = external pressure coefficients given in Figs. 6-5 through 6-7; and

(GC_{pi}) = internal pressure coefficient given in Table 6-7.

6.5.12.4.2 Buildings with $h > 60$ ft (18.3 m)

Design wind pressures on components and cladding for all buildings with $h > 60$ ft (18.3 m) shall be determined from the following equation:

$$p = q(GC_p) - q_i(GC_{pi}) \ (\text{lb/ft}^2) \ (\text{N/m}^2) \qquad \text{(Eq. 6-19)}$$

where

$q = q_z$ for windward walls calculated at height z above the ground;

$q = q_h$ for leeward walls, side walls, and roofs, evaluated at height h;

$q_i = q_h$ for windward walls, side walls, leeward walls, and roofs of enclosed buildings and for negative internal pressure evaluation in partially enclosed buildings; and

$q_i = q_z$ for positive internal pressure evaluation in partially enclosed buildings where height z is defined as the level of the highest opening in the building that could affect the positive internal pressure. For buildings sited in wind borne debris regions with glazing in the lower 60 ft (18.3 m) that is not impact resistant or protected with an impact resistant covering, the glazing shall be treated as an opening in accordance with Section 6.5.9.3. For positive internal pressure evaluation, q_i may conservatively be evaluated at height h ($q_i = q_h$);

(GC_p) = external pressure coefficient from Fig. 6-8; and

(GC_{pi}) = internal pressure coefficient given in Table 6-7.

q and q_i shall be evaluated using exposure defined in Section 6.5.6.3.2.

6.5.12.4.3 Alternative design wind pressures for components and cladding in buildings with 60 ft (18.3 m) < h < 90 ft (27.4 m)

Alternative to the requirements of Section 6.5.12.4.2, the design of components and cladding for buildings with a mean roof height greater than 60 ft (18.3 m) and less than 90 ft (27.4 m) values from Figs. 6-5–6-7 shall be used only if the height to width ratio is one or less (except as permitted by Note 6 of Fig. 6-8) and Eq. 6-18 is used.

6.5.13 Design Wind Loads on Open Buildings and Other Structures

The design wind force for open buildings and other structures shall be determined by the following formula:

$$F = q_z G C_f A_f \text{ (lb) (N)} \qquad \text{(Eq. 6-20)}$$

where

q_z = velocity pressure evaluated at height z of the centroid of area A_f using exposure defined in Section 6.5.6.3.2;

G = gust effect factor from Section 6.5.8;

C_f = net force coefficients from Tables 6-9 through 6-12; and

A_f = projected area normal to the wind except where C_f is specified for the actual surface area, ft^2 (m^2).

6.6 METHOD 3—WIND TUNNEL PROCEDURE

6.6.1 Scope

Wind tunnel tests shall be used where required by Section 6.5.2. Wind tunnel testing shall be permitted in lieu of Methods 1 and 2 for any building or structure.

6.6.2 Test Conditions

Wind tunnel tests, or similar tests employing fluids other than air, used for the determination of design wind loads for any building or other structure, shall be conducted in accordance with this section. Tests for the determination of mean and fluctuating forces and pressures shall meet all of the following conditions:

1. the natural atmospheric boundary layer has been modeled to account for the variation of wind speed with height;

2. the relevant macro (integral) length and micro length scales of the longitudinal component of atmospheric turbulence are modeled to approximately the same scale as that used to model the building or structure;

3. the modeled building or other structure and surrounding structures and topography are geometrically similar to their full-scale counterparts, except that, for low-rise buildings meeting the requirements of Section 6.5.1, tests shall be permitted for the modeled building in a single exposure site as defined in Section 6.5.6.1;

4. the projected area of the modeled building or other structure and surroundings is less than 8% of the test section cross-sectional area unless correction is made for blockage;

5. the longitudinal pressure gradient in the wind tunnel test section is accounted for;

6. Reynolds number effects on pressures and forces are minimized; and

7. response characteristics of the wind tunnel instrumentation are consistent with the required measurements.

6.6.3 Dynamic Response

Tests for the purpose of determining the dynamic response of a building or other structure shall be in accordance with Section 6.6.2. The structural model and associated analysis shall account for mass distribution, stiffness, and damping.

6.6.4 Limitations

6.6.4.1 Limitations on Wind Speeds

Variation of basic wind speeds with direction shall not be permitted unless the analysis for wind speeds conforms to the requirements of Section 6.5.4.2.

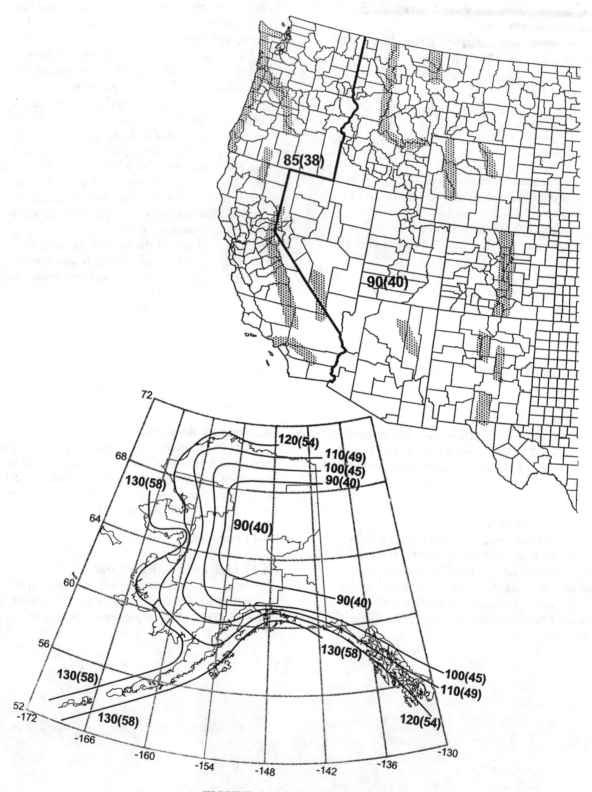

FIGURE 6-1. Basic Wind Speed

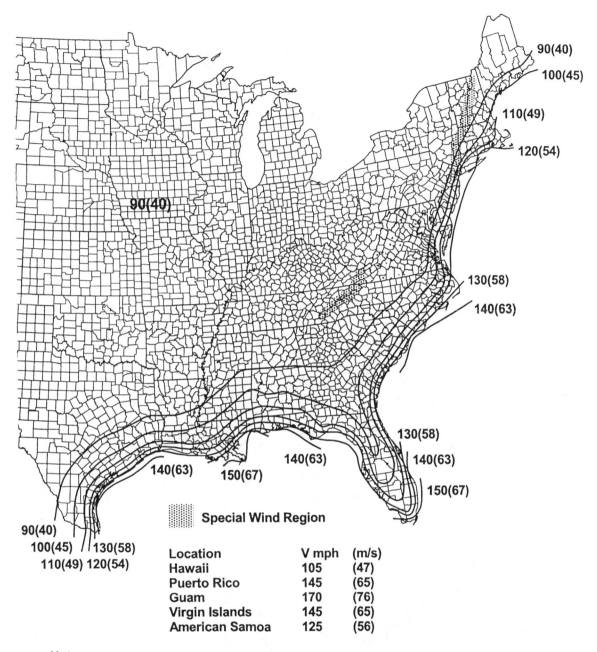

Location	V mph	(m/s)
Hawaii	105	(47)
Puerto Rico	145	(65)
Guam	170	(76)
Virgin Islands	145	(65)
American Samoa	125	(56)

Notes:
1. Values are nominal design 3-second gust wind speeds in miles per hour (m/s) at 33 ft (10 m) above ground for Exposure C category.
2. Linear interpolation between wind contours is permitted.
3. Islands and coastal areas outside the last contour shall use the last wind speed contour of the coastal area.
4. Mountainous terrain, gorges, ocean promontories, and special wind regions shall be examined for unusual wind conditions.

FIGURE 6-1. (*Continued*)

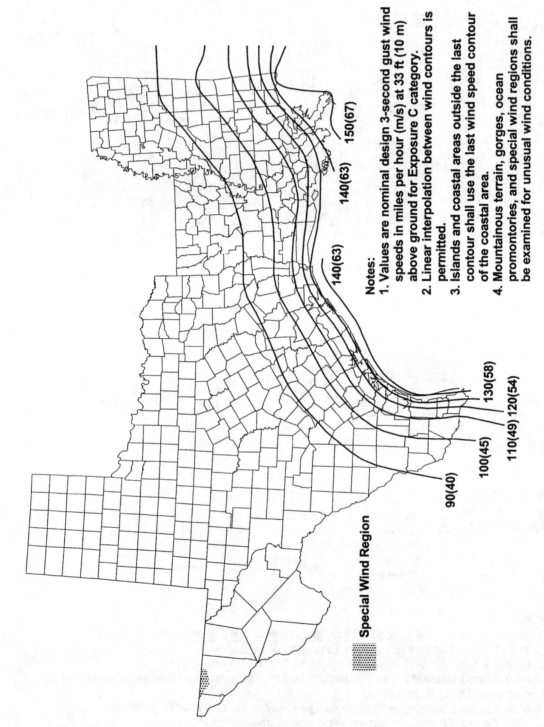

Notes:
1. Values are nominal design 3-second gust wind speeds in miles per hour (m/s) at 33 ft (10 m) above ground for Exposure C category.
2. Linear interpolation between wind contours is permitted.
3. Islands and coastal areas outside the last contour shall use the last wind speed contour of the coastal area.
4. Mountainous terrain, gorges, ocean promontories, and special wind regions shall be examined for unusual wind conditions.

150(67)

140(63)

140(63)

130(58)

120(54)

110(49)

100(45)

90(40)

Special Wind Region

FIGURE 6-1a. Basic Wind Speed—Western Gulf of Mexico Hurricane Coastline

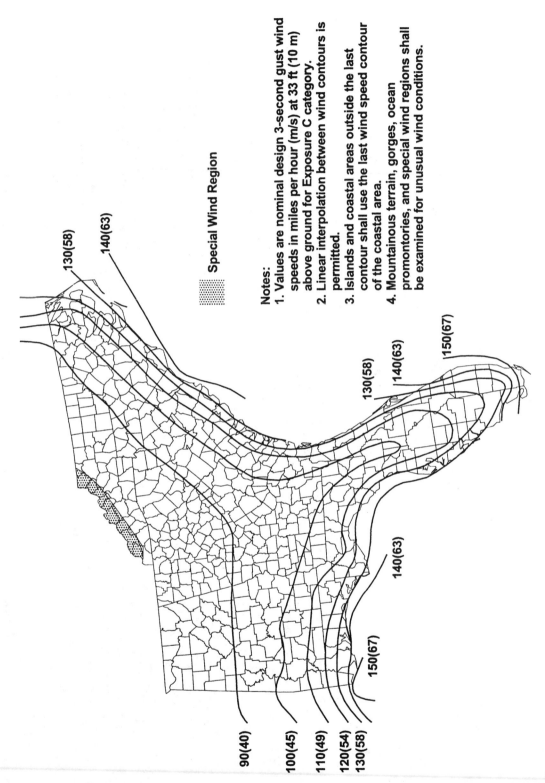

Special Wind Region

Notes:
1. Values are nominal design 3-second gust wind speeds in miles per hour (m/s) at 33 ft (10 m) above ground for Exposure C category.
2. Linear interpolation between wind contours is permitted.
3. Islands and coastal areas outside the last contour shall use the last wind speed contour of the coastal area.
4. Mountainous terrain, gorges, ocean promontories, and special wind regions shall be examined for unusual wind conditions.

130(58)
140(63)

130(58)
140(63)
150(67)

140(63)

150(67)

90(40)
100(45)
110(49)
120(54)
130(58)

FIGURE 6-1b. Basic Wind Speed—Eastern Gulf of Mexico and Southeastern U.S. Hurricane Coastline

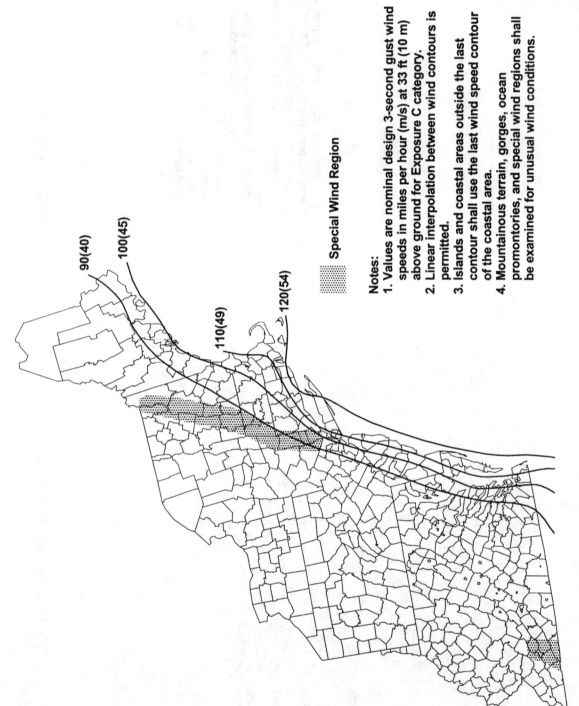

90(40)

100(45)

110(49)

120(54)

Special Wind Region

Notes:
1. Values are nominal design 3-second gust wind speeds in miles per hour (m/s) at 33 ft (10 m) above ground for Exposure C category.
2. Linear interpolation between wind contours is permitted.
3. Islands and coastal areas outside the last contour shall use the last wind speed contour of the coastal area.
4. Mountainous terrain, gorges, ocean promontories, and special wind regions shall be examined for unusual wind conditions.

FIGURE 6-1c. Basic Wind Speed—Mid and Northern Atlantic Hurricane Coastline

Topographic Factor, K_{zt}

Figure 6-2

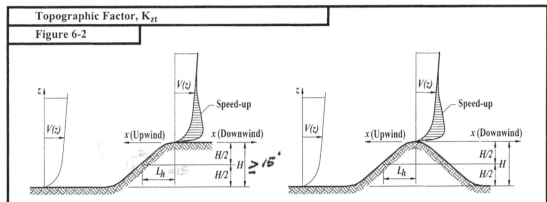

ESCARPMENT = *A CLIFF OR* 2-D RIDGE OR 3-D AXISYMMETRICAL HILL

STEEP SLOPE

Topographic Multipliers for Exposure C											
	K_1 Multiplier				K_2 Multiplier			K_3 Multiplier			
H/L_h	2-D Ridge	2-D Escarp.	3-D Axisym. Hill	x/L_h	2-D Escarp.	All Other Cases	z/L_h	2-D Ridge	2-D Escarp.	3-D Axisym. Hill
0.20	0.29	0.17	0.21	0.00	1.00	1.00	0.00	1.00	1.00	1.00
0.25	0.36	0.21	0.26	0.50	0.88	0.67	0.10	0.74	0.78	0.67
0.30	0.43	0.26	0.32	1.00	0.75	0.33	0.20	0.55	0.61	0.45
0.35	0.51	0.30	0.37	1.50	0.63	0.00	0.30	0.41	0.47	0.30
0.40	0.58	0.34	0.42	2.00	0.50	0.00	0.40	0.30	0.37	0.20
0.45	0.65	0.38	0.47	2.50	0.38	0.00	0.50	0.22	0.29	0.14
0.50	0.72	0.43	0.53	3.00	0.25	0.00	0.60	0.17	0.22	0.09
				3.50	0.13	0.00	0.70	0.12	0.17	0.06
				4.00	0.00	0.00	0.80	0.09	0.14	0.04
							0.90	0.07	0.11	0.03
							1.00	0.05	0.08	0.02
							1.50	0.01	0.02	0.00
							2.00	0.00	0.00	0.00

Notes:

1. For values of H/L_h, x/L_h and z/L_h other than those shown, linear interpolation is permitted.
2. For $H/L_h > 0.5$, assume $H/L_h = 0.5$ for evaluating K_1 and substitute 2H for L_h for evaluating K_2 and K_3.
3. Multipliers are based on the assumption that wind approaches the hill or escarpment along the direction of maximum slope.
4. Notation:
   H: Height of hill or escarpment relative to the upwind terrain, in feet (meters).
   L_h: Distance upwind of crest to where the difference in ground elevation is half the height of hill or escarpment, in feet (meters).
   K_1: Factor to account for shape of topographic feature and maximum speed-up effect.
   K_2: Factor to account for reduction in speed-up with distance upwind or downwind of crest.
   K_3: Factor to account for reduction in speed-up with height above local terrain.
   x: Distance (upwind or downwind) from the crest to the building site, in feet (meters).
   z: Height above local ground level, in feet (meters).
   μ: Horizontal attenuation factor.
   γ: Height attenuation factor.

| Topographic Factor, K_{zt} |
| Figure 6-2 (con't) |

Equations:

$$K_{zt} = (1 + K_1 K_2 K_3)^2$$

K_1 determined from table below

$$K_2 = (1 - \frac{|x|}{\mu L_h})$$

$$K_3 = e^{-\gamma z/L_h}$$

Parameters for Speed-Up Over Hills and Escarpments						
Hill Shape	**$K_1/(H/L_h)$**			γ	**μ**	
	Exposure				**Upwind of Crest**	**Downwind of Crest**
	B	**C**	**D**			
2-dimensional ridges (or valleys with negative H in $K_1/(H/L_h)$	1.30	1.45	1.55	3	1.5	1.5
2-dimensional escarpments	0.75	0.85	0.95	2.5	1.5	4
3-dimensional axisym. hill	0.95	1.05	1.15	4	1.5	1.5

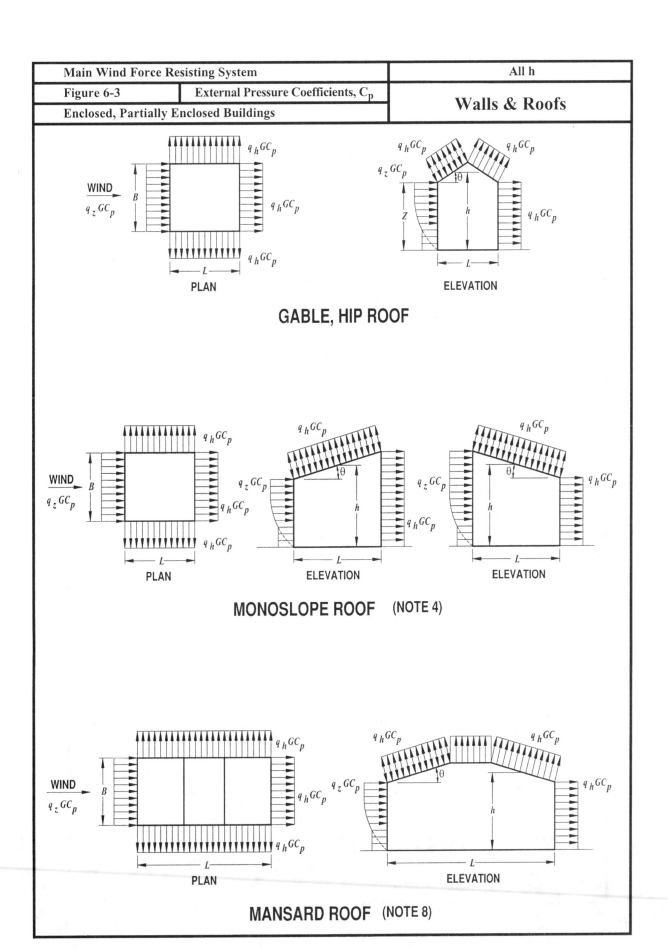

GABLE, HIP ROOF

MONOSLOPE ROOF   (NOTE 4)

MANSARD ROOF   (NOTE 8)

Main Wind Force Resisting System		All h
Figure 6-3 (con't)	**External Pressure Coefficients, C_p**	**Walls & Roofs**
Enclosed, Partially Enclosed Buildings		

Wall Pressure Coefficients, C_p			
Surface	L/B	C_p	Use With
Windward Wall	All values	0.8	q_z
Leeward Wall	0-1	-0.5	q_h
	2	-0.3	
	≥4	-0.2	
Side Wall	All values	-0.7	q_h

Roof Pressure Coefficients, C_p, for use with q_h

Wind Direction	h/L	Windward Angle, θ (degrees)								Leeward Angle, θ (degrees)		
		10	15	20	25	30	35	45	≥60#	10	15	≥20
Normal to ridge for θ ≥ 10°	≤0.25	-0.7 / 0.0*	-0.5 / 0.2	-0.3 / 0.3	-0.2 / 0.3	-0.2 / 0.4	0.0* / 0.4	0.4	0.01 θ	-0.3	-0.5	-0.6
	0.5	-0.9	-0.7	0.4 / 0.0*	-0.3 / 0.2	-0.2 / 0.2	-0.2 / 0.3	0.0* / 0.4	0.01 θ	-0.5	-0.5	-0.6
	≥1.0	-1.3**	-1.0	-0.7	-0.5 / 0.0*	-0.3 / 0.2	-0.2 / 0.2	0.0* / 0.3	0.01 θ	-0.7	-0.6	-0.6

Wind Direction		Horiz distance from windward edge	C_p	
Normal to ridge for θ < 10 and Parallel to ridge for all θ	≤ 0.5	0 to h/2	-0.9	*Value is provided for interpolation purposes.
		h/2 to h	-0.9	**Value can be reduced linearly with area over which it is applicable as follows
		h to 2 h	-0.5	
		> 2h	-0.3	
	≥ 1.0	0 to h/2	-1.3**	
		> h/2	-0.7	

Area (sq ft)	Reduction Factor
≤ 100 (9.29 sq m)	1.0
200 (23.23 sq m)	0.9
≥ 1000 (92.9 sq m)	0.8

Notes:
1. Plus and minus signs signify pressures acting toward and away from the surfaces, respectively
2. Linear interpolation is permitted for values of *L/B*, *h/L* and θ other than shown. Interpolation shall only be carried out between values of the same sign. Where no value of the same sign is given, assume 0.0 for interpolation purposes.
3. Where two values of C_p are listed, this indicates that the windward roof slope is subjected to either positive or negative pressures and the roof structure shall be designed for both conditions. Interpolation for intermediate ratios of h/L in this case shall only be carried out between C_p values of like sign.
4. For monoslope roofs, entire roof surface is either a windward or leeward surface.
5. For flexible buildings use appropriate G_f as determined by rational analysis.
6. Refer to Table 6-8 for arched roofs.
7. Notation:
   B: Horizontal dimension of building, in feet (meter), measured normal to wind direction.
   L: Horizontal dimension of building, in feet (meter), measured parallel to wind direction.
   h: Mean roof height in feet (meters), except that eave height shall be used for θ ≤ 10 degrees.
   z: Height above ground, in feet (meters).
   G: Gust effect factor.
   q_z, q_h: Velocity pressure, in pounds per square foot (N/m²), evaluated at respective height.
   θ: Angle of plane of roof from horizontal, in degrees.
8. For mansard roofs, the top horizontal surface and leeward inclined surface shall be treated as leeward surfaces from the table.

#For roof slopes greater than 80°, use $C_p = 0.8$

Main Wind Force Resisting System	h ≤ 60 ft.
Figure 6-4 **External Pressure Coefficients, GC$_{pf}$**	**Walls & Gable Roof**
Enclosed, Partially Enclosed Buildings	

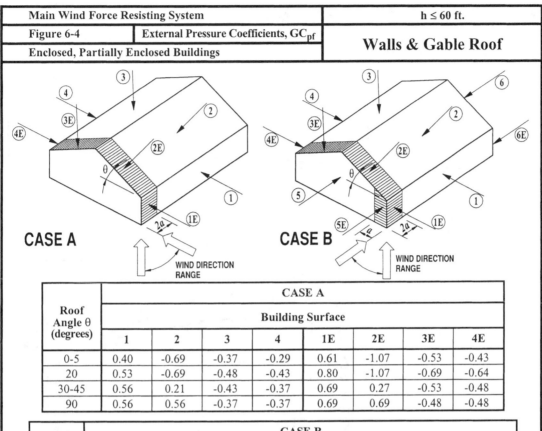

Roof Angle θ (degrees)	CASE A							
	Building Surface							
	1	**2**	**3**	**4**	**1E**	**2E**	**3E**	**4E**
0-5	0.40	-0.69	-0.37	-0.29	0.61	-1.07	-0.53	-0.43
20	0.53	-0.69	-0.48	-0.43	0.80	-1.07	-0.69	-0.64
30-45	0.56	0.21	-0.43	-0.37	0.69	0.27	-0.53	-0.48
90	0.56	0.56	-0.37	-0.37	0.69	0.69	-0.48	-0.48

Roof Angle θ (degrees)	CASE B											
	Building Surface											
	1	**2**	**3**	**4**	**5**	**6**	**1E**	**2E**	**3E**	**4E**	**5E**	**6E**
0-90	-0.45	-0.69	-0.37	-0.45	0.40	-0.29	-0.48	-1.07	-0.53	-0.48	0.61	-0.43

Notes:
1. Case A and Case B are required as two separate loading conditions to generate the wind actions, including torsion, to be resisted by the main wind-force resisting system.
2. To obtain the critical wind actions, the building shall be rotated in 90° degree increments so that each corner in turn becomes the windward corner while the loading patterns in the sketches remain fixed. For the design of structural systems providing lateral resistance in the direction parallel to the ridge line, Case A shall be based on θ = 0°.
3. Plus and minus signs signify pressures acting toward and away from the surfaces, respectively.
4. For Case A loading the following restrictions apply:
   a. The roof pressure coefficient GC_{pf}, when negative in Zone 2, shall be applied in Zone 2 for a distance from the edge of roof equal to 0.5 times the horizontal dimensions of the building measured perpendicular to the eave line or 2.5h, whichever is less; the remainder of Zone 2 extending to the ridge line shall use the pressure coefficient GC_{pf} for Zone 3.
   b. Except for moment-resisting frames, the total horizontal shear shall not be less than that determined by neglecting wind forces on roof surfaces.
5. Combinations of external and internal pressures (see Table 6-7) shall be evaluated as required to obtain the most severe loadings.
6. For values of θ other than those shown, linear interpolation is permitted.
7. Notation:
   a: 10 percent of least horizontal dimension or 0.4h, whichever is smaller, but not less than either 4% of least horizontal dimension or 3 ft (1 m).
   h: Mean roof height, in feet (meters), except that eave height shall be used for θ ≤ 10°.
   θ: Angle of plane of roof from horizontal, in degrees.

Components and Cladding		h ≤ 60 ft.
Figure 6-5A	External Pressure Coefficients, GC_p	**Walls**
Enclosed, Partially Enclosed Buildings		

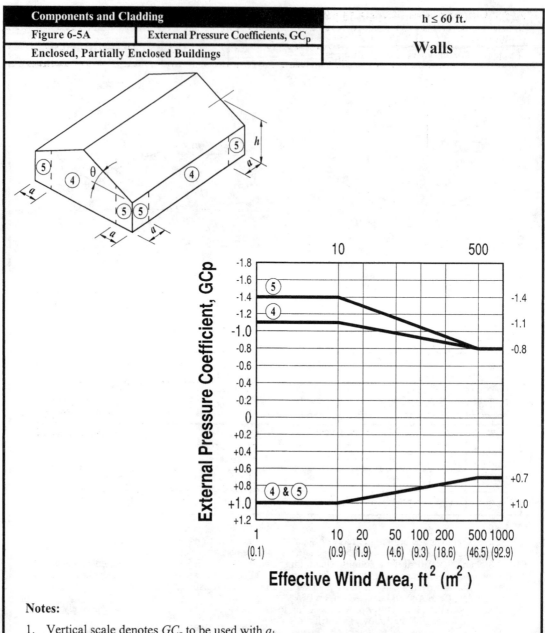

Notes:

1. Vertical scale denotes GC_p to be used with q_h.
2. Horizontal scale denotes effective wind area, in square feet (square meters).
3. Plus and minus signs signify pressures acting toward and away from the surfaces, respectively
4. Each component shall be designed for maximum positive and negative pressures.
5. Values of GC_p for walls shall be reduced by 10% when $\theta \leq 10°$.
6. Notation:
   - a: 10 percent of least horizontal dimension or 0.4h, whichever is smaller, but not less than either 4% of least horizontal dimension or 3 ft (1 m).
   - h: Mean roof height, in feet (meters), except that eave height shall be used for $\theta \leq 10°$.
   - θ: Angle of plane of roof from horizontal, in degrees.

Components and Cladding	h ≤ 60 ft.
Figure 6-5B	**External Pressure Coefficients, GC_p**	**Gable Roofs θ ≤ 10°**
Enclosed, Partially Enclosed Buildings	

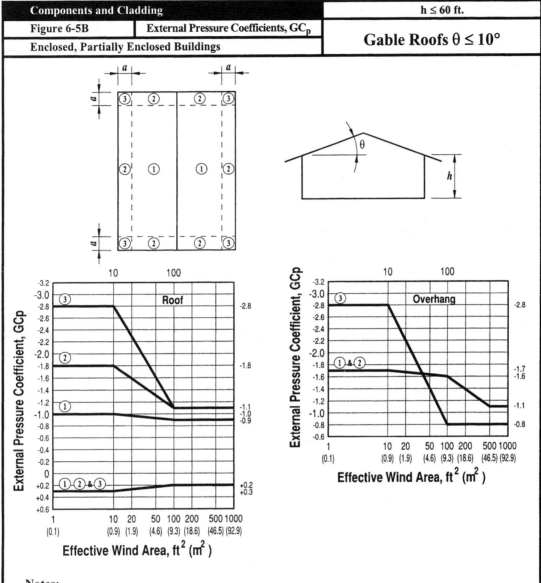

Notes:

1. Vertical scale denotes GC_p to be used with q_h.
2. Horizontal scale denotes effective wind area, in square feet (square meters).
3. Plus and minus signs signify pressures acting toward and away from the surfaces, respectively
4. Each component shall be designed for maximum positive and negative pressures.
5. If a parapet equal to or higher than 3 ft (1m) is provided around the perimeter of the roof with θ ≤ 10°, Zone 3 shall be treated as Zone 2.
6. Values of GC_p for roof overhangs include pressure contributions from both upper and lower surfaces.
7. Notation:
   a: 10 percent of least horizontal dimension or 0.4h, whichever is smaller, but not less than either 4% of least horizontal dimension or 3 ft (1 m).
   h: Eave height shall be used for θ ≤ 10°.
   θ: Angle of plane of roof from horizontal, in degrees.

Components and Cladding	h ≤ 60 ft.
Figure 6-5B (con't) External Pressure Coefficients GC$_p$	
Enclosed, Partially Enclosed Buildings	**Gable/Hip Roofs 10°< θ ≤ 30°**

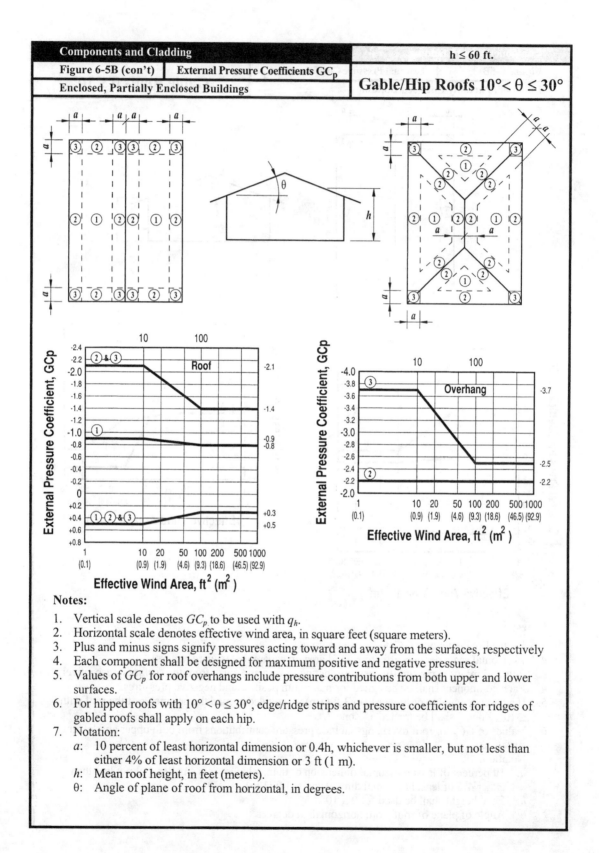

Notes:

1. Vertical scale denotes GC_p to be used with q_h.
2. Horizontal scale denotes effective wind area, in square feet (square meters).
3. Plus and minus signs signify pressures acting toward and away from the surfaces, respectively
4. Each component shall be designed for maximum positive and negative pressures.
5. Values of GC_p for roof overhangs include pressure contributions from both upper and lower surfaces.
6. For hipped roofs with $10° < θ ≤ 30°$, edge/ridge strips and pressure coefficients for ridges of gabled roofs shall apply on each hip.
7. Notation:
      a: 10 percent of least horizontal dimension or 0.4h, whichever is smaller, but not less than either 4% of least horizontal dimension or 3 ft (1 m).
      h: Mean roof height, in feet (meters).
      $θ$: Angle of plane of roof from horizontal, in degrees.

Components and Cladding		h ≤ 60 ft.
Figure 6-5B (con't)	**External Pressure Coefficients, GC$_p$**	**Gable Roofs 30°< θ ≤ 45°**
Enclosed, Partially Enclosed Buildings		

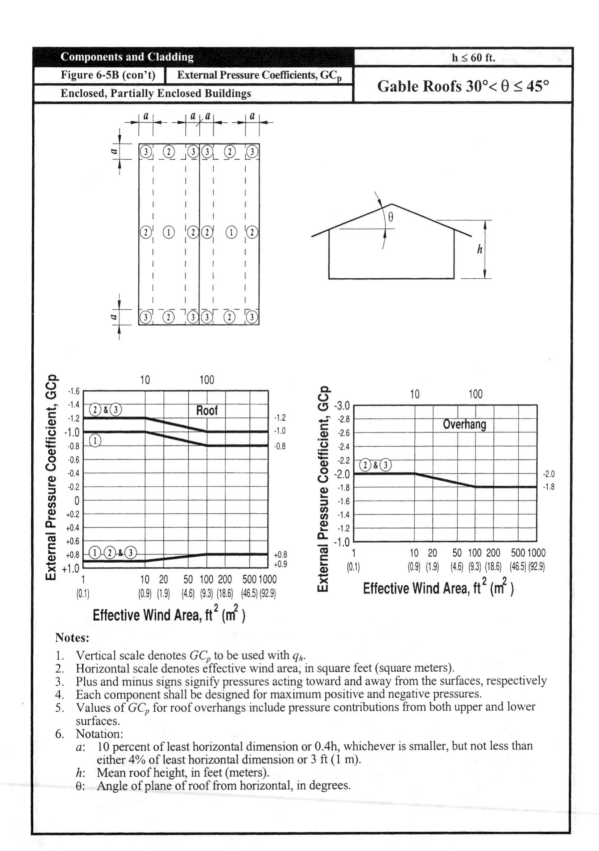

Notes:

1. Vertical scale denotes GC_p to be used with q_h.
2. Horizontal scale denotes effective wind area, in square feet (square meters).
3. Plus and minus signs signify pressures acting toward and away from the surfaces, respectively
4. Each component shall be designed for maximum positive and negative pressures.
5. Values of GC_p for roof overhangs include pressure contributions from both upper and lower surfaces.
6. Notation:
   a: 10 percent of least horizontal dimension or 0.4h, whichever is smaller, but not less than either 4% of least horizontal dimension or 3 ft (1 m).
   h: Mean roof height, in feet (meters).
   $θ$: Angle of plane of roof from horizontal, in degrees.

Components and Cladding		h ≤ 60 ft.
Figure 6-5C	External Pressure Coefficients, GC$_p$	**Stepped Roofs**
Enclosed, Partially Enclosed Buildings		

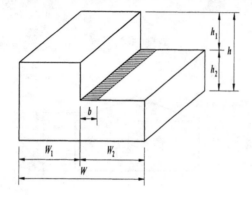

$h_1 \geq 10$ ft. (3 m)

$b = 1.5\,h_1$

$b < 100$ ft. (30.5 m)

$\dfrac{h_i}{h} = 0.3$ to 0.7

$\dfrac{W_i}{W} = 0.25$ to 0.75

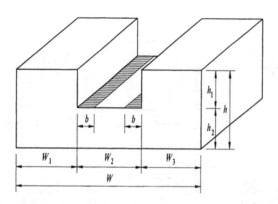

Notes:

1. On the lower level of flat, stepped roofs shown in Fig. 6-5C, the zone designations and pressure coefficients shown in Fig. 6-5B ($\theta \leq 10°$) shall apply, except that at the roof-upper wall intersection(s), Zone 3 shall be treated as Zone 2 and Zone 2 shall be treated as Zone 1. Positive values of GC_p equal to those for walls in Fig. 6-5A shall apply on the cross-hatched areas shown in Fig. 6-5C.

2. Notation:

   b: $1.5h_1$ in Fig. 6-5C, but not greater than 100 ft (30.5 m).

   h: Mean roof height, in feet (meters).

   h_i: h_1 or h_2 in Fig. 6-5C; $h = h_1 + h_2$; $h_1 \geq 10$ ft (3.1 m); $h_i/h = 0.3$ to 0.7.

   W: Building width in Fig. 6-5C.

   W_i: W_1 or W_2 or W_3 in Fig. 6-5C. $W = W_1 + W_2$ or $W_1 + W_2 + W_3$; $W_i/W = 0.25$ to 0.75.

   θ: Angle of plane of roof from horizontal, in degrees.

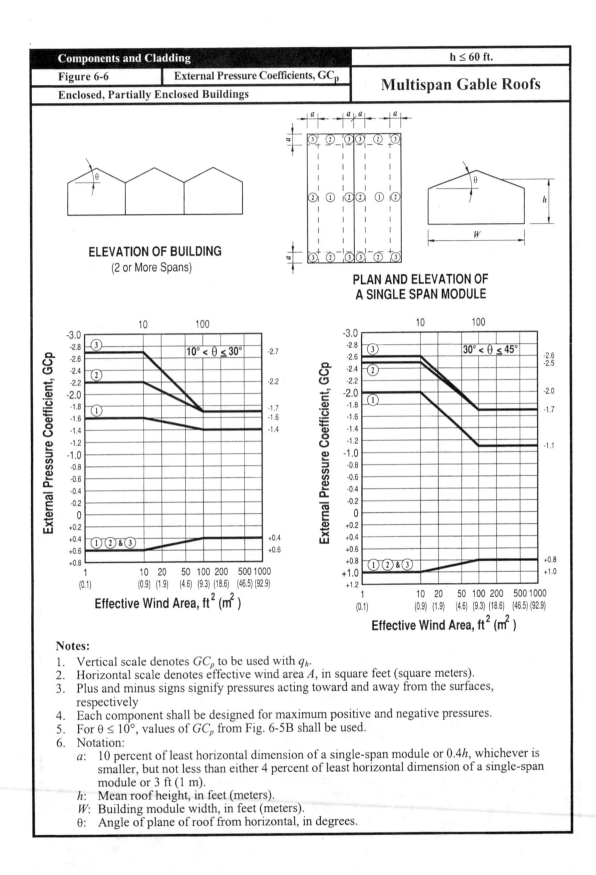

Components and Cladding		h ≤ 60 ft.
Figure 6-6	**External Pressure Coefficients, GC_p**	**Multispan Gable Roofs**
Enclosed, Partially Enclosed Buildings		

ELEVATION OF BUILDING
(2 or More Spans)

**PLAN AND ELEVATION OF
A SINGLE SPAN MODULE**

Notes:

1. Vertical scale denotes GC_p to be used with q_h.
2. Horizontal scale denotes effective wind area A, in square feet (square meters).
3. Plus and minus signs signify pressures acting toward and away from the surfaces, respectively
4. Each component shall be designed for maximum positive and negative pressures.
5. For $\theta \le 10°$, values of GC_p from Fig. 6-5B shall be used.
6. Notation:
   - a: 10 percent of least horizontal dimension of a single-span module or $0.4h$, whichever is smaller, but not less than either 4 percent of least horizontal dimension of a single-span module or 3 ft (1 m).
   - h: Mean roof height, in feet (meters).
   - W: Building module width, in feet (meters).
   - θ: Angle of plane of roof from horizontal, in degrees.

Components and Cladding		$h \le 60$ ft.
Figure 6-7A	External Pressure Coefficients, GC_p	**Monoslope Roofs**
Enclosed, Partially Enclosed Buildings		$3° < \theta \le 10°$

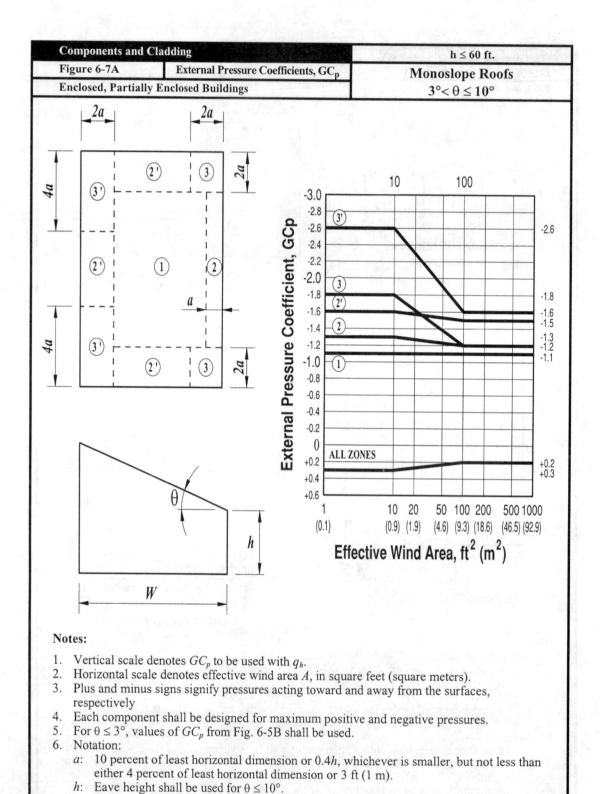

Notes:

1. Vertical scale denotes GC_p to be used with q_h.
2. Horizontal scale denotes effective wind area A, in square feet (square meters).
3. Plus and minus signs signify pressures acting toward and away from the surfaces, respectively
4. Each component shall be designed for maximum positive and negative pressures.
5. For $\theta \le 3°$, values of GC_p from Fig. 6-5B shall be used.
6. Notation:
   - a: 10 percent of least horizontal dimension or $0.4h$, whichever is smaller, but not less than either 4 percent of least horizontal dimension or 3 ft (1 m).
   - h: Eave height shall be used for $\theta \le 10°$.
   - W: Building width, in feet (meters).
   - θ: Angle of plane of roof from horizontal, in degrees.

Components and Cladding		h ≤ 60 ft.
Figure 6-7A (con't)	External Pressure Coefficients, GC$_p$	**Monoslope Roofs**
Enclosed, Partially Enclosed Buildings		10°< θ ≤ 30°

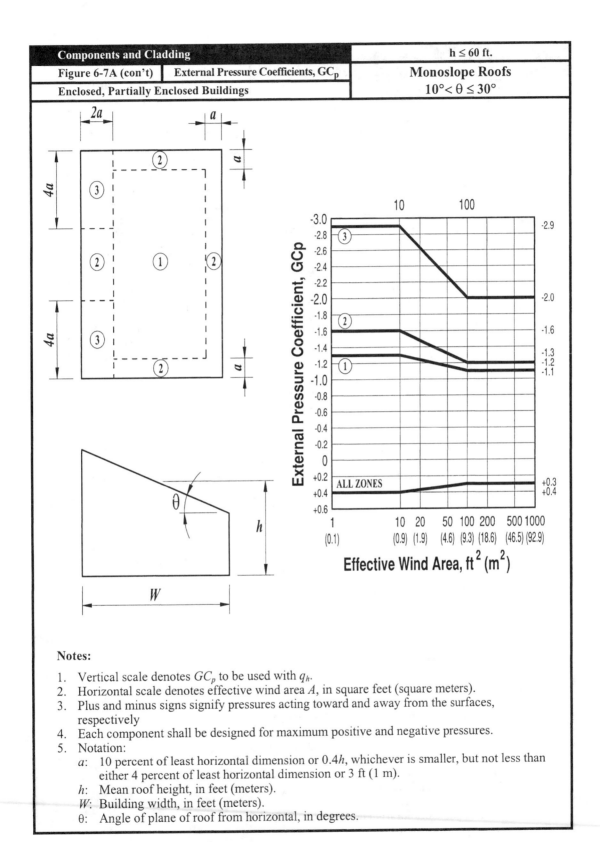

Notes:

1. Vertical scale denotes GC_p to be used with q_h.
2. Horizontal scale denotes effective wind area A, in square feet (square meters).
3. Plus and minus signs signify pressures acting toward and away from the surfaces, respectively
4. Each component shall be designed for maximum positive and negative pressures.
5. Notation:
      a: 10 percent of least horizontal dimension or 0.4h, whichever is smaller, but not less than either 4 percent of least horizontal dimension or 3 ft (1 m).
      h: Mean roof height, in feet (meters).
      W: Building width, in feet (meters).
      θ: Angle of plane of roof from horizontal, in degrees.

Components and Cladding		$h \leq 60$ ft.
Figure 6-7B	External Pressure Coefficients, GC_p	**Sawtooth Roofs**
Enclosed, Partially Enclosed Buildings		

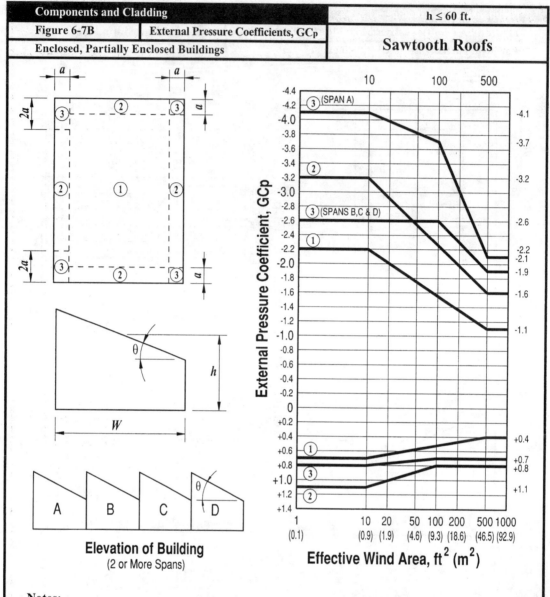

Elevation of Building
(2 or More Spans)

Notes:

1. Vertical scale denotes GC_p to be used with q_h.
2. Horizontal scale denotes effective wind area A, in square feet (square meters).
3. Plus and minus signs signify pressures acting toward and away from the surfaces, respectively
4. Each component shall be designed for maximum positive and negative pressures.
5. For $\theta \leq 10°$, values of GC_p from Fig. 6-5B shall be used.
6. Notation:
   - a: 10 percent of least horizontal dimension or $0.4h$, whichever is smaller, but not less than either 4 percent of least horizontal dimension or 3 ft (1 m).
   - h: Mean roof height, in feet (meters).
   - W: Building width, in feet (meters).
   - θ: Angle of plane of roof from horizontal, in degrees.

Components and Cladding		h >60 ft.
Figure 6-8	External Pressure Coefficients, GC$_p$	**Walls & Roofs**
Enclosed, Partially Enclosed Buildings		

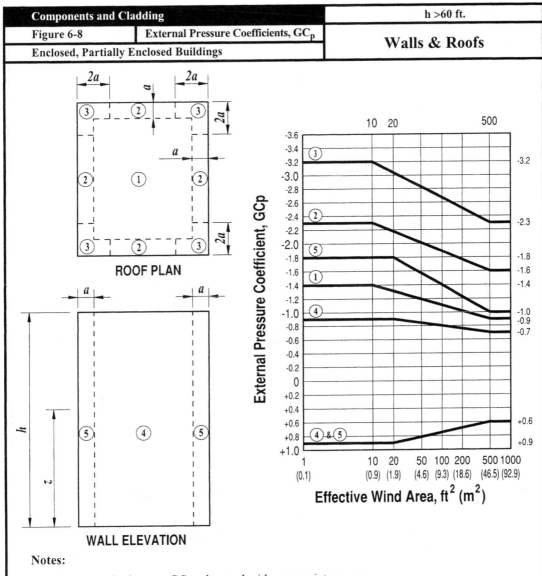

ROOF PLAN

WALL ELEVATION

Notes:

1. Vertical scale denotes GC_p to be used with appropriate q_z or q_h.
2. Horizontal scale denotes effective wind area A, in square feet (square meters).
3. Plus and minus signs signify pressures acting toward and away from the surfaces, respectively
4. Use q_z with positive values of GC_p and q_h with negative values of GC_p.
5. Each component shall be designed for maximum positive and negative pressures.
6. Coefficients are for roofs with angle $\theta \leq 10°$. For other roof angles and geometry, use GC_p values from Fig. 6-5B and attendant q_h based on exposure defined in 6.5.6.
7. If a parapet equal to or higher than 3 ft (1m) is provided around the perimeter of the roof with $\theta \leq 10°$, Zone 3 shall be treated as Zone 2.
8. Notation:
   a: 10 percent of least horizontal dimension, but not less than 3 ft (1 m).
   h: Mean roof height, in feet (meters), except that eave height shall be used for $\theta \leq 10°$.
   z: height above ground, in feet (meters).
   θ: Angle of plane of roof from horizontal, in degrees.

Main Wind Force Resisting System		h >60 ft.
Figure 6-9	Full & Partial Loadings	

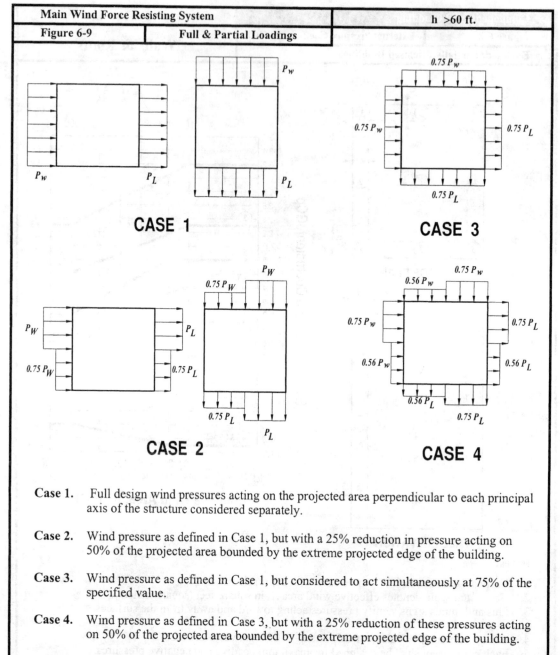

CASE 1

CASE 3

CASE 2

CASE 4

Case 1. Full design wind pressures acting on the projected area perpendicular to each principal axis of the structure considered separately.

Case 2. Wind pressure as defined in Case 1, but with a 25% reduction in pressure acting on 50% of the projected area bounded by the extreme projected edge of the building.

Case 3. Wind pressure as defined in Case 1, but considered to act simultaneously at 75% of the specified value.

Case 4. Wind pressure as defined in Case 3, but with a 25% reduction of these pressures acting on 50% of the projected area bounded by the extreme projected edge of the building.

Notes:

1. Design wind pressures for windward and leeward faces obtained in accordance with the provisions of 6.5.12 for main wind-force resisting systems for buildings with mean roof height h greater than 60 ft (18 m).
2. Diagrams show plan view of building.
3. Notation:
   P_W: Windward face design pressure.
   P_L: Leeward face design pressure.

Importance Factor, I (Wind Loads)
Table 6-1

Category	Non-Hurricane Prone Regions and Hurricane Prone Regions with V = 85-100 mph and Alaska	Hurricane Prone Regions with V > 100 mph
I	0.87	0.77
II	1.00	1.00
III	1.15	1.15
IV	1.15	1.15

Note:

1. The building and structure classification categories are listed in Table 1-1.

Main Wind Force Resisting System		h ≤ 30 ft.
Table 6-2	**Design Wind Pressure**	**Simplified Procedure**
Enclosed, Partially Enclosed Buildings		**Walls & Roofs**

DESIGN WIND PRESSURE (PSF)											
Location	**Building Classification**	**Basic Wind Speed V (MPH)**									
		85	90	100	110	120	130	140	150	160	170
Roof	Enclosed	-14	-16	-20	-24	-29	-33	-39	-45	-51	-57
	Partially Enclosed	-19	-21	-26	-31	-37	-44	-51	-58	-66	-74
Wall	Enclosed or Partially Enclosed	12	14	17	20	24	29	33	38	43	49

Notes:

1. Design wind pressures above represent the following:
   Roof – Net pressure (sum of external and internal pressures) applied normal to all roof surfaces.
   Wall – Combined net pressure (sum of windward and leeward, external and internal pressures) applied normal to all windward wall surfaces.

2. Values shown are for exposure B. For other exposures, multiply values shown by the factor below:

Exposure	Factor
C	1.40
D	1.66

3. Values shown for roof are based on a tributary area less than or equal to 100 sf. For larger tributary areas, multiply values shown by reduction factor below:

Area (SF)	Reduction Factor (Linear Interpolation Permitted)
≤ 100	1.0
250	0.9
≥ 1000	0.8

4. Values shown are for importance factor I = 1.0. For other values of I, multiply values shown by I.

5. Plus and minus signs indicate pressures acting toward and away from exterior surface, respectively.

Components and Cladding		h ≤ 30 ft.	
Table 6-3A	Design Wind Pressures	Simplified Procedure	
Enclosed Buildings		Walls & Roofs	

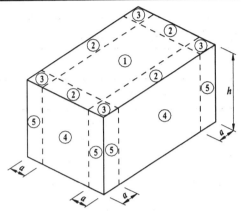

DESIGN WIND PRESSURE (PSF)

Location	Zone	Effective Wind Area (SF)	Basic Wind Speed V (MPH)									
			85	90	100	110	120	130	140	150	160	170
Roof	1	10	+10 -13	+10 -15	+10 -18	+10 -22	+11 -26	+12 -30	+14 -35	+16 -40	+19 -46	+21 -52
		20	+10 -13	+10 -14	+10 -18	+10 -21	+10 -25	+12 -30	+13 -34	+15 -39	+18 -45	+20 -51
		100	+10 -12	+10 -13	+10 -16	+10 -20	+10 -24	+10 -28	+11 -32	+13 -37	+15 -42	+17 -48
	2	10	+10 -22	+10 -24	+10 -30	+10 -36	+11 -43	+12 -51	+14 -59	+16 -68	+19 -77	+21 -87
		20	+10 -19	+10 -22	+10 -27	+10 -33	+10 -39	+12 -46	+13 -53	+15 -61	+18 -69	+20 -78
		100	+10 -14	+10 -16	+10 -19	+10 -24	+10 -28	+10 -33	+11 -38	+13 -44	+15 -50	+17 -56
	3	10	+10 -33	+10 -37	+10 -45	+10 -55	+11 -65	+12 -77	+14 -89	+16 -102	+19 -116	+21 -131
		20	+10 -27	+10 -30	+10 -37	+10 -45	+10 -54	+12 -63	+13 -73	+15 -84	+18 -96	+20 -108
		100	+10 -14	+10 -16	+10 -19	+10 -24	+10 -28	+10 -33	+11 -38	+13 -44	+15 -50	+17 -56
Walls	4	10	+13 -14	+15 -16	+18 -19	+22 -24	+26 -28	+30 -33	+35 -38	+40 -44	+46 -50	+52 -56
		50	+12 -13	+13 -14	+16 -18	+19 -22	+23 -26	+27 -30	+31 -35	+36 -40	+41 -46	+46 -51
		500	+10 -11	+11 -12	+13 -15	+16 -18	+19 -21	+23 -25	+26 -29	+30 -34	+34 -38	+39 -43
	5	10	+13 -17	+15 -19	+18 -24	+22 -29	+26 -35	+30 -41	+35 -47	+40 -54	+46 -62	+52 -70
		50	+12 -15	+13 -16	+16 -20	+19 -25	+23 -29	+27 -34	+31 -40	+36 -46	+41 -52	+46 -59
		500	+10 -11	+11 -12	+13 -15	+16 -18	+19 -21	+23 -25	+26 -29	+30 -34	+34 -38	+39 -43

Metric Conversion: 1 PSF = 47.9 pascals   1 SF = 0.0929 SM   1 MPH = 0.447 M/S

Notes:
1. Design wind pressures above represent the net pressure (sum of external and internal pressures) applied normal to all surfaces.
2. Values shown are for exposure B. For other exposures, multiply values shown by the following factor: exposure C: 1.40 and exposure D: 1.66.
3. Linear interpolation between values of tributary area is permissible.
4. Values shown are for an importance factor I = 1.0. For other values of I, multiply values shown by I.
5. Plus and minus signs signify pressure acting toward and away from the exterior surface, respectively.
6. All component and cladding elements shall be designed for both positive and negative pressures shown in the table.
7. Notation:
   a: 10 percent of least horizontal dimension or 0.4 h, whichever is smaller, but not less than 4% of least horizontal dimension or 3 ft.
   h: Mean roof height in feet (meters).

Components and Cladding				h ≤ 30 ft.	
Table 6-3B		**Net Pressure Coefficients**		**Simplified Procedure**	
Partially Enclosed Buildings				**Walls & Roofs**	

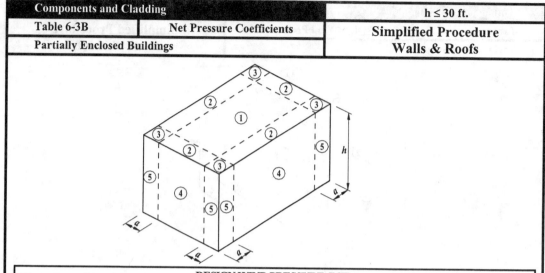

			DESIGN WIND PRESSURE (PSF)									
Location	**Zone**	**Effective Wind Area (SF)**	**Basic Wind Speed V (MPH)**									
			85	**90**	**100**	**110**	**120**	**130**	**140**	**150**	**160**	**170**
Roof	1	10	+10 -17	+10 -19	+13 -24	+16 -29	+19 -34	+22 -40	+25 -46	+29 -53	+33 -60	+37 -68
		20	+10 -17	+10 -19	+12 -23	+15 -28	+18 -33	+21 -39	+24 -45	+28 -52	+32 -59	+36 -67
		100	+10 -16	+9 -18	+11 -22	+14 -27	+16 -32	+19 -37	+22 -43	+26 -50	+29 -57	+33 -64
	2	10	+10 -26	+10 -29	+13 -36	+16 -43	+19 -52	+22 -60	+25 -70	+29 -81	+33 -92	+37 -103
		20	+10 -24	+10 -26	+12 -33	+15 -39	+18 -47	+21 -55	+24 -64	+28 -73	+32 -83	+36 -94
		100	+10 -18	+10 -20	+11 -25	+14 -30	+16 -36	+19 -42	+22 -49	+26 -57	+29 -64	+33 -73
	3	10	+10 -37	+10 -41	+13 -51	+16 -62	+19 -73	+22 -86	+25 -100	+29 -115	+33 -131	+37 -147
		20	+10 -31	+10 -35	+12 -43	+15 -52	+18 -62	+21 -73	+24 -84	+28 -97	+32 -110	+36 -125
		100	+10 -18	+10 -20	+11 -25	+14 -30	+16 -36	+19 -42	+22 -49	+26 -57	+29 -64	+33 -73
Walls	4	10	+17 -18	+19 -20	+24 -25	+29 -30	+34 -36	+40 -42	+46 -49	+53 -57	+60 -64	+68 -73
		50	+16 -17	+18 -19	+22 -23	+26 -28	+31 -34	+37 -40	+42 -46	+49 -53	+55 -60	+63 -68
		500	+14 -15	+15 -17	+19 -21	+23 -25	+27 -30	+32 -35	+37 -40	+43 -46	+49 -53	+55 -59
	5	10	+17 -21	+19 -24	+24 -30	+29 -36	+34 -43	+40 -50	+46 -58	+53 -67	+60 -76	+68 -86
		50	+16 -19	+18 -21	+22 -26	+26 -31	+31 -37	+37 -44	+42 -51	+49 -58	+55 -66	+63 -75
		500	+14 -15	+15 -17	+19 -21	+23 -25	+27 -30	+32 -35	+37 -40	+43 -46	+49 -53	+55 -59

Metric Conversion: 1 PSF = 47.9 pascals   1 SF = 0.0929 SM   1 MPH = 0.447 M/S

Notes:
1. Design wind pressures above represent the net pressure (sum of external and internal pressures) applied normal to all surfaces.
2. Values shown are for exposure B. For other exposures, multiply values shown by the following factor: exposure C: 1.40 and exposure D: 1.66.
3. Linear interpolation between values of tributary area is permissible.
4. Values shown are for an importance factor I = 1.0. For other values of I, multiply values shown by I.
5. Plus and minus signs signify pressure acting toward and away from the exterior surface, respectively.
6. All component and cladding elements shall be designed for both positive and negative pressures shown in the table.
7. Notation:
   a: 10 percent of least horizontal dimension or 0.4 h, whichever is smaller, but not less than 4% of least horizontal dimension or 3 ft.
   h: Mean roof height in feet (meters).

Terrain Exposure Constants

Table 6-4

Exposure	α	z_g (ft)	$\hat{a}$	$\hat{b}$	$\bar{\alpha}$	$\bar{b}$	c	ℓ (ft)	$\bar{\in}$	z_{min} (ft)*
A	5.0	1500	1/5	0.64	1/3.0	0.30	0.45	180	1/2.0	60
B	7.0	1200	1/7	0.84	1/4.0	0.45	0.30	320	1/3.0	30
C	9.5	900	1/9.5	1.00	1/6.5	0.65	0.20	500	1/5.0	15
D	11.5	700	1/11.5	1.07	1/9.0	0.80	0.15	650	1/8.0	7

*z_{min} = minimum height used to ensure that the equivalent height $\bar{z}$ is greater of $0.6h$ or z_{min}.
For buildings with $h \leq z_{min}$, $\bar{z}$ shall be taken as z_{min}.

Velocity Pressure Exposure Coefficients, K_h and K_z

Table 6-5

Height above ground level, z		Exposure (Note 1)					
		A		B		C	D
ft	(m)	Case 1	Case 2	Case 1	Case 2	Cases 1 & 2	Cases 1 & 2
0-15	(0-4.6)	0.68	0.32	0.70	0.57	0.85	1.03
20	(6.1)	0.68	0.36	0.70	0.62	0.90	1.08
25	(7.6)	0.68	0.39	0.70	0.66	0.94	1.12
30	(9.1)	0.68	0.42	0.70	0.70	0.98	1.16
40	(12.2)	0.68	0.47	0.76	0.76	1.04	1.22
50	(15.2)	0.68	0.52	0.81	0.81	1.09	1.27
60	(18)	0.68	0.55	0.85	0.85	1.13	1.31
70	(21.3)	0.68	0.59	0.89	0.89	1.17	1.34
80	(24.4)	0.68	0.62	0.93	0.93	1.21	1.38
90	(27.4)	0.68	0.65	0.96	0.96	1.24	1.40
100	(30.5)	0.68	0.68	0.99	0.99	1.26	1.43
120	(36.6)	0.73	0.73	1.04	1.04	1.31	1.48
140	(42.7)	0.78	0.78	1.09	1.09	1.36	1.52
160	(48.8)	0.82	0.82	1.13	1.13	1.39	1.55
180	(54.9)	0.86	0.86	1.17	1.17	1.43	1.58
200	(61.0)	0.90	0.90	1.20	1.20	1.46	1.61
250	(76.2)	0.98	0.98	1.28	1.28	1.53	1.68
300	(91.4)	1.05	1.05	1.35	1.35	1.59	1.73
350	(106.7)	1.12	1.12	1.41	1.41	1.64	1.78
400	(121.9)	1.18	1.18	1.47	1.47	1.69	1.82
450	(137.2)	1.24	1.24	1.52	1.52	1.73	1.86
500	(152.4)	1.29	1.29	1.56	1.56	1.77	1.89

Notes:

1. **Case 1:** a. All components and cladding.
   b. Main wind force resisting system in low-rise buildings designed using Figure 6-4.

   Case 2: a. All main wind force resisting systems in buildings except those in low-rise buildings designed using Figure 6-4.
   b. All main wind force resisting systems in other structures. ✓

2. The velocity pressure exposure coefficient K_z may be determined from the following formula:

   For 15 ft. $\leq z \leq z_g$ For $z < 15$ ft.

   $K_z = 2.01 (z/z_g)^{2/\alpha}$ $K_z = 2.01 (15/z_g)^{2/\alpha}$

   Note: z shall not be taken less than 100 feet for Case 1 in exposure A or less than 30 feet for Case 1 in exposure B.

3. α and z_g are tabulated in Table 6-4.

4. Linear interpolation for intermediate values of height z is acceptable.

5. Exposure categories are defined in 6.5.6.

Wind Directionality Factor, K_d

Table 6-6

Structure Type	Directionality Factor K_d*
Buildings 　　**Main Wind Force Resisting System** 　　**Components and Cladding**	0.85 0.85
Arched Roofs	0.85
Chimneys, Tanks, and Similar Structures 　　**Square** 　　**Hexagonal** 　　**Round**	0.90 0.95 0.95
Solid Signs	0.85
Open Signs and Lattice Framework	0.85
Trussed Towers 　　**Triangular, square, rectangular** 　　**All other cross sections**	0.85 0.95

*Directionality Factor K_d has been calibrated with combinations of loads specified in Section 2. This factor shall only be applied when used in conjunction with load combinations specified in 2.3 and 2.4.

Internal Pressure Coefficients for Buildings, GC$_{pi}$

Table 6-7

Enclosure Classification	GC$_{pi}$
Open Buildings	0.00
Partially Enclosed Buildings	+0.55 -0.55
Enclosed Buildings	+0.18 -0.18

Notes:

1. Plus and minus signs signify pressures acting toward and away from the internal surfaces.

2. Values of GC$_{pi}$ shall be used with q$_z$ or q$_h$ as specified in 6.5.12.

3. Two cases shall be considered to determine the critical load requirements for the appropriate condition:

    (i) a positive value of GC$_{pi}$ applied to all internal surfaces
    (ii) a negative value of GC$_{pi}$ applied to all internal surfaces

Main Wind Force Resist. Sys. / Components & Cladding		All h	
Table 6-8	**External Pressure Coefficients, C_p**	**Arched Roofs**	
Enclosed, Partially Enclosed Buildings			

Conditions	Rise-to-span ratio, r	C_p		
		Windward quarter	Center half	Leeward quarter
Roof on elevated structure	$0 < r < 0.2$	-0.9	$-0.7 - r$	-0.5
	$0.2 \leq r < 0.3$*	$1.5r - 0.3$	$-0.7 - r$	-0.5
	$0.3 \leq r \leq 0.6$	$2.75r - 0.7$	$-0.7 - r$	-0.5
Roof springing from ground level	$0 < r \leq 0.6$	$1.4r$	$-0.7 - r$	-0.5

*When the rise-to-span ratio is $0.2 \leq r \leq 0.3$, alternate coefficients given by $6r - 2.1$ shall also be used for the windward quarter.

Notes:

1. Values listed are for the determination of average loads on main windforce resisting systems.

2. Plus and minus signs signify pressures acting toward and away from the surfaces, respectively.

3. For wind directed parallel to the axis of the arch, use pressure coefficients from Fig. 6-3 with wind directed parallel to ridge.

4. For components and cladding: (1) At roof perimeter, use the external pressure coefficients in Fig. 6-5B with θ based on spring-line slope and (2) for remaining roof areas, use external pressure coefficients of this table multiplied by 0.87.

Main Wind Force Resisting System						All h	
Table 6-9		Force Coefficients, C_f				**Monoslope Roofs**	
Open Buildings							

Roof angle θ degrees	L/B						
	5	**3**	**2**	**1**	**1/2**	**1/3**	**1/5**
10	0.2	0.25	0.3	0.45	0.55	0.7	0.75
15	0.35	0.45	0.5	0.7	0.85	0.9	0.85
20	0.5	0.6	0.75	0.9	1.0	0.95	0.9
25	0.7	0.8	0.95	1.15	1.1	1.05	0.95
30	0.9	1.0	1.2	1.3	1.2	1.1	1.0

Roof angle θ degrees	Center of Pressure X/L		
	L/B		
	2 to 5	**1**	**1/5 to 1/2**
10 to 20	0.35	0.3	0.3
25	0.35	0.35	0.4
30	0.35	0.4	0.45

Notes:

1. Wind forces act normal to the surface. Two cases shall be considered: (1) wind forces directed inward; and (2) wind forces directed outward.

2. The roof angle shall be assumed to vary ± 10° from the actual angle and the angle resulting in the greatest force coefficient shall be used.

3. Notation:

   B: dimension of roof measured normal to wind direction, in feet (meters);

   L: Dimension of roof measured parallel to wind direction, in feet (meters);

   X: Distance to center of pressure from windward edge of roof, in feet (meters); and

   θ: Angle of plane of roof from horizontal, in degrees.

Other Structures			All h		
Table 6-10		Force Coefficients, C_f	Chimneys, Tanks & Similar Structures		

Cross-Section	Type of Surface	h/D		
		1	7	25
Square (wind normal to face) → ⬜	All	1.3	1.4	2.0
Square (wind along diagonal) → ◇	All	1.0	1.1	1.5
Hexagonal or octagonal → ⬡	All	1.0	1.2	1.4
Round ($D\sqrt{q_z} > 2.5$) ($D\sqrt{q_z} > 5.3$, D in m, q_z in N/m²)	Moderately smooth	0.5	0.6	0.7
	Rough (D'/D = 0.02)	0.7	0.8	0.9
	Very rough (D'/D = 0.08)	0.8	1.0	1.2
Round ($D\sqrt{q_z} \le 2.5$) ($D\sqrt{q_z} \le 5.3$, D in m, q_z in N/m²)	All	0.7	0.8	1.2

Notes:

1. The design wind force shall be calculated based on the area of the structure projected on a plane normal to the wind direction. The force shall be assumed to act parallel to the wind direction.

2. Linear interpolation is permitted for *h/D* values other than shown.

3. Notation:

   D: diameter of circular cross-section and least horizontal dimension of square, hexagonal or octagonal cross-sections at elevation under consideration, in feet (meters);

   D': depth of protruding elements such as ribs and spoilers, in feet (meters); and

   h: height of structure, in feet (meters); and

   q_z: velocity pressure evaluated at height z above ground, in pounds per square foot (N/m²).

65

Other Structures			All h	
Table 6-11		**Force Coefficients, C_f**	**Solid Freestanding Walls & Solid Signs**	

At Ground Level		Above Ground Level	
ν	C_f	M/N	C_f
≤ 3	1.2	≤ 6	1.2
5	1.3	10	1.3
8	1.4	16	1.4
10	1.5	20	1.5
20	1.75	40	1.75
30	1.85	60	1.85
≥ 40	2.0	≥ 80	2.0

Notes:

1. The term "signs" in notes below applies also to "freestanding walls".

2. Signs with openings comprising less than 30% of the gross area shall be considered as solid signs.

3. Signs for which the distance from the ground to the bottom edge is less than 0.25 times the vertical dimension shall be considered to be at ground level.

4. To allow for both normal and oblique wind directions, two cases shall be considered:

   a. resultant force acts normal to the face of the sign on a vertical line passing through the geometric center, and

   b. resultant force acts normal to the face of the sign at a distance from a vertical line passing through the geometric center equal to 0.2 times the average width of the sign.

5. Notation:

   ν: ratio of height to width;

   M: larger dimension of sign, in feet (meters); and

   N: smaller dimension of sign, in feet (meters).

Other Structures		All h
Table 6-12	**Force Coefficients, C_f**	**Open Signs &**
Open Structures		**Lattice Frameworks**

$\in$	Flat-Sided Members	Rounded Members	
		$D\sqrt{q_z} \leq 2.5$ $(D\sqrt{q_z} \leq 5.3)$	$D\sqrt{q_z} > 2.5$ $(D\sqrt{q_z} > 5.3)$
< 0.1	2.0	1.2	0.8
0.1 to 0.29	1.8	1.3	0.9
0.3 to 0.7	1.6	1.5	1.1

Notes:

1. Signs with openings comprising 30% or more of the gross area are classified as open signs.

2. The calculation of the design wind forces shall be based on the area of all exposed members and elements projected on a plane normal to the wind direction. Forces shall be assumed to act parallel to the wind direction.

3. The area A_f consistent with these force coefficients is the solid area projected normal to the wind direction.

4. Notation:

    $\in$: ratio of solid area to gross area;

    D: diameter of a typical round member, in feet (meters);

    q_z: velocity pressure evaluated at height z above ground in pounds per square foot (N/m^2).

Other Structures		All h
Table 6-13	Force Coefficients, C_f	**Trussed Towers**
Open Structures		

Tower Cross Section	C_f
Square	$4.0 \in^2 - 5.9 \in + 4.0$
Triangle	$3.4 \in^2 - 4.7 \in + 3.4$

Notes:

1. For all wind directions considered, the area A_f consistent with the specified force coefficients shall be the solid area of a tower face projected on the plane of that face for the tower segment under consideration.

2. The specified force coefficients are for towers with structural angles or similar flat-sided members.

3. For towers containing rounded members, it is acceptable to multiply the specified force coefficients by the following factor when determining wind forces on such members:

   $0.51 \in^2 + 0.57$, but not > 1.0

4. Wind forces shall be applied in the directions resulting in maximum member forces and reactions. For towers with square cross-sections, wind forces shall be multiplied by the following factor when the wind is directed along a tower diagonal:

   $1 + 0.75 \in$, but not > 1.2

5. Wind forces on tower appurtenances such as ladders, conduits, lights, elevators, etc., shall be calculated using appropriate force coefficients for these elements.

6. Loads due to ice accretion as described in Section 11 shall be accounted for.

7. Notation:

   $\in$: ratio of solid area to gross area of one tower face for the segment under consideration.

7.0 SNOW LOADS

7.1 SYMBOLS AND NOTATION

C_e = exposure factor as determined from Table 7-2;

C_s = slope factor as determined from Fig. 7-2;

C_t = thermal factor as determined from Table 7-3;

h_b = height of balanced snow load determined by dividing p_f or p_s by γ, in ft (m);

h_c = clear height from top of balanced snow load to (1) closest point on adjacent upper roof, (2) top of parapet, or (3) top of a projection on the roof, in ft (m);

h_d = height of snow drift, in ft (m);

h_e = elevation difference between the ridge line and the eaves;

h_o = height of obstruction above the surface of the roof, in ft (m);

I = importance factor as determined from Table 7-4;

l_u = length of the roof upwind of the drift, in ft (m);

L = roof length parallel to the ridge line, in ft (m);

p_d = maximum intensity of drift surcharge load, in pounds per square foot (kilonewton per square meter);

p_f = snow load on flat roofs ("flat" = roof slope $\leq 5°$), in pounds per square foot (kilonewton per square meter);

p_g = ground snow load as determined from Fig. 7-1 and Table 7-1; or a site-specific analysis, in pounds per square foot (kilonewton per square meter);

p_s = sloped-roof snow load, in pounds per square foot (kilonewton per square meter);

s = separation distance between buildings, in ft (m);

W = horizontal distance from eave to ridge, in ft (m);

w = width of snow drift, in ft (m);

β = gable roof drift parameter as determined from Eq. 7-3;

γ = snow density in pounds per cubic foot (kilonewtons per cubic meter) as determined from Eq. 7-4; and

θ = roof slope on the leeward side, in degrees.

7.2 GROUND SNOW LOADS, p_g

Ground snow loads, p_g, to be used in the determination of design snow loads for roofs shall be as set forth in Fig. 7-1 for the contiguous United States and Table 7-1 for Alaska. Site specific case studies shall be made to determine ground snow loads in areas designated CS in Fig. 7-1. Ground snow loads

for sites at elevations above the limits indicated in Fig. 7-1 and for all sites within the CS areas shall be approved by the authority having jurisdiction. Ground snow load determination for such sites shall be based on an extreme value statistical analysis of data available in the vicinity of the site using a value with a 2% annual probability of being exceeded (50-year mean recurrence interval).

Snow loads are zero for Hawaii, except in mountainous regions as determined by the authority having jurisdiction.

7.3 FLAT-ROOF SNOW LOADS, p_f

The snow load, p_f, on a roof with a slope equal to or less than 5° (1 in./ft = 4.76°) shall be calculated in pounds per square foot (kilonewton per square meter) using the following formula:

$$p_f = 0.7 C_e C_t I p_g \qquad \text{(Eq. 7-1)}$$

but not less than the following minimum values for low slope roofs as defined in Section 7.3.4: where p_g is 20 lb/ft² (0.96 kN/m²) or less, $p_f = (I)p_g$ (Importance factor times p_g) where p_g exceeds 20 lb/ft² (0.96 kN/m²), $p_f = 20(I)$ (Importance factor times 20 lb/ft²).

7.3.1 Exposure Factor, C_e
The value for C_e shall be determined from Table 7-2.

7.3.2 Thermal Factor, C_t
The value for C_t shall be determined from Table 7-3.

7.3.3 Importance Factor, I
The value for I shall be determined from Table 7-4.

7.3.4 Minimum Values of p_f for Low-Slope Roofs
Minimum values of p_f shall apply to monoslope roofs with slopes less than 15°, hip, and gable roofs with slopes less than or equal to $(70/W) + 0.5$, and curved roofs where the vertical angle from the eaves to the crown is less than 10°.

7.4 SLOPED-ROOF SNOW LOADS, p_s

Snow loads acting on a sloping surface shall be assumed to act on the horizontal projection of that surface. The sloped-roof snow load, p_s, shall be obtained by multiplying the flat-roof snow load, p_f, by the roof slope factor, C_s:

$$p_s = C_s p_f \qquad \text{(Eq. 7-2)}$$

Values of C_s for warm roofs, cold roofs, curved roofs, and multiple roofs are determined from Sections 7.4.1– 7.4.4. The thermal factor, C_t, from Table 7-3 determines if a roof is "cold" or "warm." "Slippery surface" values shall be used only where the roof's surface is unobstructed and sufficient space is available below the eaves to accept all the sliding snow. A roof shall be considered unobstructed if no objects exist on it which prevent snow on it from sliding. Slippery surfaces shall include metal, slate, glass, and bituminous, rubber and plastic membranes with a smooth surface. Membranes with an imbedded aggregate or mineral granule surface shall not be considered smooth. Asphalt shingles, wood shingles and shakes shall not be considered slippery.

7.4.1 Warm-Roof Slope Factor, C_s

For warm roofs ($C_t = 1.0$ as determined from Table 7-3) with an unobstructed slippery surface that will allow snow to slide off the eaves, the roof slope factor C_s shall be determined using the dashed line in Fig. 7-2a, provided that for non-ventilated roofs, their thermal resistance (R-value) equals or exceeds 30 ft$^2 \cdot$ h$\cdot$°F/Btu (5.3 K$\cdot$m^2/W) and for ventilated roofs, their R-value equals or exceeds 20 ft$^2 \cdot$h$\cdot$°F/Btu (3.5 K$\cdot$m^2/W). Exterior air shall be able to circulate freely under a ventilated roof from its eaves to its ridge. For warm roofs that do not meet the aforementioned conditions, the solid line in Fig. 7-2a shall be used to determine the roof slope factor C_s.

7.4.2 Cold Roof Slope Factor, C_s

Cold roofs are those with a $C_t > 1.0$ as determined from Table 7-3. For cold roofs with $C_t = 1.2$ and an unobstructed slippery surface that will allow snow to slide off the eaves, the roof slope factor C_s shall be determined using the dashed line in Fig. 7-2b. For all other cold roofs with $C_t = 1.2$, the solid line in Fig. 7-2b shall be used to determine the roof slope factor C_s. For cold roofs with $C_t = 1.1$, C_s shall be determined by taking the average of values obtained from the appropriate $C_t \le 1.0$ curve in Fig. 7-2a and the appropriate $C_t = 1.2$ curve in Fig. 7-2b.

7.4.3 Roof Slope Factor for Curved Roofs

Portions of curved roofs having a slope exceeding 70° shall be considered free of snow load, (i.e.,

$C_s = 0$). Balanced loads shall be determined from the balanced load diagrams in Fig. 7-3 with C_s determined from the appropriate curve in Fig. 7-2.

7.4.4 Roof Slope Factor for Multiple Folded Plate, Sawtooth, and Barrel Vault Roofs

Multiple folded plate, sawtooth, or barrel vault roofs shall have a $C_s = 1.0$, with no reduction in snow load because of slope (i.e., $p_s = p_f$).

7.4.5 Ice Dams and Icicles Along Eaves

Two types of warm roofs that drain water over their eaves shall be capable of sustaining a uniformly distributed load of $2p_f$ on all overhanging portions there: those that are unventilated and have an R-value less than 30 ft$^2 \cdot$ h$\cdot$°F/Btu (5.3 K$\cdot$m^2/W) and those that are ventilated and have an R-value less than 20 ft$^2 \cdot$h$\cdot$°F/Btu (3.5 K$\cdot$m^2/W). No other loads except dead loads shall be present on the roof when this uniformly distributed load is applied.

7.5 PARTIAL LOADING

The effect of having selected spans loaded with the balanced snow load and remaining spans loaded with half the balanced snow load shall be investigated as follows.

7.5.1 Continuous Beam Systems

Continuous beam systems shall be investigated for the effects of the three loadings shown in Fig. 7-4:

- Case 1: Full balanced snow load on either exterior span and half the balanced snow load on all other spans.
- Case 2: Half the balanced snow load on either exterior span and full balanced snow load on all other spans.
- Case 3: All possible combinations of full balanced snow load on any two adjacent spans and half the balanced snow load on all other spans. For this case there will be (n-1) possible combinations where n equals the number of spans in the continuous beam system.

If a cantilever is present in any of the above cases, it shall be considered to be a span.

Partial load provisions need not be applied to structural members which span perpendicular to the ridgeline in gable roofs with slopes greater than $70/W + 0.5$.

7.5.2 Other Structural Systems

Areas sustaining only half the balanced snow load shall be chosen so as to produce the greatest effects on members being analyzed.

7.6 UNBALANCED ROOF SNOW LOADS

Balanced and unbalanced loads shall be analyzed separately. Winds from all directions shall be accounted for when establishing unbalanced loads.

7.6.1 Unbalanced Snow Loads for Hip and Gable Roofs

For hip and gable roofs with a slope exceeding 70° or with a slope less than $70/W + 0.5$, unbalanced snow loads are not required to be applied. For roofs with an eave to ridge distance, W, of 20 ft or less, the structure shall be designed to resist an unbalanced uniform snow load on the leeward side equal to $1.5p_f/C_e$ for roof slopes of 5° or less, and $1.5p_s/C_e$ for roof slopes exceeding 5°. For roofs with $W > 20$ ft and with slopes (in degrees) greater than $275\ \beta p_f/\gamma W$, the structure shall be designed to resist an unbalanced uniform snow load on the leeward side equal to $1.2(1 + \beta/2)p_s/C_e$ with β given by Eq. 7-3.

$$\beta = \begin{cases} 0.5 & L/W \le 1 \\ 0.33 + 0.167L/W & 1 < L/W \le 4 \quad \text{(Eq. 7-3)} \\ 1.0 & L/W > 4 \end{cases}$$

where L is the roof length parallel to the ridgeline and W is the horizontal eave to ridge distance. For roofs with $W > 20$ ft and slopes (in degrees) equal to or less than $275\ \beta p_f/\gamma W$ the structure shall be designed to resist a linearly varying snow load on the leeward side. This linearly varying load is $1.2p_f/C_e$ at the ridge and $1.2(1 + \beta)p_f/C_e$ at the eave. However, the intensity of the surcharge at the eave, $1.2\beta p_f/C_e$, need not be taken as larger than the product of the snow density, γ, and the elevation difference between the ridgeline and the eaves, h_e.

For the unbalanced situation with $W > 20$ ft, the windward side shall have a uniform load equal to $0.3p_s$ when the angle in question is greater than $275\ \beta p_f/\gamma W$ and $0.3p_f$ when the roof slope is equal to or less than $275\ \beta p_f/\gamma W$. Balanced and unbalanced loading diagrams are presented in Fig. 7-5.

7.6.2 Unbalanced Snow Loads for Curved Roofs

Portions of curved roofs having a slope exceeding 70° shall be considered free of snow load. If the

slope of a straight line from the eaves (or the 70° point, if present) to the crown is less than 10° or greater than 60°, unbalanced snow loads shall not be taken into account.

Unbalanced loads shall be determined according to the loading diagrams in Fig. 7-3. In all cases the windward side shall be considered free of snow. If the ground or another roof abuts a Case 2 or Case 3 (see Fig. 7-3) curved roof at or within 3 ft (0.91 m) of its eaves, the snow load shall not be decreased between the 30° point and the eaves but shall remain constant at the 30° point value. This distribution is shown as a dashed line in Fig. 7-3.

7.6.3 Unbalanced Snow Loads for Multiple Folded Plate, Sawtooth, and Barrel Vault Roofs

Unbalanced loads shall be applied to folded plate, sawtooth, and barrel vaulted multiple roofs with a slope exceeding 3/8 in./ft (1.79°). According to Section 7.4.4, $C_s = 1.0$ for such roofs, and the balanced snow load equals p_f. The unbalanced snow load shall increase from one-half the balanced load at the ridge or crown (i.e., $0.5p_f$) to two times the balanced load given in Section 7.4.4 divided by C_e at the valley (i.e., $2p_f/C_e$). Balanced and unbalanced loading diagrams for a sawtooth roof are presented in Fig. 7-6. However, the snow surface above the valley shall not be at an elevation higher than the snow above the ridge. Snow depths shall be determined by dividing the snow load by the density of that snow from Eq. 7-3 which is in Section 7.7.2.

7.6.4 Unbalanced Snow Loads for Dome Roofs

Unbalanced snow loads shall be applied to domes and similar rounded structures. Snow loads, determined in the same manner as for curved roofs in Section 7.6.2, shall be applied to the downwind 90° sector in plan view. At both edges of this sector, the load shall decrease linearly to zero over sectors of 22.5° each. There shall be no snow load on the remaining 225° upwind sector.

7.7 DRIFTS ON LOWER ROOFS (AERODYNAMIC SHADE)

Roofs shall be designed to sustain localized loads from snow drifts that form in the wind shadow of: (1) higher portions of the same structure; and (2) adjacent structures and terrain features.

7.7.1 Lower Roof of a Structure

Snow that forms drifts comes from a higher roof or, with the wind from the opposite direction, from the roof on which the drift is located. These two kinds of drifts ("leeward" and "windward," respectively) are shown in Fig. 7-7. The geometry of the surcharge load due to snow drifting shall be approximated by a triangle as shown in Fig. 7-8. Drift loads shall be superimposed on the balanced snow load. If h_c/h_b is less than 0.2, drift loads are not required to be applied.

For leeward drifts the drift height h_d shall be determined directly from Fig. 7-9 using the length of the upper roof. For windward drifts the drift height shall be determined by substituting the length of the lower roof for l_u in Fig. 7-9 and using three-quarters of h_d as determined from Fig. 7-9 as the drift height. The larger of these two heights shall be used in design. If this height is equal to or less than h_c, the drift width, w, shall equal $4h_d$ and the drift height shall equal h_d. If this height exceeds h_c, the drift width, w, shall equal $4h_d^2/h_c$ and the drift height shall equal h_c. However, the drift width w shall not be greater than $8h_c$. If the drift width, w, exceeds the width of the lower roof, the drift shall be truncated at the far edge of the roof, not reduced to zero there. The maximum intensity of the drift surcharge load, p_d, equals $h_d\gamma$ where snow density, γ, is defined in Eq. 7-4:

$$\gamma = 0.13p_g + 14 \quad \text{but not more than 30 pcf}$$
$$\text{(Eq. 7-4)}$$

(in SI: $\gamma = 0.426p_g + 2.2$ but not more than 4.7 kN/ m^3).

This density shall also be used to determine h_b by dividing p_f (or p_s) by γ (in SI: also multiply by 102 to get the depth in meters).

7.7.2 Adjacent Structures and Terrain Features

The requirements in Section 7.7.1 shall also be used to determine drift loads caused by a higher structure or terrain feature within 20 ft (6.1 m) of a roof. The separation distance, s, between the roof and adjacent structure or terrain feature shall reduce applied drift loads on the lower roof by the factor $(20-s)/20$ where s is in feet $[(6.1-s)/6.1$ where s is in meters].

7.8 ROOF PROJECTIONS

The method in Section 7.7.1 shall be used to calculate drift loads on all sides of roof projections and at parapet walls. The height of such drifts shall be taken as three-quarters the drift height from Fig. 7-9 (i.e., $0.75h_d$) with l_u equal to the length of the roof upwind of the projection or parapet wall. If the side

of a roof projection is less than 15 ft (4.6 m) long, a drift load is not required to be applied to that side.

7.9 SLIDING SNOW

The extra load caused by snow sliding off a sloped roof onto a lower roof shall be determined assuming that all the snow that accumulates on the upper roof under the balanced loading condition slides onto the lower roof. The solid lines in Fig. 7-2 shall be used to determine the total extra load available from the upper roof, regardless of the surface of the upper roof.

The sliding snow load shall not be reduced unless a portion of the snow on the upper roof is blocked from sliding onto the lower roof by snow already on the lower roof or is expected to slide clear of the lower roof.

Sliding loads shall be superimposed on the balanced snow load.

7.10 RAIN-ON-SNOW SURCHARGE LOAD

For locations where p_g is 20 psf (0.96 kN/m^2) or less but not zero, all roofs with a slope less than 1/2 in./ft (2.38°), shall have a 5 psf (0.24 kN/m^2) rain-on-snow surcharge load applied to establish the design snow loads. Where the minimum flat roof design snow load from Section 7.3.4 exceeds p_f as determined by Eq. 7-1, the rain-on-snow surcharge load shall be reduced by the difference between these two values, with a maximum reduction of 5 psf (0.24 kN/m^2).

7.11 PONDING INSTABILITY

Roofs shall be designed to preclude ponding instability. For roofs with a slope less than 1/4 in./ft (1.19°), roof deflections caused by full snow loads shall be investigated when determining the likelihood of ponding instability from rain-on-snow or from snow meltwater (see Section 8.4).

7.12 EXISTING ROOFS

Existing roofs shall be evaluated for increased snow loads caused by additions or alterations. Owners or agents for owners of an existing lower roof shall be advised of the potential for increased snow loads where a higher roof is constructed within 20 ft (6.1 m) (see footnote to Table 7-2 and Section 7.7.2).

This page intentionally left blank

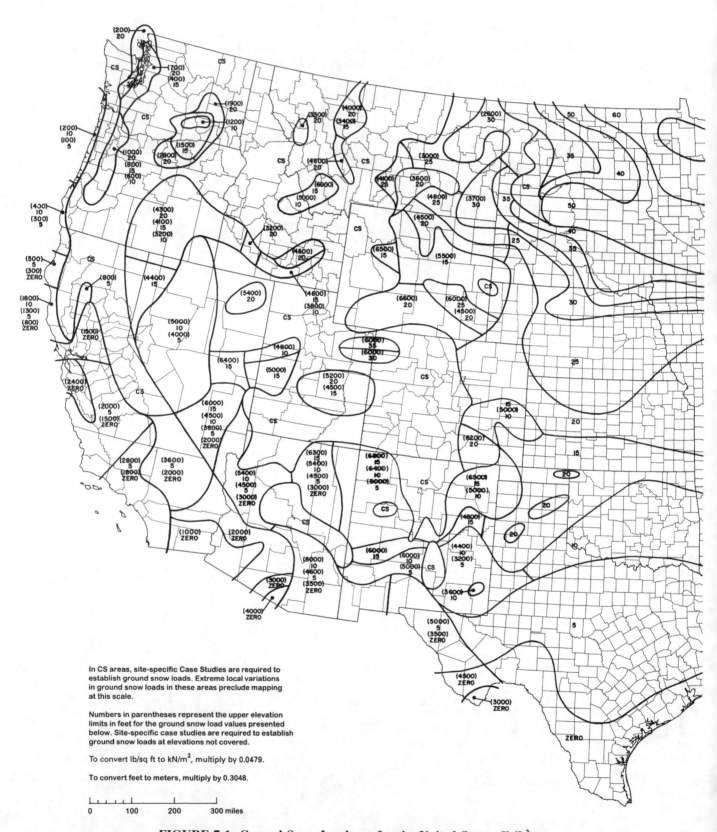

In CS areas, site-specific Case Studies are required to establish ground snow loads. Extreme local variations in ground snow loads in these areas preclude mapping at this scale.

Numbers in parentheses represent the upper elevation limits in feet for the ground snow load values presented below. Site-specific case studies are required to establish ground snow loads at elevations not covered.

To convert lb/sq ft to kN/m², multiply by 0.0479.

To convert feet to meters, multiply by 0.3048.

0   100   200   300 miles

FIGURE 7-1. Ground Snow Loads, p_g for the United States (lb/ft²)

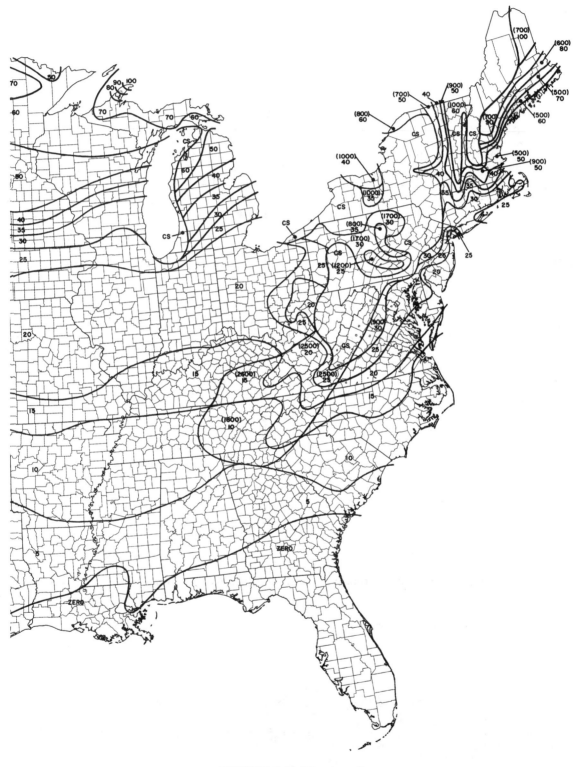

FIGURE 7-1. (*Continued*)

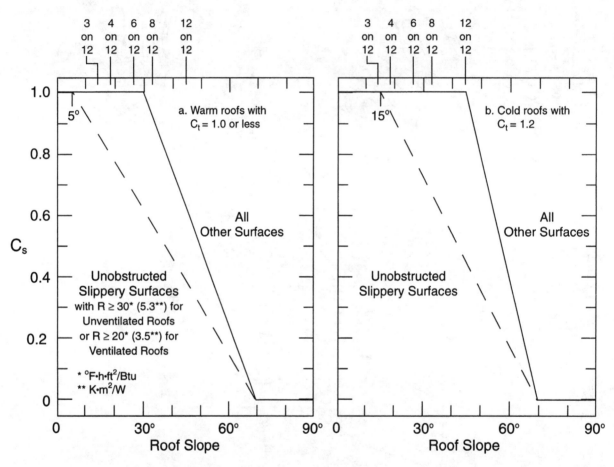

FIGURE 7-2. Graphs for Determining Roof Slope Factor C_s for Warm and Cold Roofs

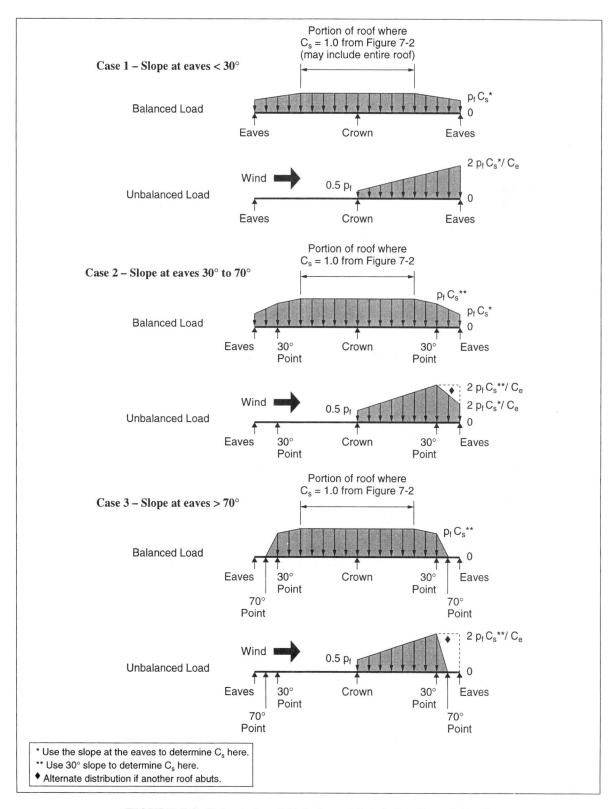

* Use the slope at the eaves to determine C_s here.
** Use 30° slope to determine C_s here.
♦ Alternate distribution if another roof abuts.

FIGURE 7-3. Balanced and Unbalanced Loads for Curved Roofs

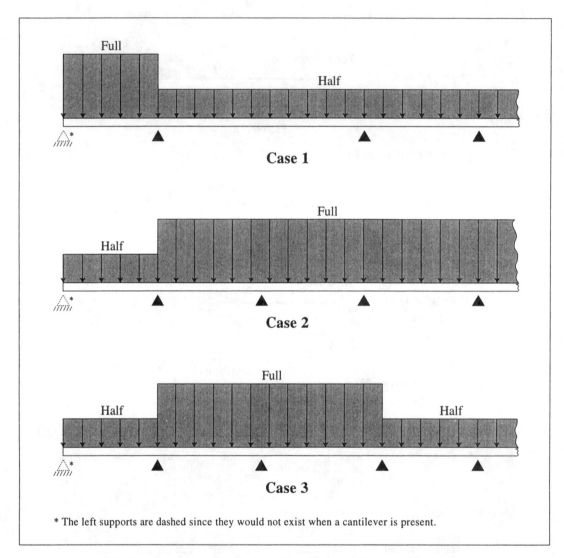

FIGURE 7-4. Partial Loading Diagrams for Continuous Beams

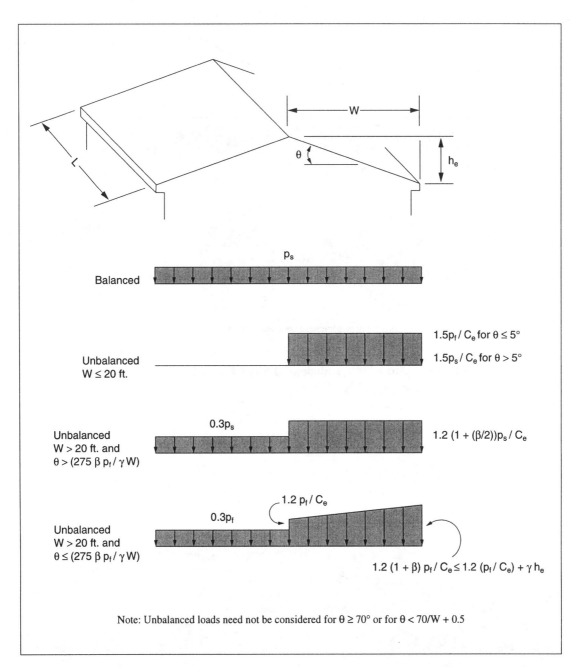

FIGURE 7-5. Balanced and Unbalanced Snow Loads for Hip and Gable Roofs

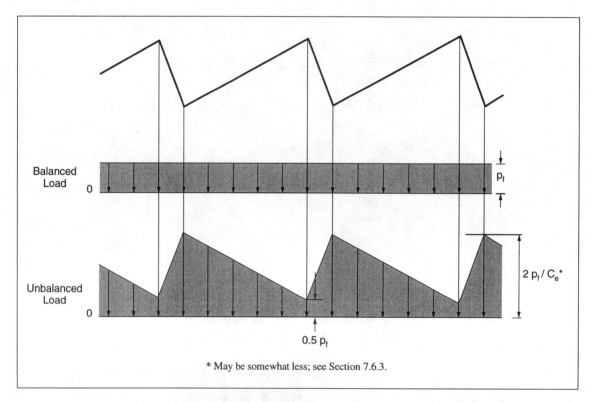

Balanced Load

0

p_f

Unbalanced Load

0

$2\,p_f\,/\,C_e{}^*$

0.5 p_f

* May be somewhat less; see Section 7.6.3.

FIGURE 7-6. Balanced and Unbalanced Snow Loads for a Sawtooth Roof

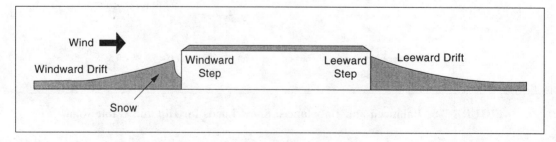

Wind

Windward Drift

Windward Step

Leeward Step

Leeward Drift

Snow

FIGURE 7-7. Drifts Formed at Windward and Leeward Steps

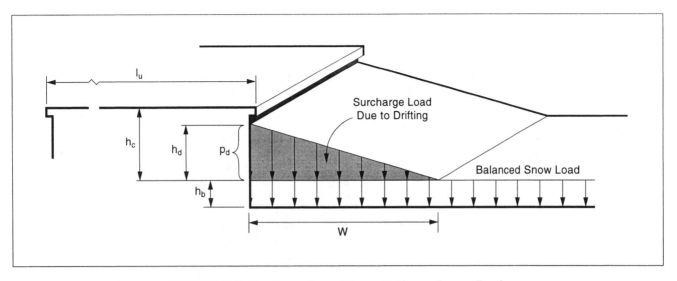

FIGURE 7-8. Configuration of Snow Drifts on Lower Roofs

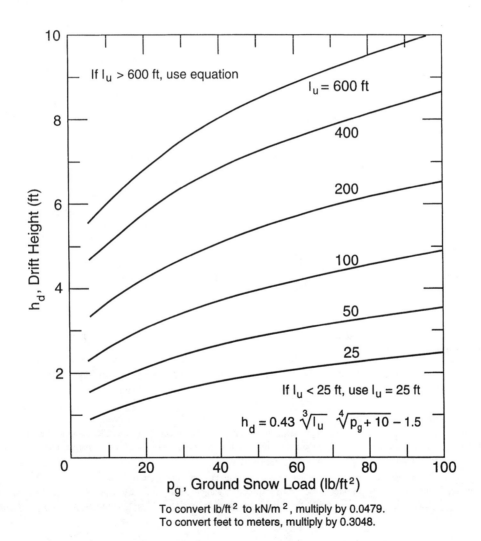

To convert lb/ft^2 to kN/m^2, multiply by 0.0479.
To convert feet to meters, multiply by 0.3048.

FIGURE 7-9. Graph and Equation for Determining Drift Height, h_d

TABLE 7-1. Ground Snow Loads, p_g, for Alaskan Locations

Location	p_g lb/ft^2	p_g (kN/m^2)	Location	p_g lb/ft^2	p_g (kN/m^2)	Location	p_g lb/ft^2	p_g (kN/m^2)
Adak	30	(1.4)	Galena	60	(2.9)	Petersburg	150	(7.2)
Anchorage	50	(2.4)	Gulkana	70	(3.4)	St Paul Islands	40	(1.9)
Angoon	70	(3.4)	Homer	40	(1.9)	Seward	50	(2.4)
Barrow	25	(1.2)	Juneau	60	(2.9)	Shemya	25	(1.2)
Barter Island	35	(1.7)	Kenai	70	(3.4)	Sitka	50	(2.4)
Bethel	40	(1.9)	Kodiak	30	(1.4)	Talkeetna	120	(5.8)
Big Delta	50	(2.4)	Kotzebue	60	(2.9)	Unalakleet	50	(2.4)
Cold Bay	25	(1.2)	McGrath	70	(3.4)	Valdez	160	(7.7)
Cordova	100	(4.8)	Nenana	80	(3.8)	Whittier	300	(14.4)
Fairbanks	60	(2.9)	Nome	70	(3.4)	Wrangell	60	(2.9)
Fort Yukon	60	(2.9)	Palmer	50	(2.4)	Yakutat	150	(7.2)

TABLE 7-2. Exposure Factor, C_e

Terrain Category	Exposure of Roof[1] Fully Exposed	Partially Exposed	Sheltered
A (see Section 6.5.3)	N/A	1.1	1.3
B (see Section 6.5.3)	0.9	1.0	1.2
C (see Section 6.5.3)	0.9	1.0	1.1
D (see Section 6.5.3)	0.8	0.9	1.0
Above the treeline in windswept mountainous areas.	0.7	0.8	N/A
In Alaska, in areas where trees do not exist within a 2-mile (3-km) radius of the site.	0.7	0.8	N/A

Notes: The terrain category and roof exposure condition chosen shall be representative of the anticipated conditions during the life of the structure. An exposure factor shall be determined for each roof of a structure.

[1]Definitions:

Partially Exposed: All roofs except as indicated below.

Fully Exposed: Roofs exposed on all sides with no shelter* afforded by terrain, higher structures or trees. Roofs that contain several large pieces of mechanical equipment, parapets which extend above the height of the balanced snow load (h_b), or other obstructions are not in this category.

Sheltered: Roofs located tight in among conifers that qualify as obstructions.

*Obstructions within a distance of $10h_o$ provide "shelter," where h_o is the height of the obstruction above the roof level. If the only obstructions are a few deciduous trees which are leafless in winter, the "fully exposed" category shall be used except for terrain Category "A." Note that these are heights above the roof. Heights used to establish the Terrain Category in Section 6.5.3 are heights above the ground.

TABLE 7-3. Thermal Factor, C_t

Thermal Condition[1]	C_t
All structures except as indicated below.	1.0
Structures kept just above freezing and others with cold, ventilated roofs in which the thermal resistance (R-value) between the ventilated space and the heated space exceeds $25°F \cdot h \cdot ft^2/Btu$ ($4.4 K \cdot m^2/W$).	1.1
Unheated structures and structures intentionally kept below freezing.	1.2
Continuously heated greenhouses[2] with a roof having a thermal resistance (R-value) less than $2.0°F \cdot h \cdot ft^2/Btu$ ($0.4 K \cdot m^2.W$).	0.85

[1]These conditions shall be representative of the anticipated conditions during winters for the life of the structure.

[2]Green houses with a constantly maintained interior temperature of 50°F (10°C) or more at any point 3 ft above the floor level during winters and having either a maintenance attendant on duty at all times or a temperature alarm system to provide warning in the event of a heating failure.

TABLE 7-4. Importance Factor, I, (Snow Loads)

Category[1]	I
I	0.8
II	1.0
III	1.1
IV	1.2

[1]See Section 1.5 and Table 1-1.

8.0 RAIN LOADS

8.1 SYMBOLS AND NOTATION

R = rain load on the undeflected roof, in pounds per square foot (kilonewtons per square meter). When the phrase "undeflected roof" is used, deflections from loads (including dead loads) shall not be considered when determining the amount of rain on the roof;

d_s = depth of water on the undeflected roof up to the inlet of the secondary drainage system when the primary drainage system is blocked (i.e., the static head), in inches (millimeters); and

d_h = additional depth of water on the undeflected roof above the inlet of the secondary drainage system at its design flow (i.e., the hydraulic head), in inches (millimeters).

8.2 ROOF DRAINAGE

Roof drainage systems shall be designed in accordance with the provisions of the code having jurisdiction. The flow capacity of secondary (overflow) drains or scuppers shall not be less than that of the primary drains or scuppers.

8.3 DESIGN RAIN LOADS

Each portion of a roof shall be designed to sustain the load of all rainwater that will accumulate on it if the primary drainage system for that portion is blocked plus the uniform load caused by water that rises above the inlet of the secondary drainage system at its design flow.

$$R = 5.2(d_s + d_h) \text{ [in SI: } R = 0.0098(d_s + d_h)]$$
(Eq. 8-1)

If the secondary drainage systems contain drain lines such lines and their point of discharge shall be separate from the primary drain lines.

8.4 PONDING INSTABILITY

"Ponding" refers to the retention of water due solely to the deflection of relatively flat roofs. Roofs with a slope less than 1/4 in./ft (1.19°) shall be investigated by structural analysis to assure that they possess adequate stiffness to preclude progressive deflection (i.e., instability) as rain falls on them or meltwater is created from snow on them. The larger of snow load or rain load shall be used in this analysis. The primary drainage system within an area subjected to ponding shall be considered to be blocked in this analysis.

8.5 CONTROLLED DRAINAGE

Roofs equipped with hardware to control the rate of drainage shall be equipped with a secondary drainage system at a higher elevation that limits accumulation of water on the roof above that elevation. Such roofs shall be designed to sustain the load of all rainwater that will accumulate on them to the elevation of the secondary drainage system plus the uniform load caused by water that rises above the inlet of the secondary drainage system at its design flow (determined from Section 8.3).

Such roofs shall also be checked for ponding instability (determined from Section 8.4).

9.0 EARTHQUAKE LOADS

[Note to user: This Section is based on the 1997 NEHRP Recommended Provisions for the Development of Seismic Regulations for New Buildings.]

9.1 GENERAL PROVISIONS

9.1.1 Purpose

Section 9 presents criteria for the design and construction of buildings and similar structures subject to earthquake ground motions. The specified earthquake loads are based upon post-elastic energy dissipation in the structure, and because of this fact, the provisions for design, detailing, and construction shall be satisfied even for structures and members for which load combinations that do not contain the earthquake effect indicate larger demands than combinations including earthquake.

9.1.2 Scope and Application

9.1.2.1 Scope

Every building, and portion thereof, shall be designed and constructed to resist the effects of earthquake motions as prescribed by these provisions. Certain nonbuilding structures, as described in Section 9.14, are within the scope and shall be designed and constructed as required for buildings. Additions to existing structures also shall be designed and constructed to resist the effects of earthquake motions as prescribed by these provisions. Existing structures and alterations to existing structures need only comply with these provisions when required by Sections 9.1.2.2 and 9.1.2.3.

Exceptions:

1. Structures located where the mapped spectral response acceleration at 1 second period, S_1, is less than or equal to 0.04g and the mapped short period spectral response acceleration, S_S, is less than or equal to 0.15g shall only be required to comply with Section 9.5.2.6.1.
2. Detached one- and two-family dwellings that are located where the mapped short period spectral response acceleration, S_S, is less than 0.4g or where the Seismic Design Category determined in accordance with Section 9.4.2 is A, B or C are exempt from the requirements of these provisions.
3. Detached one- and two-family wood frame dwellings not included in Exception 2 with not more than 2 stories above grade and satisfying the limi-

tations of Section A.9.12.10 are only required to be constructed in accordance with Section A.9.12.10.
4. Agricultural storage structures that are intended only for incidental human occupancy are exempt from the requirements of these provisions.

Special structures including, but not limited to, vehicular bridges, transmission towers, piers and wharves, hydraulic structures, and nuclear reactors require special consideration of their response characteristics and environment that is beyond the scope of these provisions.

9.1.2.2 Additions to Existing Structures

Additions shall be made to existing structures only as follows:

9.1.2.2.1

An addition that is structurally independent from an existing structure shall be designed and constructed in accordance with the seismic requirements for new structures.

9.1.2.2.2

An addition that is not structurally independent from an existing structure shall be designed and constructed such that the entire structure conforms to the seismic force resistance requirements for new structures unless the following three conditions are complied with:

1. The addition shall comply with the requirements for new structures,
2. The addition shall not increase the seismic forces in any structural element of the existing structure by more than 5% unless the capacity of the element subject to the increased forces is still in compliance with these provisions, and
3. The addition shall not decrease the seismic resistance of any structural element of the existing structure unless the reduced resistance is equal to or greater than that required for new structures.

9.1.2.3 Change of Use

When a change of use results in a structure being reclassified to a higher Seismic Use Group, the structure shall conform to the seismic requirements for new construction.

Exception: When a change of use results in a structure being reclassified from Seismic Use Group I to Seismic Use Group II and the structure is located in a seismic map area where $S_{DS} < 0.333$, compliance with these provisions is not required.

9.1.2.4 Application of Provisions

Buildings and structures within the scope of these provisions shall be designed and constructed as required by this section. When required by the authority having jurisdiction, design documents shall be submitted to determine compliance with these provisions.

9.1.2.4.1 New buildings

New buildings and structures shall be designed and constructed in accordance with the quality assurance requirements of Section 9.2.1. The analysis and design of structural systems and components, including foundations, frames, walls, floors, and roofs, shall be in accordance with the applicable requirements of Sections 9.5 and 9.7. Materials used in construction and components made of these materials shall be designed and constructed to meet the requirements of Section 9.8 through 9.12. Architectural, electrical, and mechanical systems and components including tenant improvements shall be designed in accordance with Section 9.6.

9.1.2.5 Alternate Materials and Methods of Construction

Alternate materials and methods of construction to those prescribed in these provisions shall not be used unless approved by the authority having jurisdiction. Substantiating evidence shall be submitted demonstrating that the proposed alternate, for the purpose intended, will be at least equal in strength, durability, and seismic resistance.

9.1.3 Seismic Use Groups

All structures shall be assigned to Seismic Use Group: I, II, or III, as specified in Table 9.1.3 corresponding to its Occupancy Category determined from Table 1-1:

TABLE 9.1.3. Seismic Use Group

Occupancy Category (see Table 1-1)	Seismic Use Group		
	I	II	III
I	X		
II	X		
III		X	
IV			X

9.1.3.1

High Hazard Exposure Structures: All buildings and structures assigned to Seismic Use Group III shall meet the following requirements:

9.1.3.1.1

Seismic Use Group III Structure Protected Access. Where operational access to a Seismic Use Group III structure is required through an adjacent structure, the adjacent structure shall conform to the requirements for Seismic Use Group III structures. Where operational access is less than 10 ft from the interior lot line or another structure on the same lot, protection from potential falling debris from adjacent structures shall be provided by the owner of the Seismic Use Group III structure.

9.1.3.1.2

Seismic Use Group III Function Designated seismic systems in Seismic Use Group III structures shall be provided with the capacity to function, insofar as practical, during and after an earthquake. Site-specific conditions as specified in Section 9.6.3.8 that could result in the interruption of utility services shall be considered when providing the capacity to continue to function.

9.1.3.2

This section is intentionally left blank.

9.1.3.3

This section is intentionally left blank.

9.1.3.4 Multiple Use

Structures having multiple uses shall be assigned the classification of the use having the highest Seismic Use Group except in structures having two or more portions which are structurally separated in accordance with Section 9.5.2.8, each portion shall be separately classified. Where a structurally separated portion of a structure provides access to, egress from, or shares life safety components with another portion having a higher Seismic Use Group, both portions shall be assigned the higher Seismic Use Group.

9.1.4 Occupancy Importance Factor

An occupancy importance factor, I, shall be assigned to each structure in accordance with Table 9.1.4.

9.2.1 Definitions

The definitions presented in this section provide the meaning of the terms used in these provisions.

TABLE 9.1.4. Occupancy Importance Factors

Seismic Use Group	I
I	1.0
II	1.25
III	1.5

Definitions of terms that have a specific meaning relative to the use of wood, steel, concrete, or masonry are presented in the section devoted to the material (Sections A.9.8 through A.9.12, respectively).

Active Fault: A fault determined to be active by the authority having jurisdiction, from properly substantiated geotechnical data (e.g., most recent mapping of active faults by the U.S. Geological Survey).

Addition: An increase in building area, aggregate floor area, height, or number of stories of a structure.

Adjusted Resistance (D'): The reference resistance adjusted to include the effects of all applicable adjustment factors resulting from end use and other modifying factors. Time effect factor (λ) adjustments are not included.

Alteration: Any construction or renovation to an existing structure other than an addition.

Appendage: An architectural component such as a canopy, marquee, ornamental balcony, or statuary.

Approval: The written acceptance by the authority having jurisdiction of documentation that establishes the qualification of a material, system, component, procedure, or person to fulfill the requirements of these provisions for the intended use.

Architectural Component Support: Those structural members or assemblies of members, including braces, frames, struts and attachments, that transmit all loads and forces between architectural systems, components, or elements and the structure.

Attachments: Means by which components and their supports are secured or connected to the seismic-force-resisting system of the structure. Such attachments include anchor bolts, welded connections, and mechanical fasteners.

Base: The level at which the horizontal seismic ground motions are considered to be imparted to the structure.

Base Shear: Total design lateral force or shear at the base.

Basement: A basement is any story below the lowest story above grade.

Boundary Elements: Diaphragm and shear wall boundary members to which diaphragm transfers forces. Boundary members include chords and drag struts at diaphragm and shear wall perimeters, interior openings, discontinuities, and re-entrant corners.

Boundary Members: Portions along wall and diaphragm edges strengthened by longitudinal and transverse reinforcement. Boundary members include chords and drag struts at diaphragm and shear wall perimeters, interior openings, discontinuities, and re-entrant corners.

Building: Any structure whose use could include shelter of human occupants.

Cantilevered Column System: A seismic-force-resisting system in which lateral forces are resisted entirely by columns acting as cantilevers from the foundation.

Component: A part or element of an architectural, electrical, mechanical, or structural system.

Component, Equipment: A mechanical or electrical component or element that is part of a mechanical and/or electrical system within or without a building system.

Component, Flexible: Component, including its attachments, having a fundamental period greater than 0.06 s.

Component, Rigid: Component, including its attachments, having a fundamental period less than or equal to 0.06 s.

Concrete, Plain: Concrete that is either unreinforced or contains less reinforcement than the minimum amount specified in Ref. 9.9-1 for reinforced concrete.

Concrete, Reinforced: Concrete reinforced with no less than the minimum amount required by Ref. 9.9-1, prestressed or nonprestressed, and designed on the assumption that the two materials act together in resisting forces.

Confined Region: That portion of a reinforced concrete or reinforced masonry component in which the concrete or masonry is confined by closely spaced special transverse reinforcement restraining the concrete or masonry in directions perpendicular to the applied stress.

Construction Documents: The written, graphic, electronic, and pictorial documents describing the design, locations, and physical characteristics of the project required to verify compliance with this Standard.

Container: A large-scale independent component used as a receptacle or vessel to accommodate plants, refuse, or similar uses, not including liquids.

Coupling Beam: A beam that is used to connect adjacent concrete wall elements to make them act together as a unit to resist lateral loads.

Deformability: The ratio of the ultimate deformation to the limit deformation.

High Deformability Element: An element whose deformability is not less than 3.5 when subjected to four fully reversed cycles at the limit deformation.

Limited Deformability Element: An element that is neither a low deformability or a high deformability element.

Low Deformability Element: An element whose deformability is 1.5 or less.

Deformation:

Limit Deformation: Two times the initial deformation that occurs at a load equal to 40% of the maximum strength.

Ultimate Deformation: The deformation at which failure occurs and which shall be deemed to occur if the sustainable load reduces to 80% or less of the maximum strength.

Design Earthquake Ground Motion: The earthquake effects that buildings and structures are specifically proportioned to resist as defined in Section 9.4.1.

Design Earthquake: The earthquake effects that are two-thirds of the corresponding maximum considered earthquake.

Designated Seismic Systems: The seismic force resisting system and those architectural, electrical, and mechanical systems or their components that require design in accordance with Section 9.6.1 and for which the component importance factor, I_p, is 1.0.

Diaphragm: A horizontal, or nearly horizontal, portion of the seismic force resisting system designed to transmit seismic forces to the vertical elements of the seismic force resisting system. Diaphragms which do not meet this criteria shall be considered rigid for the purposes of this standard. Diaphragms are classified as either flexible or rigid according to the requirements of Section 9.5.2.3.1.

Diaphragm, Blocked: A diaphragm in which all sheathing edges not occurring on a framing member are supported on and fastened to blocking.

Diaphragm Boundary: A location where shear is transferred into or out of the diaphragm element. Transfer is either to a boundary element or to another force-resisting element.

Diaphragm Chord: A diaphragm boundary element perpendicular to the applied load that is assumed to take axial stresses due to the diaphragm moment in a manner analogous to the flanges of a beam. Also applies to shear walls.

Displacement:

Design Displacement: The design earthquake lateral displacement, excluding additional displacement due to actual and accidental torsion, required for design of the isolation system.

Total Design Displacement: The design earthquake lateral displacement, including additional displacement due to actual and accidental torsion, required for design of the isolation system or an element thereof.

Total Maximum Displacement: The maximum considered earthquake lateral displacement, including additional displacement due to actual and accidental torsion, required for verification of the stability of the isolation system or elements thereof, design of structure separations, and vertical load testing of isolator unit prototypes.

Displacement Restraint System: A collection of structural elements that limits lateral displacement of seismically isolated structures due to the maximum considered earthquake.

Drag Strut (Collector, Tie, Diaphragm Strut): A diaphragm or shear wall boundary element parallel to the applied load that collects and transfers diaphragm shear forces to the vertical-force-resisting elements or distributes forces within the diaphragm or shear wall. A drag strut often is an extension of a boundary element that transfers forces into the diaphragm or shear wall. See Sections 9.5.2.6.3.2 and 9.5.2.6.4.2.

Effective Damping: The value of equivalent viscous damping corresponding to energy dissipated during cyclic response of the isolation system.

Effective Stiffness: The value of the lateral force in the isolation system, or an element thereof, divided by the corresponding lateral displacement.

Enclosure: An interior space surrounded by walls.

Equipment Support: Those structural members or assemblies of members or manufactured elements, including braces, frames, legs, lugs, snuggers, hangers or saddles, that transmit gravity loads and operating loads between the equipment and the structure.

Essential Facility: A structure required for post-earthquake recovery.

Factored Resistance ($\lambda\phi D$): Reference resistance multiplied by the time effect and resistance factors. This value must be adjusted for other factors such as size effects, moisture conditions, and other end-use factors.

Flexible Equipment Connections: Those connections between equipment components that permit rotational and/or translational movement without degradation of performance. Examples include universal joints, bellows expansion joints, and flexible metal hose.

Frame:

Braced Frame: An essentially vertical truss, or its equivalent, of the concentric or eccentric type that is provided in a bearing wall, building frame, or dual system to resist seismic forces.

Concentrically Braced Frame (CBF): A braced frame in which the members are subjected primarily to axial forces.

Eccentrically Braced Frame (EBF): A diagonally braced frame in which at least one end of each brace frames into a beam a short distance from a beam-column joint or from another diagonal brace.

Ordinary Concentrically Braced Frame (OCBF): A steel concentrically braced frame in which members and connections are designed in accordance with the provisions of Ref. 9.8-3 without modification.

Special Concentrically Braced Frame (SCBF): A steel or composite steel and concrete concentrically braced frame in which members and connections are designed for ductile behavior. Special concentrically braced frames shall conform to Section A.9.8.1.3.1

Moment Frame:

Intermediate Moment Frame (IMF): A moment frame in which members and joints are capable of resisting forces by flexure as well as along the axis of the members. Intermediate moment frames of reinforced concrete shall conform to Ref. 9.9-1. Intermediate moment frames of structural steel construction shall conform to Sections 9.8-3 and 10. Intermediate moment frames of composite construction shall conform to Ref. 9.10-3, Part II, Sections 6.4b, 7, 8, and 10.

Ordinary Moment Frame (OMF): A moment frame in which members and joints are capable of resisting forces by flexure as well as along the axis of the members. Ordinary moment frames shall conform to Ref. 9.9-1, exclusive of Chapter 21, Ref. 9.8-3, Section 12, or Section A.9.9.3.1.

Special Moment Frame (SMF): A moment frame in which members and joints are capable of resisting forces by flexure as well as along the axis of the members. Special moment frames shall conform to Ref. 9.8-3 or 9.9-1.

Frame System:

Building Frame System: A structural system with an essentially complete space frame providing support for vertical loads. Seismic force resistance is provided by shear walls or braced frames.

Dual Frame System: A structural system with an essentially complete space frame providing support for vertical loads. Seismic force resistance is provided by moment resisting frames and shear walls or braced frames as prescribed in Section 9.5.2.2.1.

Space Frame System: A structural system composed of interconnected members, other than bearing walls, that is capable of supporting vertical loads and, when designed for such an application, is capable of providing resistance to seismic forces.

Grade Plane: A reference plane representing the average of finished ground level adjoining the structure at all exterior walls. Where the finished ground level slopes away from the exterior walls, the reference plane shall be established by the lowest points within the area between the buildings and the lot line or, where the lot line is more than 6 ft (1,800 mm) from the structure, between the structure and a point 6 ft (1,800 mm) from the structure.

Hazardous Contents: A material that is highly toxic or potentially explosive and in sufficient quantity to pose a significant life-safety threat to the general public if an uncontrolled release were to occur.

High Temperature Energy Source: A fluid, gas, or vapor whose temperature exceeds 220°F.

Inspection, Special: The observation of the work by the special inspector to determine compliance with the approved construction documents and these standards.

Continuous Special Inspection: The full-time observation of the work by an approved special inspector who is present in the area where work is being performed.

Periodic Special Inspection: The part-time or intermittent observation of the work by an approved special inspector who is present in the area where work has been or is being performed.

Inspector, Special (who shall be identified as the Owner's Inspector): A person approved by the authority having jurisdiction to perform special inspection. The authority having jurisdiction shall have the option to approve the quality assurance personnel of a fabricator as a special inspector.

Inverted Pendulum Type Structures: Structures that have a large portion of their mass concentrated near the top and, thus, have essentially one degree of freedom in horizontal translation. The structures are usually T-shaped with a single column supporting the beams or framing at the top.

Isolation Interface: The boundary between the upper portion of the structure, which is isolated, and the lower portion of the structure, which moves rigidly with the ground.

Isolation System: The collection of structural elements that includes all individual isolator units, all structural elements that transfer force between elements of the isolation system, and all connections to other structural elements. The isolation system also includes the wind-restraint system, energy-dissipation devices, and/or the displacement restraint system if such systems and devices are used to meet the design requirements of Section 9.13.

Isolator Unit: A horizontally flexible and vertically stiff structural element of the isolation system that permits large lateral deformations under design seismic load. An isolator unit may be used either as part of or in addition to the weight-supporting system of the structure.

Joint: The geometric volume common to intersecting members.

Load:

Dead Load: The gravity load due to the weight of all permanent structural and nonstructural components of a building such as walls, floors, roofs, and the operating weight of fixed service equipment.

Gravity Load (W): The total dead load and applicable portions of other loads as defined in Section 9.5.3.2.

Live Load: The load superimposed by the use and occupancy of the building not including the wind load, earthquake load, or dead load; see Section 9.5.3.2.

Maximum Considered Earthquake Ground Motion: The most severe earthquake effects considered by these Standards as defined in Section 9.4.1.

Nonbuilding Structure: A structure, other than a building, constructed of a type included in Section 9.14 and within the limits of Section 9.14.1.1.

Occupancy Importance Factor: A factor assigned to each structure according to its Seismic Use Group as prescribed in Section 9.1.4.

Owner: Any person, agent, firm, or corporation having a legal or equitable interest in the property.

Partition: A nonstructural interior wall that spans horizontally or vertically from support to support. The supports may be the basic building frame, subsidiary structural members or other portions of the partition system.

P-Delta Effect: The secondary effect on shears and moments of structural members due to the action of the vertical loads induced by displacement of the structure resulting from various loading conditions.

Quality Assurance Plan: A detailed written procedure that establishes the systems and components subject to special inspection and testing. The type and frequency of testing and the extent and duration of special inspection are given in the quality assurance plan.

Reference Resistance (D): The resistance (force or moment as appropriate) of a member or connection computed at the reference end use conditions.

Registered Design Professional: An architect or engineer, registered or licensed to practice professional architecture or engineering, as defined by the statutory requirements of the professional registrations laws of the state in which the project is to be constructed.

Roofing Unit: A unit of roofing tile or similar material weighing more than 1 lb.

Seismic Design Category: A classification assigned to a structure based on its Seismic Use Group and the severity of the design earthquake ground motion at the site as defined in Section 9.1.2.5.

Seismic-Force-Resisting System: That part of the structural system that has been considered in the design to provide the required resistance to the seismic forces prescribed herein.

Seismic Forces: The assumed forces prescribed herein, related to the response of the structure to earthquake motions, to be used in the design of the structure and its components.

Seismic Response Coefficient: Coefficient C_s as determined from Section 9.5.3.2.1.

Seismic Use Group: A classification assigned to a structure based on its use as defined in Section 9.1.3.

Shallow Anchor: Anchors with embedment length-to-diameter ratios of less than 8.

Shear Panel: A floor, roof, or wall component sheathed to act as a shear wall or diaphragm.

Site Class: A classification assigned to a site based on the types of soils present and their engineering properties as defined in Section 9.4.1.2.

Site Coefficients: The values of F_a and F_v as indicated in Tables 9.4.1.2.4a and 9.4.1.2.4b, respectively.

Special Transverse Reinforcement: Reinforcement composed of spirals, closed stirrups, or hoops and supplementary cross-ties provided to restrain the concrete and qualify the portion of the component, where used, as a confined region.

Storage Racks: Include industrial pallet racks, moveable shelf racks, and stacker racks made of cold-formed or hot-rolled structural members. Does not include other types of racks such as drive-in and drive-through racks, cantilever racks, portable racks, or racks made of materials other than steel.

Story: The portion of a structure between the top of two successive finished floor surfaces and, for the topmost story, from the top of the floor finish to the top of the roof structural element.

Story Above Grade: Any story having its finished floor surface entirely above grade, except that a story shall be considered as a story above grade where the finished floor surface of the story immediately above is more than 6 ft (1,800 mm) above the grade plane, more than 6 ft (1,800 mm) above the finished ground level for more than 40% of the total structure perimeter, or more than 12 ft (3,600 mm) above the finished ground level at any point.

Story Drift: The difference of horizontal deflections at the top and bottom of the story as determined in Section 9.5.3.7.1.

Story Drift Ratio: The story drift, as determined in Section 9.5.3.7.1, divided by the story height.

Story Shear: The summation of design lateral seismic forces at levels above the story under consideration.

Strength:

Design Strength: Nominal strength multiplied by a strength reduction factor, ϕ.

Nominal Strength: Strength of a member or cross section calculated in accordance with the requirements and assumptions of the strength design methods of this Standard (or the referenced standards) before application of any strength reduction factors.

Required Strength: Strength of a member, cross section, or connection required to resist factored loads or related internal moments and forces in such combinations as stipulated by this Standard.

Structure: That which is built or constructed and limited to buildings and nonbuilding structures as defined herein.

Structural Observations: The visual observations performed by the registered design professional in responsible charge (or another registered design

professional) to determine that the seismic-force-resisting system is constructed in general conformance with the construction documents.

Structural-Use Panel: A wood-based panel product that meets the requirements of Ref. 9.12-3 or 9.12-5 and is bonded with a waterproof adhesive. Included under this designation are plywood, oriented strand board, and composite panels.

Subdiaphragm: A portion of a diaphragm used to transfer wall anchorage forces to diaphragm cross ties.

Testing Agency: A company or corporation that provides testing and/or inspection services. The person in charge of the special inspector(s) and the testing services shall be a registered design professional.

Tie-Down (Hold-Down): A device used to resist uplift of the boundary elements of shear walls. These devices are intended to resist load without significant slip between the device and the shear wall boundary element or be shown with cyclic testing to not reduce the wall capacity or ductility.

Time Effect Factor (λ): A factor applied to the adjusted resistance to account for effects of duration of load.

Torsional Force Distribution: The distribution of horizontal shear through a rigid diaphragm when the center of mass of the structure at the level under consideration does not coincide with the center of rigidity (sometimes referred to as diaphragm rotation).

Toughness: The ability of a material to absorb energy without losing significant strength.

Utility or Service Interface: The connection of the structure's mechanical and electrical distribution systems to the utility or service company's distribution system.

Veneers: Facings or ornamentation of brick, concrete, stone, tile, or similar materials attached to a backing.

Wall: A component that has a slope of 60° or greater with the horizontal plane used to enclose or divide space.

Bearing Wall: Any wall meeting either of the following classifications:

1. Any metal or wood stud wall that supports more than 100 lbs/linear ft (1,400 N/m) of vertical load in addition to its own weight.
2. Any concrete or masonry wall that supports more than 200 lbs/linear ft (2,900 N/m) of vertical load in addition to its own weight.

Cripple Wall: Short stud wall, less than 8 ft (2,400 mm) in height, between the foundation and the lowest framed floors with studs not less than

14 in. long—also known as a knee wall. Cripple walls can occur in both engineered structures and conventional construction.

Light Framed Wall: A wall with wood or steel studs.

Light-Framed Wood Shear Wall: A wall constructed with wood studs and sheathed with material rated for shear resistance.

Nonbearing Wall: Any wall that is not a bearing wall.

Nonstructural Wall: All walls other than bearing walls or shear walls.

Shear Wall (Vertical Diaphragm): A wall, bearing or nonbearing, designed to resist lateral seismic forces acting in the plane of the wall (sometimes referred to as a vertical diaphragm).

Wall System, Bearing: A structural system with bearing walls providing support for all or major portions of the vertical loads. Shear walls or braced frames provide seismic force resistance.

Wind-Restraint System: The collection of structural elements that provides restraint of the seismic-isolated structure for wind loads. The wind-restraint system may be either an integral part of isolator units or a separate device.

9.2.2 Symbols

The unit dimensions used with the items covered by the symbols shall be consistent throughout except where specifically noted. The symbols and definitions presented in this section apply to these provisions as indicated.

A, B, C, D, E, F = Seismic Performance Categories as defined in Tables 9.4.2.1a and 9.4.2.1b;

A, B, C, D, E, F = Site Classes as defined in Section 9.4.1.2;

A_{ch} = cross-sectional area (in.^{2} or mm^2) of a component measured to the outside of the special lateral reinforcement;

A_0 = area of the load-carrying foundation (ft^2 or m^2);

A_{sh} = total cross-sectional area of hoop reinforcement (in.^{2} or mm^2), including supplementary cross-ties, having a spacing of s_h and crossing a section with a core dimension of h_c;

A_{vd} = required area of leg (in.^{2} or mm^2) of diagonal reinforcement;

A_x = torsional amplification factor, Section 9.5.3.5.2;

a_d = incremental factor related to P-delta effects in Section 9.5.3.7.2;

a_p = amplification factor related to the response of a system or component as affected by the type of seismic attachment, determined in Section 9.6.1.3;

B_D = numerical coefficient as set forth in Table 9.13.3.3.1 for effective damping equal to β_D;

B_M = numerical coefficient as set forth in Table 9.13.3.3.1 for effective damping equal to β_M;

b = shortest plan dimension of the structure, in ft (mm) measured perpendicular to d;

C_d = deflection amplification factor as given in Table 9.5.2.2;

C_s = seismic response coefficient determined in Section 9.5.3.2 (dimensionless).

C_s = seismic response coefficient determined in Section 9.5.5.2.1 and 9.5.5.3.1 (dimensionless);

C_{sm} = modal seismic response coefficient determined in Section 9.5.4.5 (dimensionless);

C_T = building period coefficient in Section 9.5.3.3;

C_{vx} = vertical distribution factor as determined in Section 9.5.3.4;

c = distance from the neutral axis of a flexural member to the fiber of maximum compressive strain (in. or mm);

D = effect of dead load;

D_D = design displacement, in in. (mm), at the center of rigidity of the isolation system in the direction under consideration as prescribed by Eq. 9.13.3.3.1;

D'_D = design displacement, in in. (mm), at the center of rigidity of the isolation system in the direction under consideration, as prescribed by Eq. 9.13.4.2-1;

D_M = maximum displacement, in in. (mm), at the center of rigidity of the isolation system in the direction under consideration, as prescribed by Eq. 9.13.3.3.3;

D'_M = maximum displacement, in in. (mm), at the center of rigidity of the isolation system in the direction under consideration, as prescribed by Eq. 9.13.4.2-2.;

D_p = relative seismic displacement that the component must be designed to accommodate as defined in Section 9.6.1.4;

D_s = total depth of the stratum in Eq. 9.5.5.2.1.2-4 (ft or m);

D_{TD} = total design displacement, in in. (mm), of an element of the isolation system including both translational displacement at the center of rigidity and the component of torsional displacement in the direction under consideration as prescribed by Eq. 9.13.3.3.5-1;

D_{TM} = total maximum displacement, in in. (mm), of an element of the isolation system including both translational displacement at the center of rigidity and the component of torsional displacement in the direction under consideration as prescribed by Eq. 9.13.3.3.5-2;

d = overall depth of member (in. or mm) in Section 9.5;

d = longest plan dimension of the structure, in ft (mm), in Section 9.13;

d_p = longest plan dimension of the structure, in ft (mm);

E = effect of horizontal and vertical earthquake-induced forces (Sections 9.5.2.7 and 9.13);

E_{loop} = energy dissipated in kip-in. (kN-mm), in an isolator unit during a full cycle of reversible load over a test displacement range from Δ^+ to Δ^-, as measured by the area enclosed by the loop of the force-deflection curve;

e = actual eccentricity, in ft (mm), measured in plan between the center of mass of the structure above the isolation interface and the center of rigidity of the isolation system, plus accidental eccentricity, in ft (mm), taken as 5% of the maximum building dimension perpendicular to the direction of force under consideration;

F_a = acceleration-based site coefficient (at 0.3 s-period);

F^- = maximum negative force in an isolator unit during a single cycle of prototype testing at a displacement amplitude of Δ^-;

F^+ = positive force in kips (kN) in an isolator unit during a single cycle of prototype testing at a displacement amplitude of Δ^+;

F_i, F_n, F_x = portion of the seismic base shear, V, induced at Level i, n, or x, respectively, as determined in Section 9.5.3.4;

F_p = seismic force acting on a component of a structure as determined in Sections 9.5.2.6.1.2, 9.5.2.5.1.3, 9.5.2.6.1.4 or 9.6.1.3;

F_v = velocity-based site coefficient (at 1.0 s period);

F_x = total force distributed over the height of the structure above the isolation interface as prescribed by Eq. 9.13.3.5;

F_{xm} = portion of the seismic base shear, V_m, induced at Level x as determined in Section 9.5.4.6;

f'_c = specified compressive strength of concrete used in design;

f'_s = ultimate tensile strength (psi or MPa) of the bolt, stud, or insert leg wires. For A307 bolts or A108 studs, is permitted to be assumed to be 60,000 psi (415 MPa);

f_y = specified yield strength of reinforcement (psi or MPa);

f_{yh} = specified yield stress of the special lateral reinforcement (psi or kPa);

$G = \lambda\, v_s^2/g$ = average shear modulus for the soils beneath the foundation at large strain levels (psf or Pa);

$G_0 = \lambda\, v_{so}^2/g$ = average shear modulus for the soils beneath the foundation at small strain levels (psf or Pa);

g = acceleration due to gravity;

H = thickness of soil;

h = height of a shear wall measured as the maximum clear height from the foundation to the bottom of the floor or roof framing above or the maximum clear height from the top of the floor or roof framing to the bottom of the floor or roof framing above;

h = roof elevation of a structure in Section 9.6;

$\bar{h}$ = effective height of the building as determined in Section 9.5.5.2 or 9.5.5.3 (ft or m);

h_c = The core dimension of a component measured to the outside of the special lateral reinforcement (in. or mm);

h_i, h_n, h_x = height above the base Level i, n, or x, respectively;

h_{sx} = story height below Level $x = (h_x - h_{x-1})$;

I = occupancy importance factor in Section 9.1.4;

I_0 = static moment of inertia of the load-carrying foundation; see Section 9.5.5.2.1 (in.^{4} or mm^4);

I_p = component importance factor as prescribed in Section 9.6.1.5;

i = building level referred to by the subscript i; $i = 1$ designates the first level above the base;

K_p = stiffness of the component or attachment, Section 9.6.3.3;

K_y = lateral stiffness of the foundation as defined in Section 9.5.5.2.1.1 (lb/in. or N/m);

K_θ = rocking stiffness of the foundation as defined in Section 9.5.5.2.1.1 (ft-lb/degree or N-m/rad);

KL/r = lateral slenderness of a compression member measured in terms of its effective buckling length, KL, and the least radius of gyration of the member cross section, r;

K_{Dmax} = maximum effective stiffness, in kips/in. (kN/mm), of the isolation system at the design displacement in the horizontal direction under consideration as prescribed by Eq. 9.13.9.5.1-1;

K_{Dmin} = minimum effective stiffness, in kips/in. (kN/mm), of the isolation system at the design displacement in the horizontal direction under consideration as prescribed by Eq. 9.13.9.5.1-2;

K_{max} = maximum effective stiffness, in kips/in. (kN/mm), of the isolation system at the maximum displacement in the horizontal direction under consideration as prescribed by Eq. 9.13.9.5.1-3;

K_{min} = minimum effective stiffness, in kips/in. (kN/mm), of the isolation system at the maximum displacement in the horizontal direction under consideration, as prescribed by Eq. 9.13.9.5.1-4;

k = distribution exponent given in Section 9.5.3.4;

$\bar{k}$ = stiffness of the building as determined in Section 9.5.5.2.1.1 (lb/ft or N/m);

k_{eff} = effective stiffness of an isolator unit, as prescribed by Eq. 9.13.9.3-1;

L = overall length of the building (ft or m) at the base in the direction being analyzed;

L = length of bracing member (in. or mm) in Section A.9.8;

L = effect of live load in Section 9.13;

L_0 = overall length of the side of the foundation in the direction being analyzed, Section 9.5.5.2.1.2 (ft or m);

l = dimension of a diaphragm perpendicular to the direction of application of force. For open-front structures, l is the length from the edge of the diaphragm at the open front to the vertical resisting elements parallel to the direction of the applied force. For a cantilevered diaphragm, l is the length of the cantilever;

M_f = foundation overturning design moment as defined in Section 9.5.3.6 (ft-kip or kN-m);

M_0, M_{01} = overturning moment at the foundation-soil interface as determined in Sections 9.5.5.2.3 and 9.5.5.3.2 (ft-lb or N-m);

M_f = foundation overturning design moment as defined in Section 9.5.3.6;

M_t = torsional moment resulting from the location of the building masses, Section 9.5.3.5.2;

M_{ta} = accidental torsional moment as determined in Section 9.5.3.5.2;

M_x = building overturning design moment at Level x as defined in Section 9.5.3.6 or 9.5.4.7;

m = subscript denoting the mode of vibration under consideration; i.e., $m = 1$ for the fundamental mode;

N = number of stories, Section 9.5.3.3;

N = standard penetration resistance (blows/ft), ASTM D1536-84;

$\bar{N}$ = average field standard penetration resistance for the top 100 ft (30 m); see Section 9.4.1.2;

N_{ch} = average standard penetration resistance for cohesionless soil layers for the top 100 ft (30 m); see Section 9.4.1.2;

n = designates the level that is uppermost in the main portion of the building;

P_D = required axial strength on a column resulting from application of dead load, D, in Section 9.5 (kip or kN);

P_E = required axial strength on a column resulting from application of the amplified earthquake load, E', in Section 9.5 (kip or kN);

P_L = required axial strength on a column resulting from application of live load, L, in Section 9.5 (kip or kN);

P_n = nominal axial load strength (lb or N) in Section A.9.8;

P_n = algebraic sum of the shear wall and the minimum gravity loads on the joint surface acting simultaneously with the shear (lb or N);

P_u^* = required axial strength on a brace (kip or kN) in Section A.9.8;

P_x = total unfactored vertical design load at and above Level x, for use in Section 9.5.3.7.2;

PI = plasticity index, ASTM D4318-93;

Q_E = effect of horizontal seismic (earthquake-induced) forces, Section 9.5.2.7;

Q_v = load equivalent to the effect of the horizontal and vertical shear strength of the vertical segment, Section A.9.8;

R = response modification coefficient as given in Table 9.5.2.2;

R_p = component response modification factor as defined in Section 9.6.1.3;

r = characteristic length of the foundation as defined in Section 9.5.5.2.1;

r_a = characteristic foundation length as defined by Eq. 9.5.5.2.1.2-2 (ft or m);

r_m = characteristic foundation length as defined by Eq. 9.5.5.2.1.2-3 (ft or m);

r_x = ratio of the design story shear resisted by the most heavily loaded single element in the story, in direction x, to the total story shear;

S_S = mapped maximum considered earthquake, 5% damped, spectral response acceleration at short periods as defined in Section 9.4.1.2;

S_1 = mapped maximum considered earthquake, 5% damped, spectral response acceleration at a period of 1 s as defined in Section 9.4.1.2;

S_{DS} = design, 5% damped, spectral response acceleration at short periods as defined in Section 9.4.1.2;

S_{D1} = design, 5% damped, spectral response acceleration at a period of 1 s as defined in Section 9.4.1.2;

S_{MS} = maximum considered earthquake, 5% damped, spectral response acceleration at short periods adjusted for site class effects as defined in Section 9.4.1.2;

S_{M1} = maximum considered earthquake, 5% damped, spectral response acceleration at a period of 1 s adjusted for site class effects as defined in Section 9.4.1.2;

$\bar{s}_u$ = average undrained shear strength in top 100 ft (30 m); see Section 9.4.1.2, ASTM D2166-91 or ASTM D2850-87;

s_h = spacing of special lateral reinforcement (in. or mm);

T = fundamental period of the building as determined in Section 9.5.3.2.1;

T, T_1 = effective fundamental period (s) of the building as determined in Sections 9.5.5.2.1.1 and 9.5.5.3.1;

T_a = approximate fundamental period of the building as determined in Section 9.5.3.3;

T_D = effective period, in seconds (s), of the seismically isolated structure at the design displacement in the direction under consideration as prescribed by Eq. 9.13.3.3.2;

T_m = modal period of vibration of the mth mode of the building as determined in Section 9.5.4.5;

T_M = effective period, in seconds (s), of the seismically isolated structure at the maximum displacement in the direction under consideration as prescribed by Eq. 9.13.3.3.4;

T_p = fundamental period of the component and its attachment, Section 9.6.3.3;

$T_0 = 0.2 S_{D1}/S_{DS}$;

$T_S = S_{D1}/S_{DS}$;

T_M = effective period, in seconds (s), of the seismically isolated structure at the maximum displacement in the direction under consideration as prescribed by Eq. 9.13.3.3.4;

T_m = modal period of vibration (s) of the mth mode of the building as determined in Section 9.5.4.5;

T_4 = net tension in steel cable due to dead load, prestress, live load, and seismic load (Section A.9.8.5);

V = total design lateral force or shear at the base, Section 9.5.3.2;

V_b = total lateral seismic design force or shear on elements of the isolation system or elements below the isolation system as prescribed by Eq. 9.13.3.4.1;

V_s = total lateral seismic design force or shear on elements above the isolation system as prescribed by Eq. 9.13.3.4.2;

V_t = design value of the seismic base shear as determined in Section 9.5.4.8;

V_u = required shear strength (lb or N) due to factored loads in Section 9.6;

V_x = seismic design shear in Story x as determined in Section 9.5.3.5 or 9.5.4.8;

V_1 = portion of the seismic base shear, $\bar{V}$, contributed by the fundamental mode, Section 9.5.5.3 (kip or kN);

ΔV = reduction in V as determined in Section 9.5.5.2 (kip or kN);

ΔV_1 = reduction in V_1 as determined in Section 9.5.5.3 (kip or kN);

v_s = average shear wave velocity for the soils beneath the foundation at large strain levels, Section 9.5.5.2 (ft/s or m/s);

$\bar{v}_s$ = average shear wave velocity in top 100 ft (30 m); see Section 9.4.1.2;

v_{so} = average shear wave velocity for the soils beneath the foundation at small strain levels, Section 9.5.5.2 (ft/s or m/s);

W = total gravity load of the building as defined in Section 9.5.3.2. For calculation of seismic-isolated building period W is the total seismic dead load weight of the building as defined in Section 9.5.5.2 and 9.5.5.3 (kip or kN);

$\bar{W}$ = effective gravity load of the building as defined in Section 9.5.5.2 and 9.5.5.3 (kip or kN);

W_c = gravity load of a component of the building;

W_D = energy dissipated per cycle at the story displacement for the design earthquake (Section 9.13.3.2);

W_m = effective modal gravity load determined in accordance with Eq. 9.5.4.5-2;

W_p = component operating weight (lb or N);

w = moisture content (in percent), ASTM D2216-92;

w_i, w_n, w_x = portion of W that is located at or assigned to Level i, n, or x, respectively;

x = level under consideration;

$x = 1$ designates the first level above the base;

y = elevations difference between points of attachment in Section 9.6;

y = distance, in ft (mm), between the center of rigidity of the isolation system rigidity and the element of interest measured perpendicular to the direction of seismic loading under consideration Section 9.13;

z = level under consideration; $x = 1$ designates the first level above the base;

α = relative weight density of the structure and the soil as determined in Section 9.5.5.2.1;

α = angle between diagonal reinforcement and longitudinal axis of the member (degree or rad);

β = ratio of shear demand to shear capacity for the story between Level x and $x - 1$;

β = fraction of critical damping for the coupled structure-foundation system, determined in Section 9.5.5.2.1;

β_D = effective damping of the isolation system at the design displacement as prescribed by Eq. 9.13.9.5.2-1;

β_M = effective damping of the isolation system at the maximum displacement as prescribed by Eq. 9.13.5.2-2;

β_0 = foundation damping factor as specified in Section 9.5.5.2.1;

β_{eff} = effective damping of the isolation system as prescribed by Eq. 9.13.9.3-2;

γ = average unit weight of soil (lb/ft^3 or kg/m^3);

Δ = design story drift as determined in Section 9.5.3.7.1;

Δ_a = allowable story drift as specified in Section 9.5.2.8;

Δ_m = design modal story drift determined in Section 9.5.4.6;

Δ^+ = maximum positive displacement of an isolator unit during each cycle of prototype testing;

Δ^- = maximum negative displacement of an isolator unit during each cycle of prototype testing;

δ_{max} = maximum displacement at Level x, considering torsion, Section 9.5.3.5.2;

δ_{avg} = average of the displacements at the extreme points of the structure at Level x, Section 9.5.3.5.2;

δ_x = deflection of Level x at the center of the mass at and above Level x, Eq. 9.5.3.7.1;

δ_{xe} = deflection of Level x at the center of the mass at and above Level x determined by an elastic analysis, Section 9.5.3.7.1;

δ_{xem} = modal deflection of Level x at the center of the mass at and above Level x determined by an elastic analysis, Section 9.5.4.6;

δ_{xm} = modal deflection of Level x at the center of the mass at and above Level x as determined by Eq. 9.5.4.6-3 and 9.13.3.2-1;

δ_x, δ_{x1} = deflection of Level x at the center of the mass at and above Level x, Eqs. 9.5.5.2.3-1 and 9.5.5.3.2-1 (in. or mm);

θ = stability coefficient for P-delta effects as determined in Section 9.5.3.7.2;

τ = overturning moment reduction factor, Eq. 9.5.3.6;

ρ = reliability coefficient based on the extent of structural redundance present in a building as defined in Section 9.5.2.7;

ρ_s = spiral reinforcement ratio for precast prestressed piles in Sections A.9.7.4.4.5 and A.9.7.5.4.4;

ρ_x = reliability coefficient based on the extent of structural redundancy present in the seismic-force-resisting system of a building in the x direction;

λ = time effect factor;

ϕ = capacity reduction factor;

ϕ = strength reduction factor or resistance factor;

ϕ_{im} = displacement amplitude at the ith

level of the building for the fixed
base condition when vibrating in
its mth mode, Section 9.5.4.5;

Ω_0 = overstrength factor as defined in
Table 9.5.2.2;

ΣE_D = total energy dissipated, in kip-in.
(kN-mm), in the isolation system
during a full cycle of response at
the design displacement, D_D;

ΣE_M = total energy dissipated, in kip-in.
(kN-mm), in the isolation system
during a full cycle of response at
the maximum displacement, D_M;

$\Sigma|F_D^+|_{max}$ = sum, for all isolator units, of the
maximum absolute value of force,
in kips (kN), at a positive displacement equal to D_D;

$\Sigma|F_D^+|_{min}$ = sum, for all isolator units, of the
minimum absolute value of force,
in kips (kN), at a positive displacement equal to D_D;

$\Sigma|F_D^-|_{max}$ = sum, for all isolator units, of the
maximum absolute value of force,
in kips (kN), at a negative displacement equal to D_D;

$\Sigma|F_D^-|_{min}$ = sum, for all isolator units, of the
minimum absolute value of force,
in kips (kN), at a negative displacement equal to D_D;

$\Sigma|F_M^+|_{max}$ = sum, for all isolator units, of the
maximum absolute value of force,
in kips (kN), at a positive displacement equal to D_M;

$\Sigma|F_M^+|_{min}$ = sum, for all isolator units, of the
minimum absolute value of force,
in kips (kN), at a positive displacement equal to D_M;

$\Sigma|F_M^-|_{max}$ = sum, for all isolator units, of the
maximum absolute value of force,
in kips (kN), at a negative displacement equal to D_M; and

$\Sigma|F_M^-|_{min}$ = sum, for all isolator units, of the
minimum absolute value of force,
in kips (kN), at a negative displacement equal to D_M.

9.3 THIS SECTION LEFT INTENTIONALLY BLANK

9.4.1 Procedures for Determining Maximum Considered Earthquake and Design Earthquake Ground Motion Accelerations and Response Spectra

Ground motion accelerations, represented by response spectra and coefficients derived from these
spectra, shall be determined in accordance with the
general procedure of Section 9.4.1.2 or the site-specific procedure of Section 9.4.1.3. The general procedure in which spectral response acceleration parameters for the maximum considered earthquake ground
motions are derived using Maps 1 through 32, modified by site coefficients to include local site effects
and scaled to design values, are permitted to be used
for any structure except as specifically indicated in
these Provisions. The site-specific procedure also is
permitted to be used for any structure and shall be
used where specifically required by these Provisions.

9.4.1.1 Maximum Considered Earthquake Ground Motions

The maximum considered earthquake ground
motions shall be as represented by the mapped spectral response acceleration at short periods, S_S, and at
1 s, S_1, obtained from Figs. 9.4.1.1(a) through
9.4.1.1(j) of these Provisions, respectively, and adjusted for Site Class effects using the site coefficients
of Section 9.4.1.2.4. When a site-specific procedure is
used, maximum considered earthquake ground motion
shall be determined in accordance with Section
9.4.1.3.

9.4.1.2 General Procedure for Determining Maximum Considered Earthquake and Design Spectral Response Accelerations

The mapped maximum considered earthquake
spectral response acceleration at short periods (S_S)
and at 1 s (S_1) shall be determined respectively from
Spectral Acceleration Maps 1 through 32.

For buildings and structures included in the
scope of this Standard as specified in Section 9.1.2.1,
the Site Class shall be determined in accordance with
Section 9.4.1.2.1. The maximum considered earthquake spectral response accelerations adjusted for
Site Class effects, S_{MS} and S_{M1} shall be determined in
accordance with Section 9.4.1.2.4 and the design
spectral response accelerations, S_{DS} and S_{D1}, shall be
determined in accordance with Section 9.4.1.2.5. The
general response spectrum, when required by these

Provisions, shall be determined in accordance with Section 9.4.1.2.6.

9.4.1.2.1 Site Class definitions

The site shall be classified as one of the following classes:

A—Hard rock with measured shear wave velocity, $\bar{v}_s > 5,000$ ft/s (1,500 m/s)

B—Rock with 2,500 ft/s $< \bar{v}_s \leq 5,000$ (760 m/s $< \bar{v}_s \leq 1,500$ m/s)

C—Very dense soil and soft rock with 1,200 ft/s $\leq \bar{v}_s \leq 2,500$ ft/s (370 m/s $\leq \bar{v}_s \leq 760$ m/s) or $\bar{N}$ or $\bar{N}_{ch} > 50$ or $\bar{s}_u \geq 2,000$ psf (100 kPa)

D—Stiff soil with 600 ft/s $\leq \bar{v}_s \leq 1,200$ ft/s (180 m/s $\leq \bar{v}_s \leq 370$ m/s) or with $15 \leq \bar{N}$ or $\bar{N}_{ch} \leq 50$ or 1,000 psf $\leq \bar{s}_u \leq 2,000$ psf (50 kPa $\leq \bar{s}_u \leq 100$ kPa)

E—A soil profile with $\bar{v}_s < 600$ ft/s (180 m/s) or any profile with more than 10 ft (3 m) of soft clay. Soft clay is defined as soil with $PI > 20$, $w \geq 40\%$, and $s_u < 500$ psf (25 kPa)

F—Soils requiring site-specific evaluations:
1. Soils vulnerable to potential failure or collapse under seismic loading such as liquefiable soils, quick and highly sensitive clays, collapsible weakly cemented soils.
2. Peats and/or highly organic clays ($H > 10$ ft [3 m] of peat and/or highly organic clay where H = thickness of soil)
3. Very high plasticity clays ($H > 25$ ft [7.6 m] with $PI > 75$)
4. Very thick soft/medium stiff clays ($H > 120$ ft [37 m])

Exception: When the soil properties are not known in sufficient detail to determine the Site Class, Class D shall be used. Site Class E shall be used when the authority having jurisdiction determines that Site Class E is present at the site or in the event that Site E is established by geotechnical data.

The following standards are referenced in the provisions for determining the seismic coefficients:

Reference 9.4.1.2-1   ASTM D1586-84, Test Method for Penetration Test and Split-Barrell Sampling of Soils, 1984.

Reference 9.4.1.2-2   ASTM D4318-93, Test Method for Liquid Limit, Plastic Limit, and Plasticity Index of Soils, 1993.

Reference 9.4.1.2-3   ASTM D2216-92, Test Method for Laboratory Determination of Water (Moisture) Content of Soil and Rock, 1992.

Reference 9.4.1.2-4   ASTM D2166-91, Test Method for Unconfined Compressive Strength of Cohesive Soil, 1991.

Reference 9.4.1.2-5   ASTM D2850-87, Test Method for Unconsolidated, Undrained Compressive Strength of Cohesive Soils in Triaxial Compression, 1987.

9.4.1.2.2 Steps for classifying a site

The Site Class of a site shall be determined using the following steps:

Step 1: Check for the four categories of Site Class F requiring site-specific evaluation. If the site corresponds to any of these categories, classify the site as Site Class F and conduct a site-specific evaluation.

Step 2: Check for the existence of a total thickness of soft clay > 10 ft (3 m), where a soft clay layer is defined by: $s_u < 500$ psf (25 kPa), $w \geq 40\%$, and $PI > 20$. If this criterion is satisfied, classify the site as Site Class E.

Step 3: Categorize the site using one of the following three methods with $\bar{v}_s$, $\bar{N}$, and $\bar{s}_u$ computed in all cases as specified by the definitions in Section 9.4.1.2.3:

a. The $\bar{v}_s$ method: Determine $\bar{v}_s$ for the top 100 ft (30 m) of soil. Compare the value of $\bar{v}_s$ with those given in Section 9.4.1.2 and Table 9.4.1.2 and assign the corresponding Site Class.

$\bar{v}_s$ for rock, Site Class B, shall be measured on site or estimated by a geotechnical engineer or engineering geologist/seismologist for competent rock with moderate fracturing and weathering.

$\bar{v}_s$ for softer and more highly fractured and weathered rock shall be measured on site or shall be classified as Site Class C.

The classification of hard rock, Site Class A, shall be supported by on-site measurements of $\bar{v}_s$ or on profiles of the same rock type in the same formation with an equal or greater degree of weathering and fracturing. Where hard rock conditions are known to be continuous to a depth of at least 100 ft (30 m), surficial measurements of v_s are not prohibited from being extrapolated to assess $\bar{v}_s$.

The rock categories, Site Classes A

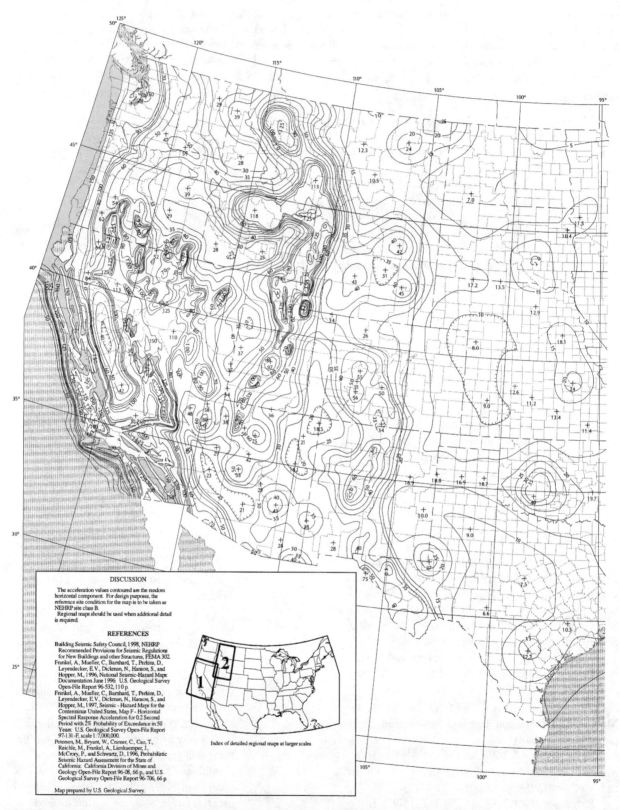

FIGURE 9.4.1.1 (a). Maximum Considered Earthquake Ground Motion for Conterminous United States. Of 0.2 s Spectral Response Acceleration (5% of Critical Damping), Site Class B

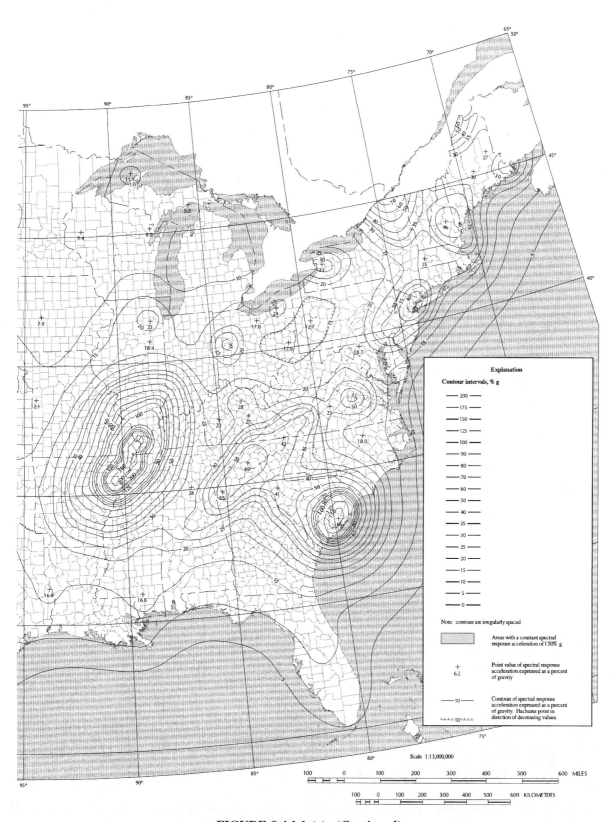

FIGURE 9.4.1.1 (a). (*Continued*)

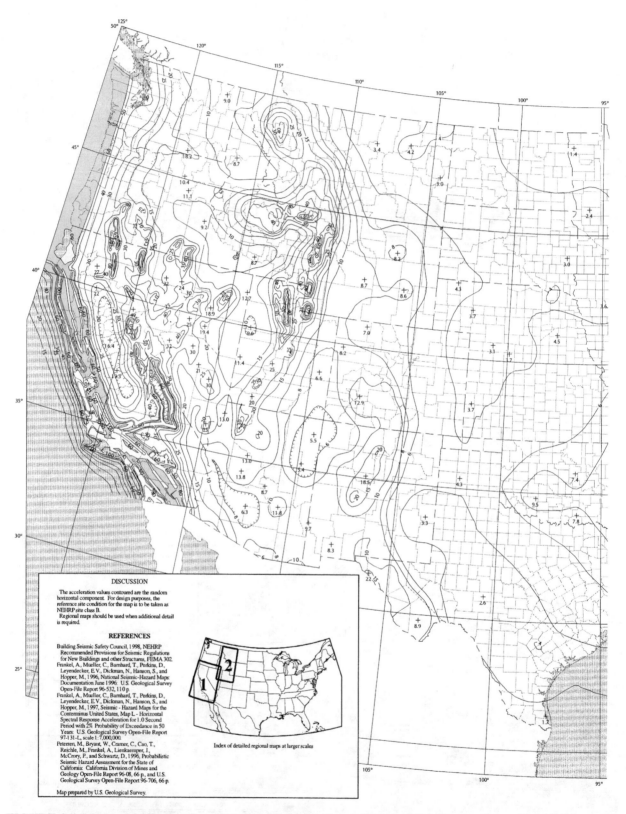

FIGURE 9.4.1.1 (b). Maximum Considered Earthquake Ground Motion for Conterminous United States. Of 1.0 s Spectral Response Acceleration (5% of Critical Damping), Site Class B

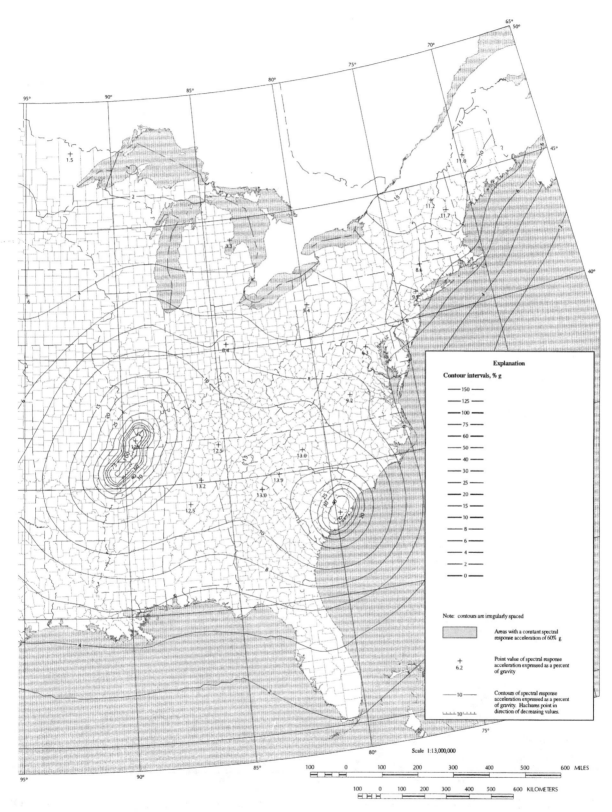

FIGURE 9.4.1.1 (b). (*Continued*)

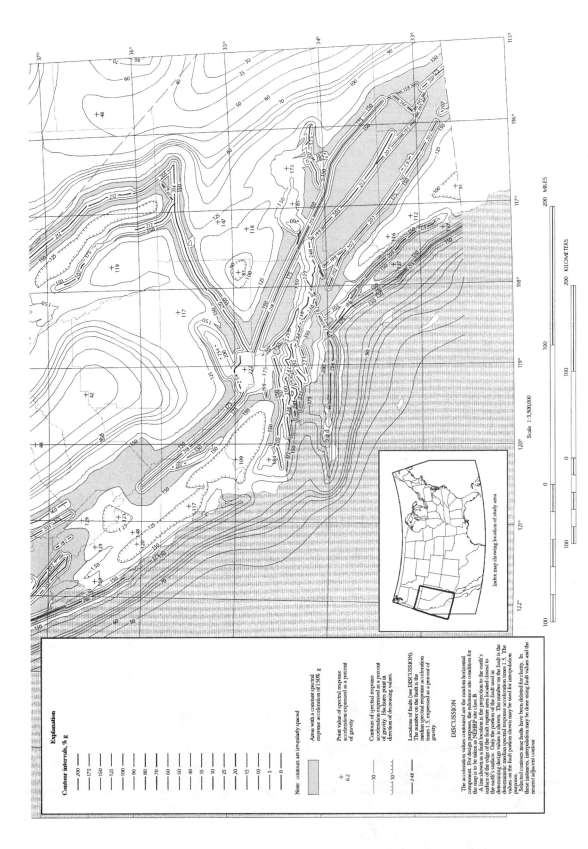

FIGURE 9.4.1.1 (c). Maximum Considered Earthquake Ground Motion for Region 1 of 0.2 s Spectral Response Acceleration (5% of Critical Damping), Site Class B

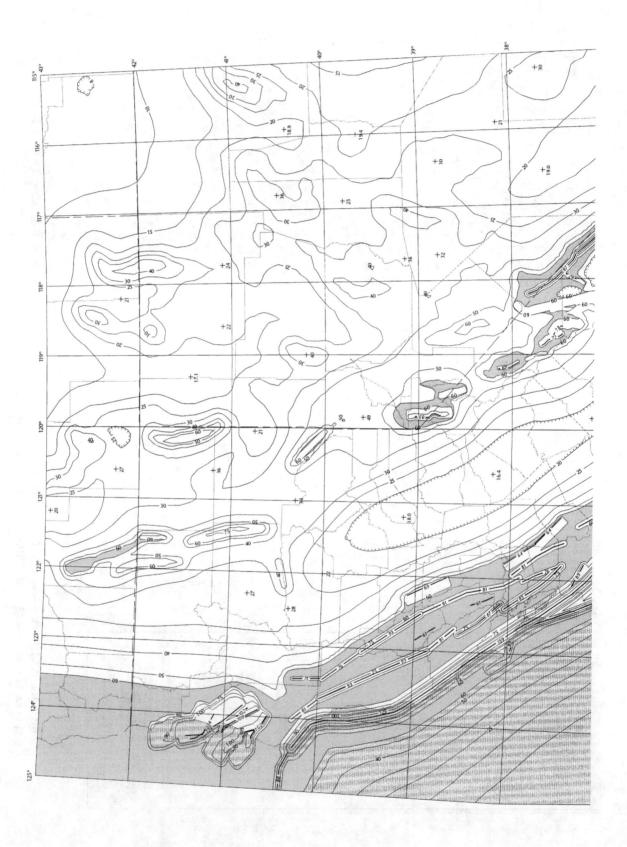

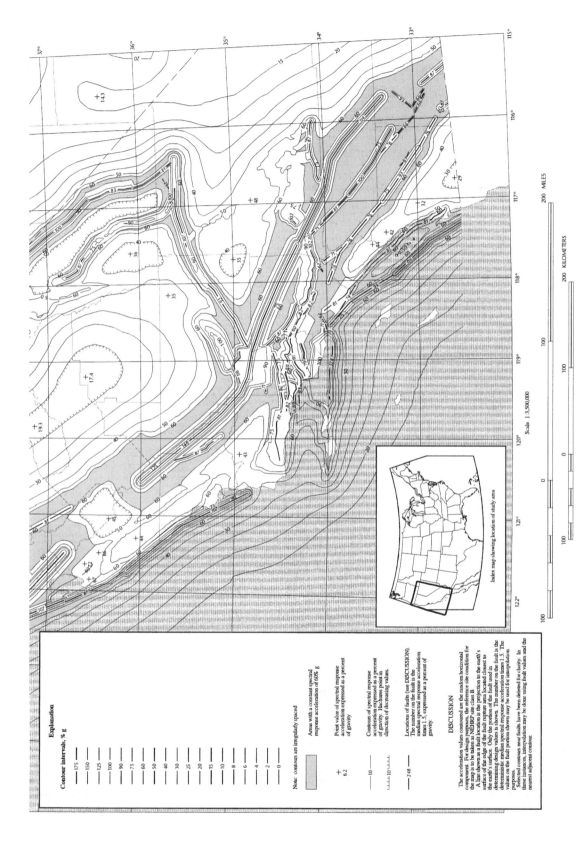

FIGURE 9.4.1.1 (d). Maximum Considered Earthquake Ground Motion for Region 1 of 1.0 s Spectral Response Acceleration (5% of Critical Damping), Site Class B

107

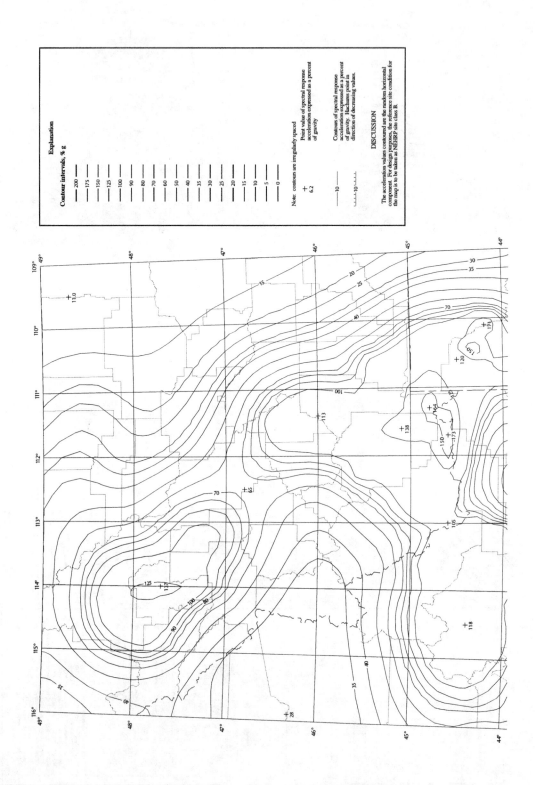

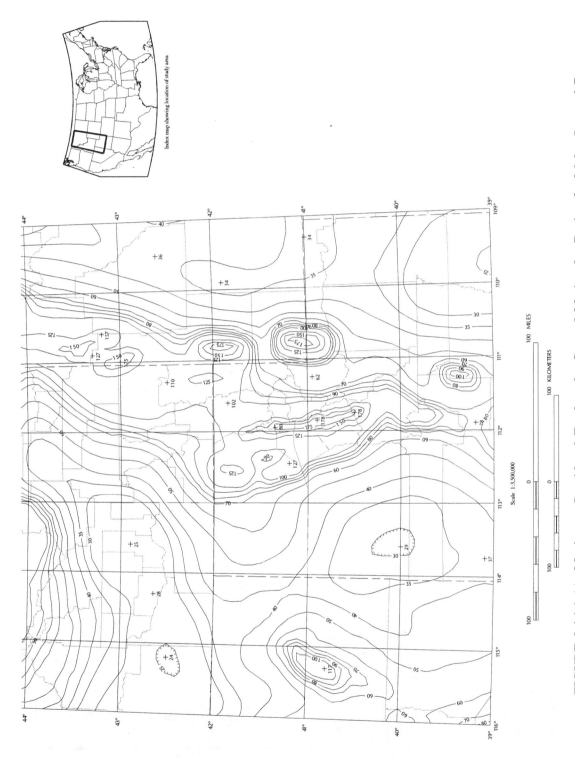

FIGURE 9.4.1.1 (e). Maximum Considered Earthquake Ground Motion for Region 2 of 0.2 s Spectral Response Acceleration (5% of Critical Damping), Site Class B

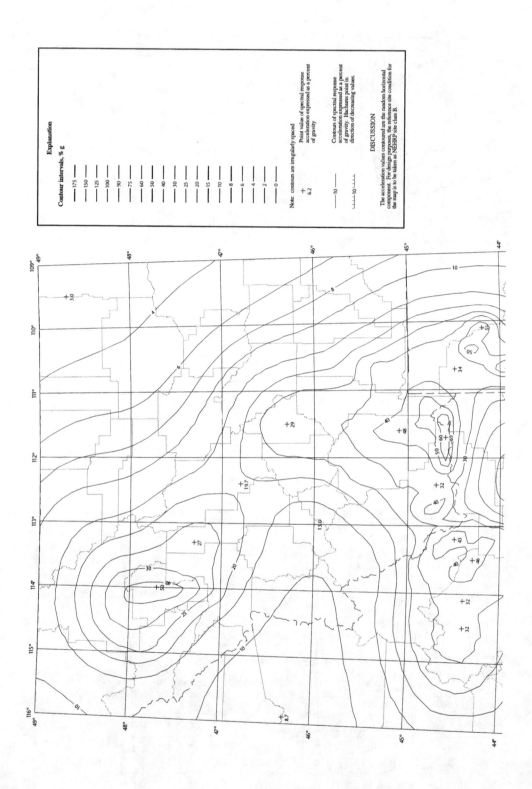

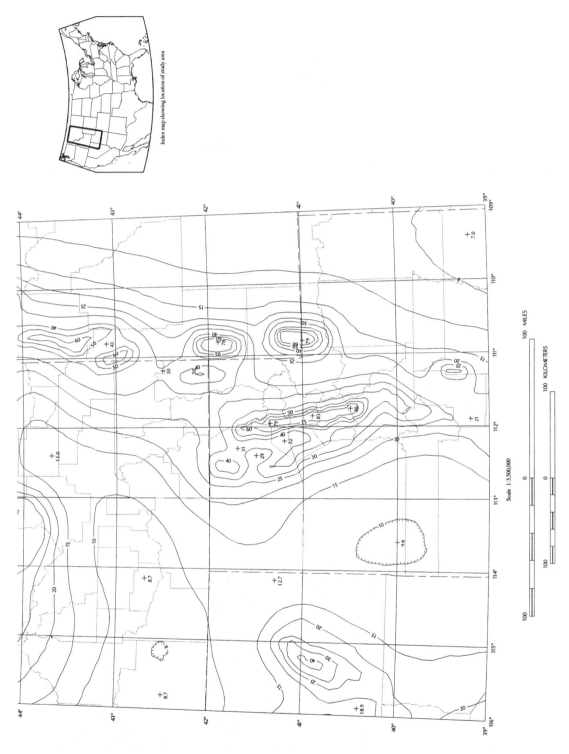

FIGURE 9.4.1.1 (f). Maximum Considered Earthquake Ground Motion for Region 2 of 1.0 s Spectral Response Acceleration (5% of Critical Damping), Site Class B

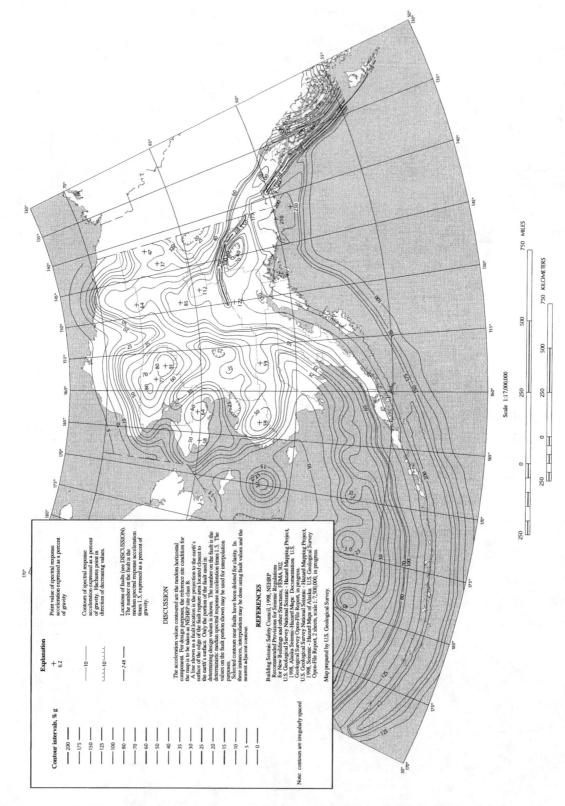

FIGURE 9.4.1.1 (g-1). Maximum Considered Earthquake Ground Motion for Alaska of 0.2 s Spectral Response Acceleration (5% of Critical Damping), Site Class B

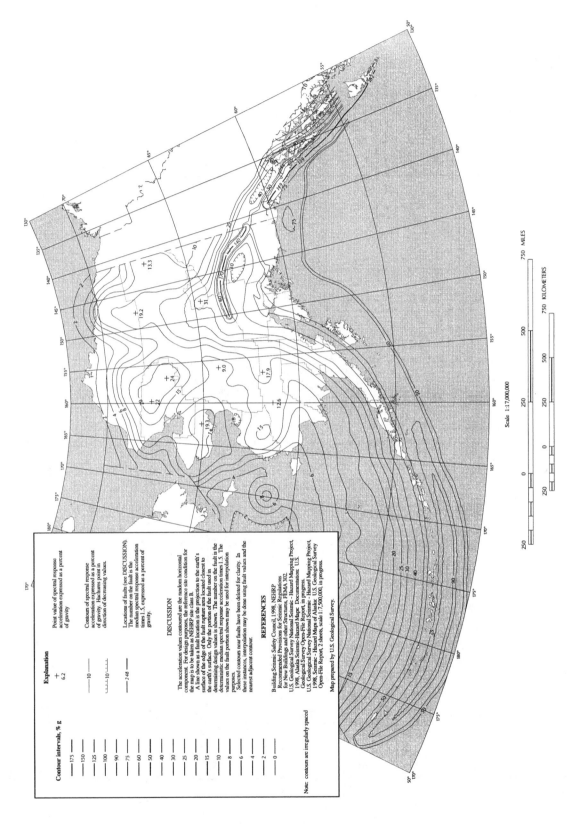

FIGURE 9.4.1.1 (g-2). Maximum Considered Earthquake Ground Motion for Alaska of 1.0 s Spectral Response Acceleration (5% of Critical Damping), Site Class B

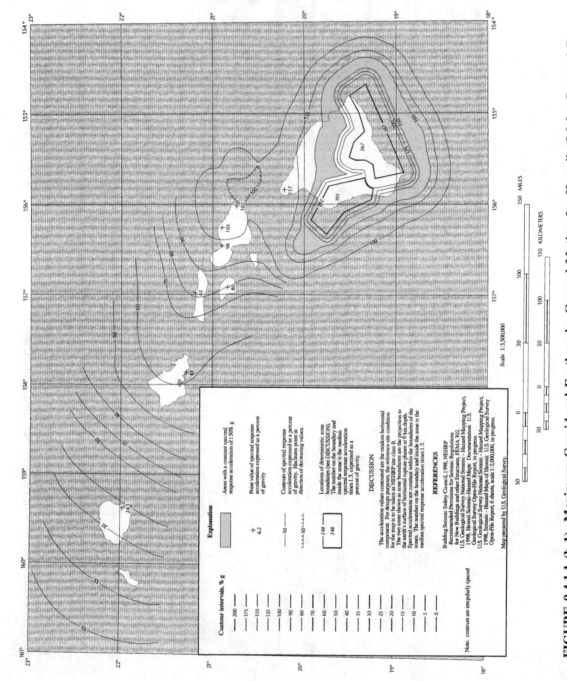

FIGURE 9.4.1.1 (h-1). Maximum Considered Earthquake Ground Motion for Hawaii of 0.2 s Spectral Response Acceleration (5% of Critical Damping), Site Class B

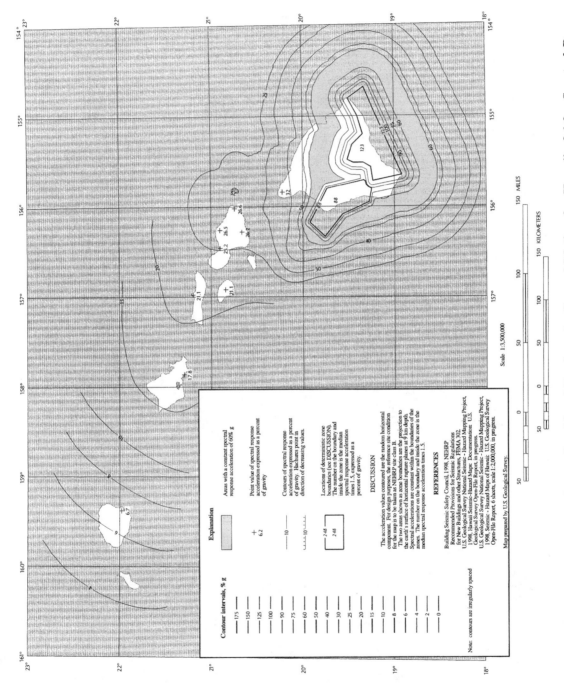

FIGURE 9.4.1.1 (h-2). Maximum Considered Earthquake Ground Motion for Hawaii of 1.0 s Spectral Response Acceleration (5% of Critical Damping), Site Class B

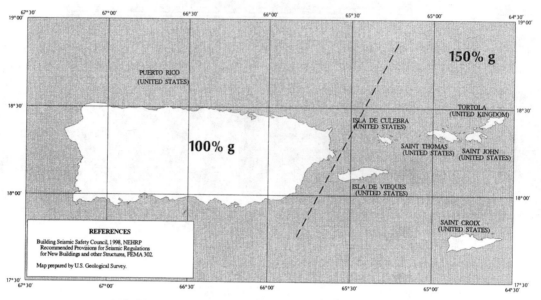

0.2 SEC SPECTRAL RESPONSE ACCELERATION (5% OF CRITICAL DAMPING)

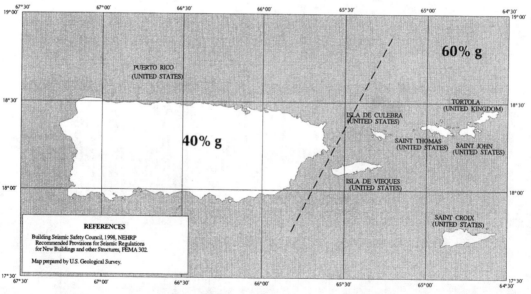

1.0 SEC SPECTRAL RESPONSE ACCELERATION (5% OF CRITICAL DAMPING)

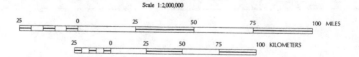

FIGURE 9.4.1.1 (i). Maximum Considered Earthquake Ground Motion for Puerto Rico, Culebra, Vieques, St. Thomas, St. John, and St. Croix of 0.2 and 1.0 s Spectral Response Acceleration (5% of Critical Damping), Site Class B

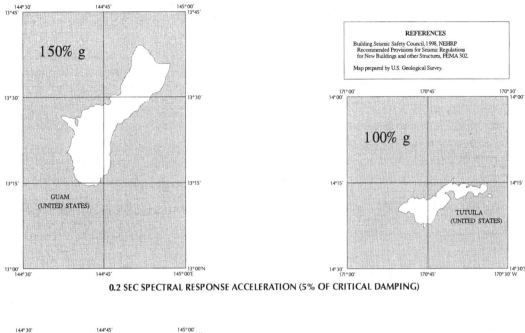

0.2 SEC SPECTRAL RESPONSE ACCELERATION (5% OF CRITICAL DAMPING)

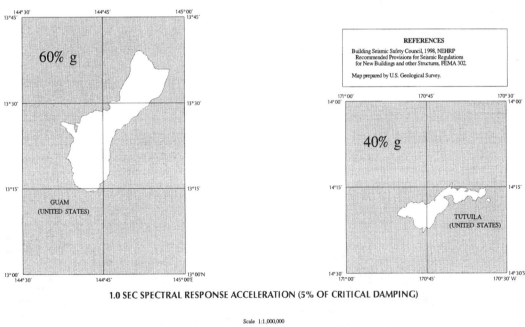

1.0 SEC SPECTRAL RESPONSE ACCELERATION (5% OF CRITICAL DAMPING)

Scale 1:1,000,000

25   0   25 MILES

25   0   25 KILOMETERS

FIGURE 9.4.1.1 (j). Maximum Considered Earthquake Ground Motion for Guam and Tutuilla of 0.2 and 1.0 s Spectral Response Acceleration (5% of Critical Damping), Site Class B

and B, shall not be assigned to a site if there is more than 10 ft (3 m) of soil between the rock surface and the bottom of the spread footing or mat foundation.

b. The $\bar{N}$ method: Determine $\bar{N}$ for the top 100 ft (30 m) of soil. Compare the value of $\bar{N}$ with those given in Section 9.4.1.2 and Table 9.4.1.2 and assign the corresponding Site Class.

c. The s_u method: For cohesive soil layers, determine $\bar{s}_u$ for the top 100 ft (30 m) of soil. For cohesionless soil layers determine $\bar{N}_{ch}$ for the top 100 ft (30 m) of soil. Cohesionless soil is defined by a $PI < 20$, where cohesive soil is defined by a $PI > 20$. Compare the values of s_u and N_{ch} with those given in Section 9.4.1.2 and Table 9.4.1.2 and assign the corresponding Site Class. When the $\bar{N}_{ch}$ and $\bar{s}_u$ criteria differ, assign the category with the softer soil (Site Class E soil is softer than D).

9.4.1.2.3 Definitions of Site Class parameters

The definitions presented below apply to the upper 100 ft (30 m) of the site profile. Profiles containing distinctly different soil layers shall be subdivided into those layers designated by a number that ranges from 1 to n at the bottom where there are a total of n distinct layers in the upper 100 ft (30 m). Where some of the n layers are cohesive and others are not,

k is the number of cohesive layers and m is the number of cohesionless layers. The symbol i refers to any one of the layers between 1 and n.

v_{si} is the shear wave velocity in ft/s (m/s).

d_i is the thickness of any layer between 0 and 100 ft (30 m).

$\bar{v}_s$ is:

$$\bar{v}_s = \frac{\sum_{i=1}^{n} d_i}{\sum_{i=1}^{n} \frac{d_i}{v_{si}}} \qquad \text{(Eq. 9.4.1.2-1)}$$

whereby $\sum_{i=1}^{n} d_i$ is equal to 100 ft (30 m).

$\bar{N}_i$ is the Standard Penetration Resistance (ASTM D1586-84) not to exceed 100 blows/ft as directly measured in the field without corrections.

$\bar{N}$ is:

$$\bar{N} = \frac{\sum_{i=1}^{n} d_i}{\sum_{i=1}^{n} \frac{d_i}{N_i}} \qquad \text{(Eq. 9.4.1.2-2)}$$

TABLE 9.4.1.2. Site Classification

Site Class	$\bar{v}_s$	$\bar{N}$ or $\bar{N}_{ch}$	$\bar{s}_u$
A Hard rock	>5,000 fps (>1,500 m/s)	not applicable	not applicable
B Rock	2,500 to 5,000 fps (760 to 1,500 m/s)	not applicable	not applicable
C Very dense soil and soft rock	1,200 to 2,500 fps (370 to 760 m/s)	>50	>2,000 psf (>100 kPa)
D Stiff soil	600 to 1,200 fps (180 to 370 m/s)	15 to 50	1,000 to 2,000 psf (50 to 100 kPa)
E Soil	<600 fps (<180 m/s)	<15	<1,000 psf (<50 kPa)
F—Soils requiring site specific evaluation		1. Soils vulnerable to potential failure or collapse 2. Peats and/or highly organic clays 3. Very high plasticity clays 4. Very thick soft/medium clays	

$\bar{N}_{ch}$ is:

$$\bar{N}_{ch} = \frac{d_s}{\sum_{i=1}^{m} \frac{d_i}{N_i}} \qquad \text{(Eq. 9.4.1.2-3)}$$

whereby $\sum_{i=1}^{m} d_i = d_s$. (Use only d_i and N_i for cohesionless soils.)

d_s is the total thickness of cohesionless soil layers in the top 100 ft (30 m).

s_{ui} is the undrained shear strength in psf (kPa), not to exceed 5,000 psf (240 kPa), ASTM D2166-91 or D2850-87.

$\bar{s}_i$ is:

$$\bar{s}_u = \frac{d_c}{\sum_{i=1}^{k} \frac{d_i}{s_{ui}}} \qquad \text{(Eq. 9.4.1.2-4)}$$

whereby $\sum_{i=1}^{k} d_i = d_c$.

d_c is the total thickness $(100 - d_s)$ of cohesive soil layers in the top 100 ft (30 m).

PI is the plasticity index, ASTM D4318-93.

w is the moisture content in percent, ASTM D2216-92.

9.4.1.2.4 Site coefficients and adjusted maximum considered earthquake spectral response acceleration parameters

The maximum considered earthquake spectral response acceleration for short periods (S_{MS}) and at 1 s (S_{M1}), adjusted for site class effects, shall be determined by Eqs. 9.4.1.2.4-1 and 9.4.1.2.4-2, respectively:

$$S_{MS} = F_a S_s \qquad \text{(Eq. 9.4.1.2.4-1)}$$

$$S_{M1} = F_v S_1 \qquad \text{(Eq. 9.4.1.2.4-2)}$$

where

S_1 = mapped maximum considered earthquake spectral response acceleration at a period of 1 s as determined in accordance with Section 9.4.1;

S_s = mapped maximum considered earthquake spectral response acceleration at short periods as determined in accordance with Section 9.4.1; and

where site coefficients F_a and F_v are defined in Tables 9.4.1.2.4a and b, respectively.

9.4.1.2.5 Design spectral response acceleration parameters

Design earthquake spectral response acceleration at short periods, S_{DS}, and at 1 s period, S_{D1}, shall be determined from Eqs. 9.4.1.2.5-1 and 9.4.1.2.5-2, respectively:

$$S_{DS} = \frac{2}{3} S_{MS} \qquad \text{(Eq. 9.4.1.2.5-1)}$$

$$S_{D1} = \frac{2}{3} S_{M1} \qquad \text{(Eq. 9.4.1.2.5-2)}$$

TABLE 9.4.1.2.4a. Values of F_a as a Function of Site Class and Mapped Short-Period Maximum Considered Earthquake Spectral Acceleration

Site Class	Mapped Maximum Considered Earthquake Spectral Response Acceleration at Short Periods				
	$S_S \leq 0.25$	$S_S = 0.5$	$S_S = 0.75$	$S_S = 1.0$	$S_S \geq 1.25$[b]
A	0.8	0.8	0.8	0.8	0.8
B	1.0	1.0	1.0	1.0	1.0
C	1.2	1.2	1.1	1.0	1.0
D	1.6	1.4	1.2	1.1	1.0
E	2.5	1.7	1.2	0.9	—[a]
F	—[a]	—[a]	—[a]	—[a]	—[a]

Note: Use straight line interpolation for intermediate values of S_S.

[a]Site specific geotechnical investigation and dynamic site response analyses shall be performed.

[b]Site specific studies required per Section 9.4.1.2.4 may result in higher values of S_s than included on the hazard maps, as may the provisions of Section 9.13.

TABLE 9.4.1.2.4b. Values of F_v as a Function of Site Class and Mapped 1-Second Period Maximum Considered Earthquake Spectral Acceleration

Site Class	Mapped Maximum Considered Earthquake Spectral Response Acceleration at 1-Second Periods				
	$S_1 \leq 0.1$	$S_1 = 0.2$	$S_1 = 0.3$	$S_1 = 0.4$	$S_1 \geq 0.5^b$
A	0.8	0.8	0.8	0.8	0.8
B	1.0	1.0	1.0	1.0	1.0
C	1.7	1.6	1.5	1.4	1.3
D	2.4	2.0	1.8	1.6	1.5
E	3.5	3.2	2.8	2.4	—^{a}
F	—^{a}	—^{a}	—^{a}	—^{a}	—^{a}

Note: Use straight line interpolation for intermediate values of S_1.

^{a}Site specific geotechnical investigation and dynamic site response analyses shall be performed.

^{b}Site specific studies required per Section 9.4.1.2.4 may result in higher values of S_1 than included on the hazard maps, as may the provisions of Section 9.13.

9.4.1.2.6 General procedure response spectrum

Where a design response spectrum is required by these Provisions and site-specific procedures are not used, the design response spectrum curve shall be developed as indicated in Fig. 9.4.1.2.6 and as follows:

1. For periods less than or equal to T_0, the design spectral response acceleration, S_a, shall be taken as given by Eq. 9.4.1.2.6-1:

$$S_a = S_{DS} \left(0.4 + 0.6 \frac{T}{T_0} \right) \quad \text{(Eq. 9.4.1.2.6-1)}$$

2. For periods greater than or equal to T_0 and less than or equal to T_S, the design spectral response acceleration, S_a, shall be taken as equal to S_{DS}.

3. For periods greater than T_s, the design spectral response acceleration, S_a, shall be taken as given by Eq. 9.4.1.2.6-2:

$$S_a = \frac{S_{D1}}{T} \quad \text{(Eq. 9.4.1.2.6-2)}$$

where

S_{DS} = design spectral response acceleration at short periods;

S_{D1} = design spectral response acceleration at 1 s period, in units of g-s;

T = fundamental period of the structure (s);

$T_0 = 0.2 S_{D1}/S_{DS}$; and

$T_S = S_{D1}/S_{DS}$.

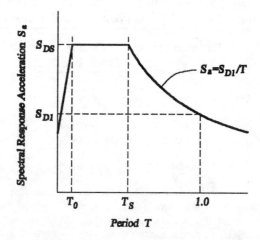

FIGURE 9.4.1.2.6. Design Response Spectrum

9.4.1.3 Site-Specific Procedure for Determining Ground Motion Accelerations

A site-specific study shall account for the regional seismicity and geology, the expected recurrence rates and maximum magnitudes of events on known faults and source zones, the location of the site with respect to these, near source effects if any, and the characteristics of subsurface site conditions.

9.4.1.3.1 Probabilistic maximum considered earthquake

When site-specific procedures are utilized, the maximum considered earthquake ground motion shall be taken as that motion represented by a 5% damped acceleration response spectrum having a 2% probability of exceedance within a 50 year period. The maximum considered earthquake spectral response acceleration at any period S_{aM}, shall be taken from that spectrum.

Exception: Where the spectral response ordinates or a 5% damped spectrum having a 2% probability of exceedance within a 50 year period at periods of 0.2 s or 1 s exceed the corresponding ordinate of the deterministic limit of Section 9.4.1.3.2, the maximum considered earthquake ground motion shall be taken as the lesser of the probabilistic maximum considered earthquake ground motion or the deterministic maximum considered earthquake ground motion of Section 9.4.1.3.3 but shall not be taken less than the deterministic limit ground motion of Section 9.4.1.3.2.

9.4.1.3.2 Deterministic limit on maximum considered earthquake ground motion

The deterministic limit on maximum considered earthquake ground motion shall be taken as the response spectrum determined in accordance with Fig. 9.4.1.3.2, where F_a and F_v are determined in accordance with Section 9.4.1.2.4, with the value of S_s taken as 1.5g and the value of S_1 taken as 0.6g.

9.4.1.3.3 Deterministic maximum considered earthquake ground motion

The deterministic maximum considered earthquake ground motion response spectrum shall be calculated as 150% of the median spectral response accelerations (S_{aM}) at all periods resulting from a characteristic earthquake on any known active fault within the region.

9.4.1.3.4 Site-specific design ground motion

Where site-specific procedures are used to determine the maximum considered earthquake ground motion response spectrum, the design spectral response acceleration at any period shall be determined from Eq. 9.4.1.3.4:

$$S_a = \frac{2}{3} S_{aM} \qquad \text{(Eq. 9.4.1.3.4)}$$

and shall be greater than or equal to 80% of the S_a determined by the general response spectrum in Section 9.4.1.2.6.

9.4.2 Seismic Design Category

Structures shall be assigned a Seismic Design Category in accordance with Section 9.4.2.1.

9.4.2.1 Determination of Seismic Design Category

All structures shall be assigned to a Seismic Design Category based on their Seismic Use Group and

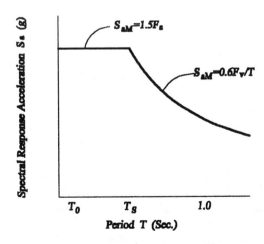

FIGURE 9.4.1.3.2. Deterministic Limit on Maximum Considered Earthquake Response Spectrum

TABLE 9.4.2.1a. Seismic Design Category Based on Short Period Response Accelerations

Value of S_{DS}	Seismic Use Group		
	I	II	III
$S_{DS} < 0.167g$	A	A	A
$0.167g \leq S_{DS} < 0.33g$	B	B	C
$0.33g \leq S_{DS} < 0.50g$	C	C	D
$0.50g \leq S_{DS}$	D[a]	D[a]	D[a]

[a]Seismic Use Group I and II structures located on sites with mapped maximum considered earthquake spectral response acceleration at 1 s period, S_1, equal to or greater than $0.75g$ shall be assigned to Seismic Design Category E and Seismic Use Group III structures located on such sites shall be assigned to Seismic Design Category F.

the design spectral response acceleration coefficients, S_{DS} and S_{D1}, determined in accordance with Section 9.4.1.2.5. Each building and structure shall be assigned to the most severe Seismic Design Category in accordance with Table 9.4.2.1a or 9.4.2.1b, irrespective of the fundamental period of vibration of the structure, T.

9.4.2.2 Site Limitation for Seismic Design Categories E and F

A structure assigned to Category E or F shall not be sited where there is a known potential for an active fault to cause rupture of the ground surface at the structure.

Exception: Detached one- and two-family dwellings of light-frame construction.

TABLE 9.4.2.1b. Seismic Design Category Based on 1 s Period Response Accelerations

Value of S_{D1}	Seismic Use Group		
	I	II	III
$S_{D1} < 0.067g$	A	A	A
$0.067g \leq S_{D1} < 0.133g$	B	B	C
$0.133g \leq S_{D1} < 0.20g$	C	C	D
$0.20g \leq S_{D1}$	D[a]	D[a]	D[a]

[a]Seismic Use Group I and II structures located on sites with mapped maximum considered earthquake spectral response acceleration at 1 s period, S_1, equal to or greater than $0.75g$ shall be assigned to Seismic Design Category E and Seismic Use Group III structures located on such sites shall be assigned to Seismic Design Category F.

9.4.3 Quality Assurance

The performance required of structures in Seismic Design Categories C, D, E, or F, requires that special attention be paid to quality assurance during construction. Refer to Section A.9.3 for supplementary provisions.

9.5 STRUCTURAL DESIGN CRITERIA, ANALYSIS & PROCEDURES

9.5.1 This section has been intentionally left blank

9.5.2 Structural Design Requirements

9.5.2.1 Design Basis

The seismic analysis and design procedures to be used in the design of structures and their components shall be as prescribed in this chapter. The design ground motions can occur along any horizontal direction of a structure. The design seismic forces, and their distribution over the height of the structure, shall be established in accordance with the procedures in Sections 9.5.3 and 9.5.4 and the corresponding internal forces in the members of the structure shall be determined using a linearly elastic model. An approved alternative procedure shall not be used to establish the seismic forces and their distribution unless the corresponding internal forces and deformations in the members are determined using a model consistent with the procedure adopted.

Individual members shall be provided with adequate strength to resist the shears, axial forces, and moments determined in accordance with these provisions, and connections shall develop the strength of the connected members or the forces indicated above. The deformation of the structure shall not exceed the prescribed limits when the structure is subjected to the design seismic forces.

A continuous load path, or paths, with adequate strength and stiffness shall be provided to transfer all forces from the point of application to the final point of resistance. The foundation shall be designed to resist the forces developed and accommodate the movements imparted to the structure by the design ground motions. The dynamic nature of the forces, the expected ground motion, and the design basis for strength and energy dissipation capacity of the structure shall be included in the determination of the foundation design criteria.

9.5.2.2 Basic Seismic-Force-Resisting Systems

The basic lateral and vertical seismic-force-resisting system shall conform to one of the types indicated in Table 9.5.2.2. Each type is subdivided by the types of vertical element used to resist lateral seismic forces. The structural system used shall be in accordance with the Seismic Design Category and height limitations indicated in Table 9.5.2.2. The appropriate response modification, coefficient R, system overstrength factor, Ω_0, and the deflection amplification factor (C_d) indicated in Table 9.5.2.2 shall be used in determining the base shear, element design forces, and design story drift. Special framing requirements are indicated in Sections 9.5.2.6, 9.8, 9.9, 9.10, 9.11 and 9.12 for structures assigned to the various Seismic Design Categories.

Seismic-force-resisting systems that are not contained in Table 9.5.2.2 shall be permitted if analytical and test data are submitted that establish the dynamic characteristics and demonstrate the lateral force resistance and energy dissipation capacity to be equivalent to the structural systems listed in Table 9.5.2.2 for equivalent response modification coefficient, R, system overstrength coefficient, Ω_0, and deflection amplification factor, C_d, values.

9.5.2.2.1 Dual system

For a dual system, the moment frame shall be capable of resisting at least 25% of the design seismic forces. The total seismic force resistance is to be provided by the combination of the moment frame and the shear walls or braced frames in proportion to their rigidities.

9.5.2.2.2 Combinations of framing systems

Different seismic-force-resisting systems are permitted along the two orthogonal axes of the structure. Combinations of seismic-force-resisting systems shall comply with the requirements of this section.

9.5.2.2.2.1 R and Ω_0 factors:

The response modification, coefficient, R in the direction under consideration at any story shall not exceed the lowest response modification coefficient, R, for the seismic force resisting system in the same direction considered above that story excluding penthouses. For other than dual systems where a combination of different structural systems is utilized to resist lateral forces in the same direction, the value of R used in that direction shall not be greater than the least value of any of the systems utilized in the same direction. If a system other than a dual system with a response modification coefficient, R, with a value of less than 5 is used as part of the seismic-force-resisting system in any direction of the structure, the lowest such value shall be used for the entire structure. The system overstrength factor, Ω_0, in the direction under consideration at any story shall not be less than the largest value of this factor for the seismic-force-resisting system in the same direction considered above that story.

Exceptions:

1. The limit does not apply to supported structural systems with a weight equal to or less than 10% of the weight of the structure.
2. Detached one- and two-family dwellings of light-frame construction.

9.5.2.2.2.2 Combination framing detailing requirements:

The detailing requirements of Section 9.5.2.6 required by the higher response modification coefficient, R, shall be used for structural components common to systems having different response modification coefficients.

9.5.2.2.3 Seismic design Categories A, B and C

The structural framing system for structures assigned to Seismic Design Categories A, B and C shall comply with the structure height and structural limitations in Table 9.5.2.2.

9.5.2.2.4 Seismic design Categories D and E

The structural framing system for a structure assigned to Seismic Design Categories D and E shall comply with Section 9.5.2.2.3 and the additional provisions of this section.

9.5.2.2.4.1 Increased building height limit:

The height limits in Table 9.5.2.2 is permitted to be increased to 240 ft (75 m) in buildings that have steel braced frames or concrete cast-in-place shear walls and that meet the requirements of this section. In such buildings the braced frames or cast-in-place special reinforced concrete shear walls in any one plane shall resist no more than 60% of the total seismic forces in each direction, neglecting torsional effects. The seismic force in any braced frame or shear wall in any one plane resulting from torsional effects shall not exceed 20% of the total seismic force in that braced frame or shear wall.

9.5.2.2.4.2 Interaction effects:

Moment resisting frames that are enclosed or adjoined by more rigid elements not considered to be part of the seismic-force-resisting system shall be designed so that the

TABLE 9.5.2.2. Design Coefficients and Factors for Basic Seismic-Force-Resisting Systems

Basic Seismic Force Resisting System	Response Modification Coefficient, R^a	System Over-strength Factor, $\Omega_0{}^g$	Deflection Amplification Factor, $C_d{}^b$	Structural System Limitations and Building Height (ft) Limitations[c]				
				Seismic Design Category				
				A & B	C	D^d	E^e	F^e
Bearing Wall Systems								
Ordinary steel concentrically braced frames	4	2	3 1/2	NL	NL	160	160	160
Special reinforced concrete shear walls	5	2 1/2	5	NL	NL	160	160	160
Ordinary reinforced concrete shear walls	4	2 1/2	4	NL	NL	NP	NP	NP
Detailed plain concrete shear walls	2 1/2	2 1/2	2	NL	NP	NP	NP	NP
Ordinary plain concrete shear walls	1 1/2	2 1/2	1 1/2	NL	NP	NP	NP	NP
Special reinforced masonry shear walls	5	2 1/2	3 1/2	NL	NL	160	160	100
Intermediate reinforced masonry shear walls	3 1/2	2 1/2	2 1/4	NL	NL	NP	NP	NP
Ordinary reinforced masonry shear walls	2	2 1/2	1 3/4	NL	160	NP	NP	NP
Detailed plain masonry shear walls	2	2 1/2	1 3/4	NL	NP	NP	NP	NP
Ordinary plain masonry shear walls	1 1/2	2 1/2	1 1/4	NL	NP	NP	NP	NP
Light-framed walls sheathed with wood structural panels rated for shear resistance	6	3	4	NL	NL	65	65	65
Light-framed walls with shear panels of all other materials	2	2 1/2	2	NL	NL	35	NP	NP
Building Frame Systems								
Steel eccentrically braced frames, moment resisting, connections at columns away from links	8	2	4	NL	NL	160	160	100
Steel eccentrically braced frames, non-moment resisting, connections at columns away from links	7	2	4	NL	NL	160	160	100
Special steel concentrically braced frames	6	2	5	NL	NL	160	160	100
Ordinary steel concentrically braced frames	5	2	4 1/2	NL	NL	160	100	100
Special reinforced concrete shear walls	6	2 1/2	5	NL	NL	160^h	160	100
Ordinary reinforced concrete shear walls	5	2 1/2	4 1/2	NL	NL	NP	NP	NP
Detailed plain concrete shear walls	3	2 1/2	2 1/2	NL	NP	NP	NP	NP
Ordinary plain concrete shear walls	2	2 1/2	2	NL	NP	NP	NP	NP
Composite eccentrically braced frames	8	2	4	NL	NL	160	160	100

TABLE 9.5.2.2. Design Coefficients and Factors for Basic Seismic-Force-Resisting Systems (*Continued*)

Basic Seismic Force Resisting System	Response Modification Coefficient, R^a	System Over-strength Factor, $\Omega_0{}^g$	Deflection Amplification Factor, $C_d{}^b$	Structural System Limitations and Building Height (ft) Limitations[c]				
				Seismic Design Category				
				A & B	C	D^d	E^e	F^e
Composite concentrically braced frames	5	2	4 1/2	NL	NL	160	160	100
Ordinary composite braced frames	3	2	3	NL	NL	NP	NP	NP
Composite steel plate shear walls	6 1/2	2 1/2	5 1/2	NL	NL	160	160	100
Special composite reinforced concrete shear walls with steel elements	6	2 1/2	5	NL	NL	160	160	100
Ordinary composite reinforced concrete shear walls with steel elements	5	2 1/2	4 1/2	NL	NL	NP	NP	NP
Special reinforced masonry shear walls	5 1/2	2 1/2	4	NL	NL	160	160	100
Intermediate reinforced masonry shear walls	4	2 1/2	4	NL	NL	NP	NP	NP
Ordinary reinforced masonry shear walls	2 1/2	2 1/2	2 1/4	NL	160	NP	NP	NP
Detailed plain masonry shear walls	2 1/2	2 1/2	2 1/4	NL	NP	NP	NP	NP
Ordinary plain masonry shear walls	1 1/2	2 1/2	1 1/4	NL	NP	NP	NP	NP
Light-framed walls sheathed with wood structural panels rated for shear resistance	6 1/2	2 1/2	4 1/2	NL	NL	65	65	65
Light-framed walls with shear panels of all other materials	2 1/2	2 1/2	2 1/2	NL	NL	35	NP	NP
Moment Resisting Frame Systems								
Special steel moment frames	8	3	5 1/2	NL	NL	NL	NL	NL
Special steel truss moment frames	7	3	5 1/2	NL	NL	160	100	NP
Intermediate steel moment frames	6	3	5	NL	NL	160	100	NP[h]
Ordinary steel moment frames	4	3	3 1/2	NL	NL	35[h]	NP[h,i]	NP[h,i]
Special reinforced concrete moment frames	8	3	5 1/2	NL	NL	NL	NL	NL
Intermediate reinforced concrete moment frames	5	3	4 1/2	NL	NL	NP	NP	NP
Ordinary reinforced concrete moment frames	3	3	2 1/2	NL	NP	NP	NP	NP
Special composite moment frames	8	3	5 1/2	NL	NL	NL	NL	NL
Intermediate composite moment frames	5	3	4 1/2	NL	NL	NP	NP	NP
Composite partially restrained moment frames	6	3	5 1/2	160	160	100	NP	NP
Ordinary composite moment frames	3	3	2 1/2	NL	NP	NP	NP	NP
Special masonry moment frames	5 1/2	3	5	NL	NL	160	160	100

TABLE 9.5.2.2. Design Coefficients and Factors for Basic Seismic-Force-Resisting Systems (*Continued*)

Basic Seismic Force Resisting System	Response Modification Coefficient, R^a	System Over-strength Factor, $\Omega_0{}^g$	Deflection Amplification Factor, $C_d{}^b$	Structural System Limitations and Building Height (ft) Limitations[c]				
				Seismic Design Category				
				A & B	C	D^d	E^e	F^e
Dual Systems with Special Moment Frames Capable of Resisting at Least 25% of Prescribed Seismic Forces								
Steel eccentrically braced frames, moment resisting connections, at columns away from links	8	2 1/2	4	NL	NL	NL	NL	NL
Steel eccentrically braced frames, non-moment resisting connections, at columns away from links	7	2 1/2	4	NL	NL	NL	NL	NL
Special steel concentrically braced frames	8	2 1/2	6 1/2	NL	NL	NL	NL	NL
Ordinary Steel concentrically braced frames	6	2 1/2	5	NL	NL	NL	NL	NL
Special reinforced concrete shear walls	8	2 1/2	6 1/2	NL	NL	NL	NL	NL
Ordinary reinforced concrete shear walls	7	2 1/2	6	NL	NL	NP	NP	NP
Composite eccentrically braced frames	8	2 1/2	4	NL	NL	NL	NL	NL
Composite concentrically braced frames	6	2 1/2	5	NL	NL	NL	NL	NL
Composite steel plate shear walls	8	2 1/2	6 1/2	NL	NL	NL	NL	NL
Special composite reinforced concrete shear walls with steel elements	8	2 1/2	6 1/2	NL	NL	NL	NL	NL
Ordinary composite reinforced concrete shear walls with steel elements	7	2 1/2	6	NL	NL	NP	NP	NP
Special reinforced masonry shear walls	7	3	6 1/2	NL	NL	NL	NL	NL
Intermediate reinforced masonry shear walls	6 1/2	3	5 1/2	NL	NL	NP	NP	NP
Dual Systems with Intermediate Moment Frames Capable of Resisting at Least 25% of Prescribed Seismic Forces								
Special steel concentrically braced frames[f]	6	2 1/2	5	NL	NL	160	100	NP
Ordinary steel concentrically braced frames[f]	5	2 1/2	4 1/2	NL	NL	160	100	NP
Special reinforced concrete shear walls	6	2 1/2	5	NL	NL	160	100	100
Ordinary reinforced concrete shear walls	5 1/2	2 1/2	4 1/2	NL	NL	NP	NP	NP
Ordinary reinforced masonry shear walls	3	3	2 1/2	NL	160	NP	NP	NP

TABLE 9.5.2.2. Design Coefficients and Factors for Basic Seismic-Force-Resisting Systems (*Continued*)

Basic Seismic Force Resisting System	Response Modification Coefficient, R^a	System Over-strength Factor, $\Omega_0{}^g$	Deflection Amplification Factor, $C_d{}^b$	Structural System Limitations and Building Height (ft) Limitations[c]				
				Seismic Design Category				
				A & B	C	D^d	E^e	F^e
Intermediate reinforced masonry shear walls	5	3	4 1/2	NL	NL	NP	NP	NP
Composite concentrically braced frames	5	2 1/2	4 1/2	NL	NL	160	100	NP
Ordinary composite braced frames	4	2 1/2	3	NL	NL	NP	NP	NP
Ordinary composite reinforced concrete shear walls with steel elements	5	3	4 1/2	NL	NL	NP	NP	NP
Inverted Pendulum Systems and Cantilevered Column Systems								
Special steel moment frames	2 1/2	2	2 1/2	NL	NL	NL	NL	NL
Ordinary steel moment frames	1 1/4	2	2 1/2	NL	NL	NP	NP	NP
Special reinforced concrete moment frames	2 1/2	2	1 1/4	NL	NL	NL	NL	NL
Structural Steel Systems Not Specifically Detailed for Seismic Resistance	3	3	3	NL	NL	NP	NP	NP

[a]Response modification coefficient, R, for use throughout the Standard. Note R reduces forces to a strength level, not an allowable stress level.
[b]Deflection amplification factor, C_d, for use in Secs. 9.5.3.7.1 and 9.5.3.7.2
[c]NL = Not Limited and NP = No Permitted. For metric units use 30 m for 100 ft and use 50 m for 160 ft. Heights are measured from the base of the structure as defined in Sec. 9.2.1.
[d]See Sec. 9.5.2.2.4.1 for a description of building systems limited to buildings with a height of 240 feet (75 m) or less.
[e]See Sec. 9.5.2.2.4 and 9.5.2.2.4.5 for building systems limited to buildings with a height of 160 feet (50 m) or less.
[f]Ordinary moment frame is permitted to be used in lieu of Intermediate moment frame in Seismic Design Categories B and C.
[g]The tabulated value of the overstrength factor, Ω_0, may be reduced by subtracting 1/2 for structures with flexible diaphragms but shall not be taken as less than 2.0 for any structure.
[h]Steel ordinary moment frames and Intermediate moment frames are permitted in single story buildings up to a height of 60 feet, when the moment joints of field connections are bolted end plates and the dead load of the roof does not exceed 15 psf.
[i]Steel ordinary moment frames are permitted in buildings up to a height of 35 feet, where the dead load of the walls, floors, and roofs does not exceed 15 psf.

action or failure of those elements will not impair the vertical load and seismic force-resisting capability of the frame. The design shall provide for the effect of these rigid elements on the structural system at structure deformations corresponding to the design story drift (Δ) as determined in Section 9.5.3.7. In addition, the effects of these elements shall be considered when determining whether a structure has one or more of the irregularities defined in Section 9.5.2.3.

9.5.2.2.4.3 Deformational compatibility: Every structural component not included in the seismic force-resisting system in the direction under consideration shall be designed to be adequate for the vertical load-carrying capacity and the induced moments and shears resulting from the design story drift (Δ) as determined in accordance with Section 9.5.3.7 (also see Section 9.5.2.8).

Exception: Reinforced concrete frame members not designed as part of the seismic force resisting system shall comply with Section 21.9 of Ref. 9.9-1.

When determining the moments and shears induced in components that are not included in the seismic-force-resisting system in the direction under consideration, the stiffening effects of adjoining rigid structural and nonstructural elements shall be considered and a rational value of member and restraint stiffness shall be used.

9.5.2.2.4.4 Special moment frames: A special moment frame that is used but not required by Table 9.5.2.2 shall not be discontinued and supported by a more rigid system with a lower response modification coefficient (*R*) unless the requirements of Sections 9.5.2.6.2.4 and 9.5.2.6.4.3 are met. Where a special moment frame is required by Table 9.5.2.2, the frame shall be continuous to the foundation.

9.5.2.2.5 Seismic design Category F

The framing systems of structures assigned to Seismic Design Category F shall conform to the requirements of Section 9.5.2.2.4 for Seismic Design Categories D and E and to the additional requirements and limitations of this section. The increased height limit of Section 9.5.2.2.4.1 for braced frame or shear wall systems shall be reduced from 240 ft (75 m) to 160 ft (50 m).

9.5.2.3 Structure Configuration

Structures shall be classified as regular or irregular based upon the criteria in this section. Such classification shall be based on the plan and vertical configuration.

9.5.2.3.1 Diaphragm flexibility

Diaphragms shall be considered flexible when the maximum lateral deformation of the diaphragm is more than two times the average story drift of the associated story. The loadings used for this calculation shall be those prescribed by Section 9.5.3.

9.5.2.3.2 Plan irregularity

Structures having one or more of the irregularity types listed in Table 9.5.2.3.2 shall be designated as having plan structural irregularity. Such structures assigned to the Seismic Design Categories listed in Table 9.5.2.3.2 shall comply with the requirements in the sections referenced in that table.

9.5.2.3.3 Vertical irregularity

Structures having one or more of the irregularity types listed in Table 9.5.2.3.3 shall be designated as having vertical irregularity. Such structures assigned to the Seismic Design Categories listed in Table 9.5.2.3.3 shall comply with the requirements in the sections referenced in that table.

Exceptions:

1. Vertical structural irregularities of Types 1a, 1b, or 2 in Table 9.5.2.3.3 do not apply where no story drift ratio under design lateral seismic force is greater than 130% of the story drift ratio of the next story above. Torsional effects need not be considered in the calculation of story drifts. The story drift ratio relationship for the top 2 stories of the structure are not required to be evaluated.
2. Irregularities of Types 1a, 1b, and 2 of Table 9.5.2.3.3 are not required to be considered for one-story buildings in any Seismic Design Category or for two-story buildings in Seismic Design Categories A, B, C, or D.

9.5.2.4 Redundancy

A reliability factor, ρ, shall be assigned to all structures in accordance with this Section, based on the extent of structural redundancy inherent in the lateral-force-resisting system.

9.5.2.4.1 Seismic design Categories A, B, and C

For structures in Seismic Design Categories A, B, and C, the value of ρ is 1.0.

9.5.2.4.2 Seismic design Category D

For structures in Seismic Design Category D, ρ shall be taken as the largest of the values of ρ_x calculated at each story of the structure "*x*" in accordance with Eq. 9.5.2.4.2 as follows:

$$\rho_x = 2 - \frac{20}{r_{max_x} \sqrt{A_x}} \quad \text{(Eq. 9.5.2.4.2)}$$

where

r_{max_x} = ratio of the design story shear resisted by the single element carrying the most shear force in the story to the total story shear, for a given direction of loading. For braced frames, the value of r_{max_x} is equal to the lateral force component in the most heavily loaded brace element divided by the story shear. For moment frames r_{max_x} shall be taken as the maximum of the sum of the shears in any two adjacent columns in the plane of a moment frame divided by the story shear. For columns common to two bays with moment resisting connections on opposite sides at the level under consideration, 70% of the shear in that column may be used in the column shear summation. For shear walls, r_{max_x} shall be taken equal to the shear in the most heavily loaded wall or wall pier multiplied by $10/l_w$ (the metric coefficient is $3.3/l_w$), where l_w is the wall or wall pier length in ft (m) divided by the story shear. For dual systems, r_{max_x} shall be taken as the maximum value as defined above considering all

TABLE 9.5.2.3.2. Plan Structural Irregularities

Irregularity Type and Description	Reference Section	Seismic Design Category Application
1a. Torsional Irregularity	9.5.2.6.4.3	D, E, and F
Torsional irregularity is defined to exist where the maximum story drift, computed including accidental torsion, at one end of the structure transverse to an axis is more than 1.2 times the average of the story drifts at the two ends of the structure along the axis being considered. Torsional irregularity requirements in the reference sections apply only to structures in which the diaphragms are rigid or semi-rigid.	9.5.3.5.2	C, D, E, and F
1b. Extreme Torsional Irregularity	9.5.2.6.4.3	D
Extreme torsional irregularity is defined to exist where the maximum story drift, computed including accidental torsion, at one end of the structure transverse to an axis is more than 1.4 times the average of the story drifts at the two ends of the structure along the axis being considered. Extreme torsional irregularity requirements in the reference sections apply only to structures in which the diaphragms are rigid or semi-rigid.	9.5.3.5.2 9.5.2.6.5.1	C and D E and F
2. Re-entrant Corners	9.5.2.6.4.3	D, E, and F
Plan configurations of a structure and its lateral force-resisting system contain re-entrant corners, where both projections of the structure beyond a re-entrant corner are greater than 15% of the plan dimension of the structure in the given direction.		
3. Diaphragm Discontinuity	9.5.2.6.4.3	D, E, and F
Diaphragms with abrupt discontinuities or variations in stiffness, including those having cutout or open areas greater than 50% of the gross enclosed diaphragm area, or changes in effective diaphragm stiffness of more than 50% from one story to the next.		
4. Out-of-Plane Offsets	9.5.2.6.4.3	D, E, and F
Discontinuities in a lateral force resistance path, such as out-of-plane offsets of the vertical elements.	9.5.2.6.2.11	B, C, D, E, or F
5. Nonparallel Systems	9.5.2.6.3.1	C, D, E, and F
The vertical lateral force-resisting elements are not parallel to or symmetric about the major orthogonal axes of the lateral force-resisting system.		

lateral load resisting elements in the story. The lateral loads shall be distributed to elements based on relative rigidities considering the interaction of the dual system. For dual systems, the value of ρ need not exceed 80% of the value calculated above; and

A_x = floor area in square feet of the diaphragm level immediately above the story.

The value of ρ need not exceed 1.5, which may be used for any structure. The value of ρ shall not be taken as less than 1.0.

Exception: For structures with lateral-force-resisting systems in any direction consisting solely of special moment frames, the lateral-force-resisting system shall be configured such that the value of ρ calculated in accordance with this section does not exceed 1.25.

The metric equivalent of Eq. 9.5.2.4.2 is:

$$\rho_x = 2 - \frac{6.1}{r_{\max_x}\sqrt{A_x}}$$

where A_x is in square meters.

9.5.2.4.3 Seismic design Categories E and F

For structures in Seismic Design Categories E and F, the value of ρ shall be calculated as indicated in Section 9.5.2.4.2, above.

Exception: For structures with lateral-force-resisting systems in any direction consisting solely of special moment frames, the lateral-force-resisting system shall be configured such that the value of ρ calculated in accordance with Section 9.5.2.4.2 does not exceed 1.1

TABLE 9.5.2.3.3. Vertical Structural Irregularities

Irregularity Type and Description	Reference Section	Seismic Design Category Application
1a. Stiffness Irregularity: Soft Story A soft story is one in which the lateral stiffness is less than 70% of that in the story above or less than 80% of the average stiffness of the three stories above.	9.5.2.5.3	D, E, and F
1b. Stiffness Irregularity: Extreme Soft Story An extreme soft story is one in which the lateral stiffness is less than 60% of that in the story above or less than 70% of the average stiffness of the three stories above.	9.5.2.5.3 9.5.2.6.5.1	D, E, and F
2. Weight (Mass) Irregularity Mass irregularity shall be considered to exist where the effective mass of any story is more than 150% of the effective mass of an adjacent story. A roof that is lighter than the floor below need not be considered.	9.5.2.5.3	D, E, and F
3. Vertical Geometric Irregularity Vertical geometric irregularity shall be considered to exist where the horizontal dimension of the lateral force-resisting system in any story is more than 130% of that in an adjacent story.	9.5.2.5.3	D, E, and F
4. In-Plane Discontinuity in Vertical Lateral Force-Resisting Elements In-plane discontinuity in vertical lateral force-resisting elements shall be considered to exist where an in-plane offset of the lateral force-resisting elements is greater than the length of those elements or there exists a reduction in stiffness of the resisting element in the story below.	9.5.2.5.3 9.5.2.6.2.11	B, C, D, E, and F
5. Discontinuity in Lateral Strength: Weak Story A weak story is one in which the story lateral strength is less than 80% of that in the story above. The story strength is the total strength of all seismic-resisting elements sharing the story shear for the direction under consideration.	9.5.2.6.2.4 9.5.2.5.3 9.5.2.6.5.1	B, C, D, E, and F D, E, and F E and F

9.5.2.5 Analysis Procedures

A structural analysis shall be made for all structures in accordance with the requirements of this section. This section prescribes the minimum analysis procedure to be followed. Use of the procedure in Section 9.5.4 or, with the approval of the regulatory agency, an alternate generally accepted procedure, including the use of an approved site-specific spectrum, is permitted for any structure. The limitations on the base shear stated in Section 9.5.4 apply to any such analysis that is force-based.

A simplified analysis, in accordance with Section 9.5.3.8 shall be permitted to be used for any structure in Seismic Use Group I.

Exceptions:

1. Buildings of light-framed construction not exceeding three stories in height, excluding basements,
2. Buildings of any construction other than light-framed, not exceeding two stories in height.

9.5.2.5.1 Seismic design Category A

Regular or irregular structures assigned to Category A shall be analyzed for minimum lateral forces given by Eq. 9.5.2.5.1, applied independently, in each of two orthogonal directions:

$$F_x = 0.01w_x \qquad \text{(Eq. 9.5.2.5.1)}$$

where

F_x = design lateral force applied at Story x; and

w_x = portion of the total gravity load of the structure W, located or assigned to Level x where W is as defined in Section 9.5.3.2.

9.5.2.5.2 Seismic design Categories B and C

The analysis procedures in Section 9.5.3 shall be used for regular or irregular structures assigned to Category B or C or a more rigorous analysis shall be made.

9.5.2.5.3 Seismic design Categories D, E, and F

The analysis procedures identified in Table 9.5.2.5.3 shall be used for structures assigned to Categories D, E, and F or a more rigorous analysis shall be made. For regular structures 5 stories or less in height and having a period, T, of 0.5 s or less, the design spectral response accelerations, S_{DS} and S_{D1}, need not exceed the values calculated in accordance with Section 9.4.1.2.5 using the values of the site adjusted maximum considered earthquake spectral response accelerations S_{MS} and S_{M1} given by Eq. 9.5.2.5.3-1 and 9.5.2.5.3-2, respectively:

$$S_{MS} = 1.5F_a \qquad \text{(Eq. 9.5.2.5.3-1)}$$

$$S_{M1} = 0.6F_v \qquad \text{(Eq. 9.5.2.5.3-2)}$$

where F_a and F_v are determined in accordance with Section 9.4.1.2.5 using values of S_S and S_1, respectively, of $1.5g$ and $0.6g$. For the purpose of this section, structures may be considered regular if they do not have plan irregularities 1a, 1b, or 4 of Table 9.5.2.3.2 or vertical irregularities 1a, 1b, 4, or 5 of Table 9.5.2.3.3.

9.5.2.5.4 Diaphragms

The deflection in the plane of the diaphragm shall not exceed the permissible deflection of the attached elements. Permissible deflection shall be that deflection that will permit the attached elements to maintain structural integrity under the individual loading and continue to support the prescribed loads. Floor and roof diaphragms shall be designed to resist design seismic forces determined in accordance with Eq. 9.5.2.5.4 as follows:

$$F_{px} = \frac{\sum\limits_{i=x}^{n} F_i}{\sum\limits_{i=x}^{n} w_i} \, w_{px} \qquad \text{(Eq. 9.5.2.5.4)}$$

where

F_{px} = diaphragm design force;
F_i = design force applied to Level i;
w_i = weight tributary to Level i; and
w_{px} = weight tributary to the diaphragm at Level x.

The force determined from Eq. 9.5.2.5.4 need not exceed $0.4S_{DS}Iw_{px}$ but shall not be less than $0.2S_{DS}Iw_{px}$. When the diaphragm is required to transfer design seismic force from the vertical resisting elements above the diaphragm to other vertical resisting elements below the diaphragm due to offsets in the placement of the elements or to changes in relative lateral stiffness in the vertical elements, these forces shall be added to those determined from Eq. 9.5.2.5.4.

9.5.2.6 Design, Detailing Requirements, and Structural Component Load Effects

The design and detailing of the components of the seismic force-resisting system shall comply with the requirements of this section. Foundation design shall conform to the applicable requirements of Section 9.7. The materials and the systems composed of those materials shall conform to the requirements and limitations of Sections 9.8 through 9.12 for the applicable category.

TABLE 9.5.2.5.3. Analysis Procedures for Seismic Design Categories D, E, and F

Structure Description	Reference and Procedures
1. Structures designated as regular and up to 240 ft (75 m) in height	Section 9.5.3
2. Structures that have only vertical structural irregularities of Type 1a, 1b, 2, or 3 in Table 9.5.2.3.2 or plan irregularities of Type 1a or 1b of Table 9.5.2.3.1 and have a height exceeding 5 stories or 65 ft (20 m) in height	Section 9.5.4
3. All other structures designated as having plan or vertical structural irregularities	Section 9.5.3 plus the effect of the irregularity on the dynamic response shall be analyzed.
4. Structures in areas with S_{D1} of 0.2 and greater with a period greater than T_0 located on Site Class F or Site Class E soils.	A site-specific response spectra shall be used but the design base shear shall not be less than that determined from Section 9.5.3.2; also, the modal seismic design coefficient, C_{sm}, shall not be limited per Section 9.5.4.5.

9.5.2.6.1 Seismic design Category A

The design and detailing of structures assigned to Category A shall comply with the requirements of this section.

9.5.2.6.1.1 Component load effects: All structure components shall be provided with strengths sufficient to resist the effects of the seismic forces prescribed herein. Loads shall be combined as prescribed in Section 9.5.2.7. The direction of application of seismic forces used in design shall be that which will produce the most critical load effect in each component. The design seismic forces may be applied separately in each of two orthogonal horizontal directions and combined orthogonal effects may be neglected.

9.5.2.6.1.2 Load path connections: All parts of the structure between separation joints shall be interconnected to form a continuous path to the seismic force-resisting system, and the connections shall be capable of transmitting the seismic force (F_p) induced by the parts being connected. Any smaller portion of the structure shall be tied to the remainder of the structure with elements having a design strength capable of transmitting a seismic force of 0.133 times the short period design spectral response acceleration coefficient, S_{DS}, times the weight of the smaller portion or 5% of the portion's weight, whichever is greater. This connection force does not apply to the overall design of the lateral force resisting system. Connection design forces need not exceed the maximum forces that the structural system can deliver to the connection where the maximum system capacity is computed with a Ω_0 factor of 3.

A positive connection for resisting a horizontal force acting parallel to the member shall be provided for each beam, girder, or truss to its support. The connection shall have a minimum strength of 5% of the dead plus live load reaction. One means to provide the strength is to use connecting elements such as slabs.

9.5.2.6.1.3 Anchorage of concrete or masonry walls: Concrete and masonry walls shall be anchored to the roof and all floors and members that provide lateral support for the wall or which are supported by the wall. The anchorage shall provide a direct connection between the walls and the roof or floor construction. The connections shall be capable of resisting the greater of a seismic lateral force (F_p) induced by the wall or 400 times S_{DS} in pounds per linear foot of wall (5,840 times S_{DS} N/m) multiplied by the occu-

pancy importance factor, *I*. Walls shall be designed to resist bending between anchors where the anchor spacing exceeds 4 ft (1.2 m).

9.5.2.6.2 Seismic design Category B

Structures assigned to Seismic Design Category B shall conform to the requirements of Section 9.5.2.6.1 for Seismic Design Category A and the requirements of this section.

9.5.2.6.2.1 Component load effects: Seismic load effects on components shall be determined from the load analysis as required by Section 9.5.2.5, by other portions of Sections 9.5.2.6.2 and 9.5.2.7. The second-order effects shall be included where applicable. Where these seismic load effects exceed the minimum load path connection forces given in Section 9.5.2.6.1, they shall govern. Components shall satisfy the load combinations in Section 2.

9.5.2.6.2.2 Openings: Where openings occur in shear walls, diaphragms, or other plate-type elements, reinforcement at the edges of the openings shall be designed to transfer the stresses into the structure. The edge reinforcement shall extend into the body of the wall or diaphragm a distance sufficient to develop the force in the reinforcement. The extension must be sufficient in length to allow dissipation or transfer of the force without exceeding the shear and tension capacity of the diaphragm or the wall.

9.5.2.6.2.3 Direction of seismic load: The direction of application of seismic forces used in design shall be that which will produce the most critical load effect in each component. This requirement will be deemed satisfied if the design seismic forces are applied separately and independently in each of two orthogonal directions.

9.5.2.6.2.4 Discontinuities in vertical system: Structures with a discontinuity in lateral capacity, vertical irregularity Type 5 as defined in Table 9.5.2.3.3, shall not be over 2 stories or 30 ft (9 m) in height where the "weak" story has a calculated strength of less than 65% of the story above.

Exception: The limit does not apply where the "weak" story is capable of resisting a total seismic force equal to 75% of the deflection amplification factor, C_d, times the design force prescribed in Section 9.5.3.

9.5.2.6.2.5 Nonredundant systems: The design of a structure shall consider the potentially adverse effect that the failure of a single member, connection, or component of the seismic force-resisting system will have on the stability of the structure. See Section 1.3.

9.5.2.6.2.6 Collector elements: Collector elements shall be provided that are capable of transferring the seismic forces originating in other portions of the structure to the element providing the resistance to those forces.

9.5.2.6.2.7 Diaphragms: The deflection in the plane of the diaphragm, as determined by engineering analysis, shall not exceed the permissible deflection of the attached elements. Permissible deflection shall be that deflection which will permit the attached element to maintain its structural integrity under the individual loading and continue to support the prescribed loads.

At diaphragm discontinuities, such as openings and re-entrant corners, the design shall assure that the dissipation or transfer of edge (chord) forces combined with other forces in the diaphragm is within shear and tension capacity of the diaphragm.

Diaphragms shall provide for both the shear and bending stresses resulting from these forces. Diaphragms shall have ties or struts to distribute the wall anchorage forces into the diaphragm. Diaphragm connections shall be positive, mechanical or welded type connections.

9.5.2.6.2.8 Bearing walls: Exterior and interior bearing walls and their anchorage shall be designed for a force normal to the surface equal to 40% of the short period design spectral response acceleration, S_{DS}, times the weight of wall (W_c) associated with the anchor, with a minimum force of 10% of the weight of the wall. Interconnection of wall elements and connections to supporting framing systems shall have sufficient ductility, rotational capacity, or sufficient strength to resist shrinkage, thermal changes, and differential foundation settlement when combined with seismic forces. The connections shall also satisfy Section 9.5.2.6.1.3.

9.5.2.6.2.9 Inverted pendulum-type structures: Supporting columns or piers of inverted pendulum-type structures shall be designed for the bending moment calculated at the base determined using the procedures given in Section 9.5.3 and varying uniformly to a moment at the top equal to one-half the calculated bending moment at the base.

9.5.2.6.2.10 Anchorage of nonstructural systems: When required by Section 9.6, all portions or components of the structure shall be anchored for the seismic force, F_p, prescribed therein.

9.5.2.6.2.11 Columns supporting discontinuous walls or frames: Columns supporting discontinuous walls or frames of structures having plan irregularity Type 4 of Table 9.5.2.3.2 or vertical irregularity Type 4 of Table 9.5.2.3.3 shall have the design strength to resist the maximum axial force that can develop in accordance with the special combination of loads of Section 9.5.2.7.

9.5.2.6.3 Seismic design Category C

Structures assigned to Category C shall conform to the requirements of Section 9.5.2.6.2 for Category B and to the requirements of this section.

9.5.2.6.3.1 Direction of seismic load: For structures that have plan structural irregularity Type 5 in Table 9.5.2.3.2 the critical direction requirement of Section 9.5.2.6.2.3 shall not be deemed to be satisfied unless components and their foundations are designed for the following orthogonal combination of prescribed loads: 100% of the forces for one direction plus 30% of the forces for the perpendicular direction. The combination requiring the maximum component strength shall be used.

9.5.2.6.3.2 Collector elements: Collector elements shall be provided that are capable of transferring the seismic forces originating in other portions of the structure to the element providing the resistance to those forces. Collector elements, splices, and their connections to resisting elements shall resist the special load combinations of Section 9.5.2.7.

Exception: In structures or portions thereof braced entirely by light frame shear walls, collector elements, splices and connections to resisting elements need only be designed to resist forces in accordance with Section 9.5.2.5.4.

The quantity $\Omega_0 E$ in Eq. 9.5.2.7-3 need not exceed the maximum force that can be transferred to the collector by the diaphragm and other elements of the lateral-force-resisting system.

9.5.2.6.3.3 Anchorage of concrete or masonry walls: Concrete or masonry walls shall be anchored to all floors, roofs, and members that provide out-of-plane lateral support for the wall or that are supported by the wall. The anchorage shall provide a positive direct connection between the wall and floor, roof, or supporting member capable of resisting horizontal forces specified in this section for structures with flexible diaphragms or of Section 9.6.1.3 for structures with diaphragms that are not flexible.

Anchorage of walls to flexible diaphragms shall have the strength to develop the out-of-plane force given by Eq. 9.5.2.6.3.3:

$$F_p = 1.2S_{DS}/W_p \qquad \text{(Eq. 9.5.2.6.3.3)}$$

where

F_p = design force in the individual anchors;
S_{DS} = design spectral response acceleration at short periods per Section 9.4.1.2.5;
I = occupancy importance factor per Section 9.1.4; and
W_p = weight of the wall tributary to the anchor.

Diaphragms shall be provided with continuous ties or struts between diaphragm chords to distribute these anchorage forces into the diaphragms. Added chords may be used to form subdiaphragms to transmit the anchorage forces to the main continuous cross ties. The maximum length to width ratio of the structural subdiaphragm shall be 2-1/2 to 1. Connections and anchorages capable of resisting the prescribed forces shall be provided between the diaphragm and the attached components. Connections shall extend into the diaphragm a sufficient distance to develop the force transferred into the diaphragm.

In wood diaphragms, the continuous ties shall be in addition to the diaphragm sheathing. Anchorage shall not be accomplished by use of toe nails or nails subject to withdrawal nor shall wood ledgers of framing be used in cross-grain bending or cross-grain tension. The diaphragm sheathing shall not be considered effective as providing the ties or struts required by this section.

In metal deck diaphragms, the metal deck shall not be used as the continuous ties required by this section in the direction perpendicular to the deck span.

Diaphragm to wall anchorage using embedded straps shall be attached to or hooked around the reinforcing steel or otherwise terminated so as to effectively transfer forces to the reinforcing steel.

9.5.2.6.4 Seismic design Category D

Structures assigned to Category D shall conform to the requirements of Section 9.5.2.6.3 for Category C and to the requirements of this section.

9.5.2.6.4.1 Direction of seismic load: the independent orthogonal procedure given in Section 9.5.2.6.2.3 will not be satisfactory for the critical direction requirement for any structure. The orthogonal combination procedure in Section 9.5.2.6.3.1 will be deemed satisfactory for any structure.

9.5.2.6.4.2 Collector elements: In addition to the requirements of Section 9.5.2.6.3.2, collector elements, splices, and their connections to resisting elements shall resist the forces determined in accordance with Section 9.5.2.5.4.

9.5.2.6.4.3 Plan or vertical irregularities: When the ratio of the strength provided in any story to the strength required is less than two-thirds of that ratio for the story immediately above, the potentially adverse effect shall be analyzed and the strengths shall be adjusted to compensate for this effect.

For structures having a plan structural irregularity of Type 1, 2, 3, or 4 in Table 9.5.2.3.2 or a vertical structural irregularity of Type 4 in Table 9.5.2.3.3, the design forces determined from Section 9.5.3.2 shall be increased 25% for connections of diaphragms to vertical elements and to collectors and for connections of collectors to the vertical elements.

9.5.2.6.4.4 Vertical seismic forces: The vertical component of earthquake ground motion shall be considered in the design of horizontal cantilever and horizontal prestressed components. The load combinations used in evaluating such components shall include E as defined by Eqs. 9.5.2.7-1 and 9.5.2.7-2. Horizontal cantilever structural components shall be designed for a minimum net upward force of 0.2 times the dead load in addition to the applicable load combinations of Section 9.5.2.7.

9.5.2.6.5 Seismic design Categories E and F

Structures assigned to Seismic Design Categories E and F shall conform to the requirements of Section 9.5.2.6.4 for Seismic Design Category D and to the requirements of this section.

9.5.2.6.5.1 Plan or vertical irregularities: Structures having plan irregularity Type 1b of Table 9.5.2.3.1 or vertical irregularities Type 1b or 5 of Table 9.5.2.3.3 shall not be permitted.

9.5.2.7 Combination of Load Effects

The effects on the structure and its components due to seismic forces shall be combined with gravity loads in accordance with the combination of load effects given in Section 2. For use with those combinations, the earthquake-induced force effect shall include vertical and horizontal effects as given by Eq. 9.5.2.7-1 or, as applicable, Eq. 9.5.2.7-2, 9.5.2.7-3, or 9.5.2.7-4. The vertical seismic effect term $0.2 S_{DS}D$ need not be included where S_{DS} is equal to or less than 0.125 in Eqs. 9.5.2.7-1, 9.5.2.7-2, 9.5.2.7-3, and 9.5.2.7-4.

for Eq. 5 in Section 2.3.2 or Eq. 3 in Section 2.4.1:

$$E = \rho Q_E + 0.2 S_{DS}D \quad \text{(Eq. 9.5.2.7-1)}$$

for Eq. 7 in Section 2.3.2 or Eq. 5 in Section 2.4.1:

$$E = \rho Q_E - 0.2 S_{DS}D \quad \text{(Eq. 9.5.2.7-2)}$$

where

E = effect of horizontal and vertical earthquake-induced forces;

S_{DS} = design spectral response acceleration at short periods obtained from Section 9.4.1.2.5;

D = effect of dead load, D;

Q_E = effect of horizontal seismic (earthquake-induced) forces; and

ρ = reliability factor.

For columns supporting discontinuous lateral-force-resisting elements, the axial compression in the columns shall be computed using the following load in Eq. 5 in Section 2.3.2 or Eq. 3 in Section 2.4.1:

$$E = \Omega_0 Q_E + 0.2 S_{DS}D \quad \text{(Eq. 9.5.2.7-3)}$$

The axial forces in such columns need not exceed the capacity of other elements of the structure to transfer such loads to the column.

For brittle materials, systems and connections the following load also shall be used in Eq. 7 in Section 2.3.2 or Eq. 5 in Section 2.4.1:

$$E = \Omega_0 Q_E - 0.2 S_{DS}D \quad \text{(Eq. 9.5.2.7-4)}$$

The factor Ω_0 shall be greater than or equal to 2 and less than or equal to 3.

9.5.2.8 Deflection and Drift Limits

The design story drift (Δ) as determined in Section 9.5.3.7 or 9.5.4.6, shall not exceed the allowable story drift (Δ_a) as obtained from Table 9.5.2.8 for any story. For structures with significant torsional deflections, the maximum drift shall include torsional effects. All portions of the structure shall be designed and constructed to act as an integral unit in resisting seismic forces unless separated structurally by a distance sufficient to avoid damaging contact under total deflection (δ_x) as determined in Section 9.5.3.7.1.

9.5.3 Equivalent Lateral Force Procedure

9.5.3.1 General

Section 9.5.3 provides required minimum standards for the equivalent lateral force procedure of

TABLE 9.5.2.8. Allowable Story Drift, Δ_a[a]

Structure	Seismic Use Group		
	I	II	III
Structures, other than masonry shear wall or masonry wall frame structures, 4 stories or less with interior walls, partitions, ceilings and exterior wall systems that have been designed to accommodate the story drifts	$0.025h_{sx}$[b]	$0.020h_{sx}$	$0.015h_{sx}$
Masonary cantilever shear wall structures[c]	$0.010h_{sx}$	$0.010h_{sx}$	$0.010h_{sx}$
Other masonry shear wall structures	$0.007h_{sx}$	$0.007h_{sx}$	$0.007h_{sx}$
Masonry wall frame structures	$0.013h_{sx}$	$0.013h_{sx}$	$0.010h_{sx}$
All other structures	$0.020h_{sx}$	$0.015h_{sx}$	$0.010h_{sx}$

[a]h_{sx} is the story height below Level x.
[b]There shall be no drift limit for single-story structures with interior walls, partitions, ceilings, and exterior wall systems that have been designed to accommodate the story drifts. The structure separation requirement of Section 9.5.2.8 is not waived.
[c]Structures in which the basic structural system consists of masonry shear walls designed as vertical elements cantilevered from their base or foundation support which are so constructed that moment transfer between shear walls (coupling) is negligible.

seismic analysis of structures. For purposes of analysis, the structure is considered to be fixed at the base. See Section 9.5.2.5 for limitations on the use of this procedure.

9.5.3.2 Seismic Base Shear

The seismic base shear (V) in a given direction shall be determined in accordance with the following equation:

$$V = C_s W \qquad \text{(Eq. 9.5.3.2-1)}$$

where

C_s = seismic response coefficient determined in accordance with Section 9.5.3.2.1; and

W = total dead load and applicable portions of other loads listed below.

1. In areas used for storage, a minimum of 25% of the floor live load shall be applicable. The 50 psf floor live load for passenger cars in parking garages need not be considered.
2. Where an allowance for partition load is included in the floor load design, the actual partition weight or a minimum weight of 10 psf of floor area, whichever is greater, shall be applicable.
3. Total operating weight of permanent equipment and the effective contents of vessels.

Where the flat roof snow load (see Section 7.3) exceeds 30 lbs/ft^2, the design snow load shall be included in W. Where siting and load duration conditions warrant and the authority having jurisdiction approves, the amount of the snow load included in W may be reduced to no less than 20% of the design snow load.

9.5.3.2.1 Calculation of seismic response coefficient

When the fundamental period of the structure is computed, the seismic design coefficient (C_s) shall be determined in accordance with the following equation:

$$C_s = \frac{S_{DS}}{R/I} \qquad \text{(Eq. 9.5.3.2.1-1)}$$

where

S_{DS} = design spectral response acceleration in the short period range as determined from Section 9.4.1.2.5;

R = response modification factor in Table 9.5.2.2; and

I = occupancy importance factor determined in accordance with Section 9.1.4.

A soil-structure interaction reduction shall not be used unless Section 9.5.5 or another generally accepted procedure approved by the authority having jurisdiction is used.

Alternatively, the seismic response coefficient, (C_s), need not be greater than the following equation:

$$C_s = \frac{S_{D1}}{T(R/I)} \qquad \text{(Eq. 9.5.3.2.1-2)}$$

but shall not be taken less than:

$$C_s = 0.044 I S_{DS} \qquad \text{(Eq. 9.5.3.2.1-3)}$$

nor for buildings and structures in Seismic Design Categories E and F:

$$C_s = \frac{0.5 S_1}{R/I} \qquad \text{(Eq. 9.5.3.2.1-4)}$$

where I and R are as defined above and

S_{D1} = design spectral response acceleration at a period of 1.0 s, in units of g-s, as determined from Section 9.4.1.2.5;

T = fundamental period of the structure (s) determined in Section 9.5.3.3; and

S_1 = mapped maximum considered earthquake spectral response acceleration determined in accordance with Section 9.4.1.

A soil-structure interaction reduction is permitted when determined using Section 9.5.5.

9.5.3.3 Period Determination

The fundamental period of the structure (T) in the direction under consideration shall be established using the structural properties and deformational characteristics of the resisting elements in a properly substantiated analysis. The fundamental period (T) shall not exceed the product of the coefficient for upper limit on calculated period (C_u) from Table 9.5.3.3 and the approximate fundamental period (T_a) determined from Eq. 9.5.3.3-1.

$$T_a = C_T h_n^{3/4} \qquad \text{(Eq. 9.5.3.3-1)}$$

TABLE 9.5.3.3. Coefficient for Upper Limit on Calculated Period

Design Spectral Response Acceleration at 1 s, S_{D1}	Coefficient C_u
≥0.4	1.2
0.3	1.3
0.2	1.4
0.15	1.5
0.1	1.7
0.05	1.7

where

C_T = 0.035 for structures in which the lateral force-resisting system consists of moment resisting frames of steel providing 100% of the required lateral force resistance and such frames are not enclosed or adjoined by more rigid components tending to prevent the frames from deflecting when subjected to seismic forces (the metric coefficient is 0.0853);

C_T = 0.030 for structures in which the lateral force-resisting system consists of moment resisting frames of reinforced concrete providing 100% of the required lateral force resistance and such frames are not enclosed or adjoined by more rigid components tending to prevent the frames from deflecting when subjected to seismic forces (the metric coefficient is 0.0731);

C_T = 0.030 for structures in which the lateral force-resisting system consists of steel eccentrically-braced frames acting along or with moment resisting frames (the metric coefficient is 0.0731);

C_T = 0.020 for all other structures (the metric coefficient is 0.0488); and

h_n = the height in feet above the base to the highest level of the structure.

Alternately, it shall be permitted to determine the approximate fundamental period (T_a), in seconds, from the following equation for structures not exceeding 12 stories in height in which the lateral force-resisting system consists entirely of concrete or steel moment resisting frames and the story height is at least 10 ft (3 m):

$$T_a = 0.1N \qquad \text{(Eq. 9.5.3.3-2)}$$

where N = number of stories.

9.5.3.4 Vertical Distribution of Seismic Forces

The lateral seismic force (F_x) (kip or kN) induced at any level shall be determined from the following equations:

$$F_x = C_{vx}V \qquad \text{(Eq. 9.5.3.4-1)}$$

and

$$C_{vx} = \frac{w_x h_x^k}{\sum_{i=1}^{n} w_i h_i^k} \qquad \text{(Eq. 9.5.3.4-2)}$$

where

C_{vx} = vertical distribution factor;

V = total design lateral force or shear at the base of the structure, (kip or kN);

w_i and w_x = portion of the total gravity load of the structure (W) located or assigned to Level i or x;

h_i and h_x = height (ft or m) from the base to Level i or x; and

k = exponent related to the structure period as follows: for structures having a period of 0.5 s or less, k = 1; for structures having a period of 2.5 s or more, k = 2; and for structures having a period between 0.5 and 2.5 s, k shall be 2 or shall be determined by linear interpolation between 1 and 2.

9.5.3.5 Horizontal Shear Distribution and Torsion

The seismic design story shear in any story (V_x) (kip or kN) shall be determined from the following equation:

$$V_x = \sum_{i=1}^{n} F_i \qquad \text{(Eq. 9.5.3.5)}$$

where F_i = the portion of the seismic base shear (V) (kip or kN) induced at Level i.

9.5.3.5.1 Direct shear

The seismic design story shear (V_x) (kip or kN) shall be distributed to the various vertical elements of the seismic force-resisting system in the story under consideration based on the relative lateral stiffness of the vertical resisting elements and the diaphragm.

9.5.3.5.2 Torsion

The design shall include the torsional moment (M_t) (kip or kN) resulting from the location of the structure masses plus the accidental torsional moments (M_{ta}) (kip or kN) caused by assumed displacement of the mass each way from its actual location by a distance equal to 5% of the dimension of the structure perpendicular to the direction of the applied forces.

Structures of Seismic Design Categories C, D, E, and F, where Type 1 torsional irregularity exists as defined in Table 9.5.2.3.2 shall have the effects accounted for by multiplying the sum of M_t plus M_{ta} at each level by a torsional amplification factor (A_x) determined from the following equation:

$$A_x = \left(\frac{\delta_{max}}{1.2\delta_{avg}} \right)^2 \qquad \text{(Eq. 9.5.3.5.2)}$$

where

δ_{max} = the maximum displacement at Level x (in. or mm); and

δ_{avg} = average of the displacements at the extreme points of the structure at Level x (in. or mm).

The torsional amplification factor (A_x) is not required to exceed 3.0. The more severe loading for each element shall be considered for design.

9.5.3.6 Overturning

The structure shall be designed to resist overturning effects caused by the seismic forces determined in Section 9.5.3.4 At any story, the increment of overturning moment in the story under consideration shall be distributed to the various vertical force-resisting elements in the same proportion as the distribution of the horizontal shears to those elements.

The overturning moments at Level x (M_x) (kip·ft or kN·m) shall be determined from the following equation:

$$M_x = \tau \sum_{i=x}^{n} F_i(h_i - h_x) \qquad \text{(Eq. 9.5.3.6)}$$

where

F_i = portion of the seismic base shear (V) induced at Level i;

h_i and h_x = height (in ft or m) from the base to Level i or x;

τ = overturning moment reduction factor, determined as follows: for the top 10 stories, $\tau = 1.0$; for the 20th story from the top and below, $\tau = 0.8$; and for stories between the 20th and 10th stories below the top, a value between 1.0 and 0.8 determined by a straight line interpolation.

The foundations of structures, except inverted pendulum-type structures, shall be designed for the foundation overturning design moment (M_f) (kip·ft or kN·m) at the foundation-soil interface determined using the equation for the overturning moment at Level x (M_x) (kip·ft or kN·m) above with an overturning moment reduction factor (τ) of 0.75 for all structure heights.

9.5.3.7 Drift Determination and P-Delta Effects

Story drifts and, where required, member forces and moments due to P-delta effects shall be determined in accordance with this section. Determination of story drifts shall be based on the application of the design seismic forces to a mathematical model of the physical structure. The model shall include the stiffness and strength of all elements that are significant to the distribution of forces and deformations in the structure and shall represent the spatial distribution of the mass and stiffness of the structure. In addition, the model shall comply with the following:

1. Stiffness properties of reinforced concrete and masonry elements shall consider the effects of cracked sections; and
2. For steel moment resisting frame systems, the contribution of panel zone deformations to overall story drift shall be included.

9.5.3.7.1 Story drift determination

The design story drift (Δ) shall be computed as the difference of the deflections at the top and bottom of the story under consideration.

Exception: For structures of Seismic Design Categories C, D, E and F having plan irregularity Types 1a or 1b of Table 9.5.2.3.2, the design story drift, Δ, shall be computed as the largest difference of the deflections along any of the edges of the structure at the top and bottom of the story under consideration.

The deflections of Level x at the center of the mass (δ_x) (in. or mm) shall be determined in accordance with the following equation:

$$\delta_x = C_d \delta_{xe}/I \qquad \text{(Eq. 9.5.3.7.1)}$$

where

C_d = deflection amplification factor in Table 9.5.2.2;
δ_{xe} = deflections determined by an elastic analysis; and
I = importance factor determined in accordance with Section 9.1.4.

The elastic analysis of the seismic force-resisting system shall be made using the prescribed seismic design forces of Section 9.5.3.4.

For determining compliance with the story drift limitation of Section 9.5.2.8, the deflections of Level x at the center of mass (δ_x) (in. or mm) shall be calculated as required in this section. For the purposes of this drift analysis only, the upper-bound limitation specified in Section 9.5.3.3 does not apply for computing forces and displacements.

Where applicable, the design story drift (Δ) (in. or mm) shall be increased by the incremental factor relating to the P-delta effects as determined in Section 9.5.3.7.2.

9.5.3.7.2 P-Delta effects

P-delta effects on story shears and moments, the resulting member forces and moments, and the story drifts induced by these effects are not required to be considered when the stability coefficient (θ) as determined by the following equation is equal to or less than 0.10:

$$\theta = \frac{P_x \Delta}{V_x h_{sx} C_d} \qquad \text{(Eq. 9.5.3.7.2-1)}$$

where

P_x = total vertical design load at and above Level x. (kip or kN); when computing P_x, no individual load factor need exceed 1.0;
Δ = design story drift occurring simultaneously with V_x, (in. or mm);
V_x = seismic shear force acting between Levels x and $x - 1$, (kip or kN);
h_{sx} = story height below Level x, (in or mm); and
C_d = deflection amplification factor in Table 9.5.2.2.

The stability coefficient (θ) shall not exceed θ_{max} determined as follows:

$$\theta_{max} = \frac{0.5}{\beta C_d} \leq 0.25 \qquad \text{(Eq. 9.5.3.7.2-2)}$$

where β is the ratio of shear demand to shear capacity for the story between Level x and $x - 1$. This ratio may be conservatively taken as 1.0.

When the stability coefficient (θ) is greater than 0.10 but less than or equal to θ_{max}, the incremental factor related to P-delta effects (a_d) shall be determined by rational analysis. To obtain the story drift for including the P-delta effect, the design story drift determined in Section 9.5.3.7.1 shall be multiplied by $1.0/(1 - \theta)$.

When θ is greater than θ_{max}, the structure is potentially unstable and shall be redesigned.

When the P-delta effect is included in an automated analysis, Eq. 9.5.3.7.2-2 must still be satisfied, however, the value of θ computed from Eq. 9.5.3.7.2-1 using the results of the P-delta analysis may be divided by $(1 + \theta)$ before checking Eq. 9.5.3.7.2-2.

9.5.3.8 Simplified Analysis Procedure for Seismic Design of Buildings

See Section 9.5.2.5 for limitations on the use of this procedure. For purposes of this analysis procedure, a building is considered to be fixed at the base.

9.5.3.8.1 Seismic base shear

The seismic base shear, V, in a given direction shall be determined in accordance with the following formula:

$$V = \frac{1.2 S_{DS}}{R} W \qquad \text{(Eq. 9.5.3.8.1)}$$

where

S_{DS} = design elastic response acceleration at short period as determined in accordance with Section 9.4.1.2.5;
R = response modification factor from Table 9.5.2.2;
W = effective seismic weight of the structure, including the total dead load and other loads listed below.

1. In areas used for storage, a minimum of 25% of the reduced floor live load (floor live load in public arages and open parking structures need not be included)
2. Where an allowance for partition load is included in the floor load design, the actual partition weight or a minimum weight of 10 psf of floor area, whichever is greater
3. Total weight of permanent operating equipment
4. 20% of flat roof snow load where flat snow load exceeds 30 psf (1.44 kN/m²).

9.5.3.8.2 Vertical distribution

The forces at each level shall be calculated using the following formula:

$$F_x = \frac{1.2S_{DS}}{R} w_x \qquad \text{(Eq. 9.5.3.8.2)}$$

where

w_x = portion of the effective seismic weight of the structure, W, at level x.

9.5.4 Modal Analysis Procedure

9.5.4.1 General

Section 9.5.4 provides required standards for the modal analysis procedure of seismic analysis of structures. See Section 9.5.2.5 for requirements for use of this procedure. The symbols used in this method of analysis have the same meaning as those for similar terms used in Section 9.5.3, with the subscript m denoting quantities in the mth mode.

9.5.4.2 Modeling

A mathematical model of the structure shall be constructed that represents the spatial distribution of mass and stiffness throughout the structure.

For regular structures with independent orthogonal seismic-force-resisting systems, independent two-dimensional models are permitted to be constructed to represent each system. For irregular structures or structures without independent orthogonal systems, a three-dimensional model incorporating a minimum of three dynamic degrees of freedom consisting of translation in two orthogonal plan directions and torsional rotation about the vertical axis shall be included at each level of the structure. Where the diaphragms are not rigid compared to the vertical elements of the lateral-force-resisting system, the model should include representation of the diaphragm's flexibility and such additional dynamic degrees of freedom as are required to account for the participation of the diaphragm in the structure's dynamic response. In addition, the model shall comply with the following:

1. Stiffness properties of concrete and masonry elements shall consider the effects of cracked sections and
2. For steel moment frame systems, the contribution of panel zone deformations to overall story drift shall be included.

9.5.4.3 Modes

An analysis shall be conducted to determine the natural modes of vibration for the structure including the period of each mode, the modal shape vector Φ, the modal participation factor, and modal mass. The analysis shall include a sufficient number of modes to obtain a combined modal mass participation of at least 90% of the actual mass in each of two orthogonal directions.

9.5.4.4 Periods

The required periods, mode shapes, and participation factors of the structure in the direction under consideration shall be calculated by established methods of structural analysis for the fixed base condition using the masses and elastic stiffnesses of the seismic force-resisting system.

9.5.4.5 Modal Base Shear

The portion of the base shear contributed by the mth mode (V_m) shall be determined from the following equations:

$$V_m = C_{sm}W_m \qquad \text{(Eq. 9.5.4.5-1)}$$

$$W_m = \frac{\left(\sum\limits_{i=1}^{n} w_i \varphi_{im}\right)^2}{\sum\limits_{i=1}^{n} w_i \varphi_{im}^2} \qquad \text{(Eq. 9.5.4.5-2)}$$

where

C_{sm} = modal seismic design coefficient determined below;
W_m = effective modal gravity load;
w_i = portion of the total gravity load of the structure at Level i; and
ϕ_m = displacement amplitude at the ith level of the structure when vibrating in its mth mode.

The modal seismic design coefficient (C_{sm}) shall be determined in accordance with the following equation:

$$C_{sm} = \frac{S_{am}}{R|I} \qquad \text{(Eq. 9.5.4.5-3)}$$

where

S_{am} = design spectral response acceleration at period T_m determined from either the general design response spectrum of Section 9.4.1.2.6 or a site-specific response spectrum per Section 9.4.1.3;

R = response modification factor determined from Table 9.5.2.2;

I = occupancy importance factor determined in accordance with Section 9.1.4; and

T_m = modal period of vibration (in seconds) of the mth mode of the structure.

Exceptions:

1. When the general design response spectrum of Section 9.4.1.2.6 is used for structures where any modal period of vibration (T_m) exceeds 4.0 s, the modal seismic design coefficient (C_{sm}) for that mode shall be determined by the following equation:

$$C_{sm} = \frac{4S_{D1}}{(R|I)T_m^2} \qquad \text{(Eq. 9.5.4.5-4)}$$

The reduction due to soil-structure interaction as determined in Section 9.5.5.3 is permitted to be used.

9.5.4.6 Modal Forces, Deflections, and Drifts

The modal force (F_{xm}) at each level shall be determined by the following equations:

$$F_{xm} = C_{vxm}V_m \qquad \text{(Eq. 9.5.4.6-1)}$$

and

$$C_{vxm} = \frac{w_x\varphi_{xm}}{\displaystyle\sum_{i=1}^{n} w_i\varphi_{im}} \qquad \text{(Eq. 9.5.4.6-2)}$$

where

C_{vxm} = vertical distribution factor in the mth mode;

V_m = total design lateral force or shear at the base in the mth mode;

w_i and w_x = portion of the total gravity load of the structure (W) located or assigned to Level i or x;

ϕ_{xm} = displacement amplitude at the xth level of the structure when vibrating in its mth mode; and

ϕ_{im} = displacement amplitude at the ith level of the structure when vibrating in its mth mode.

The modal deflection at each level (δ_{xm}) shall be determined by the following equations:

$$\delta_{xm} = \frac{C_d\delta_{xem}}{I} \qquad \text{(Eq. 9.5.4.6-3)}$$

and

$$\delta_{xem} = \left(\frac{g}{4\pi^2}\right)\left(\frac{T_m^2 F_{xm}}{w_x}\right) \qquad \text{(Eq. 9.5.4.6-4)}$$

where

C_d = deflection amplification factor determined from Table 9.5.2.2;

δ_{xem} = deflection of Level x in the mth mode at the center of the mass at Level x determined by an elastic analysis;

g = acceleration due to gravity (ft/s^2);

I = occupancy importance factor determined in accordance with Section 9.1.4;

T_m = modal period of vibration, in seconds, of the mth mode of the structure;

F_{xm} = portion of the seismic base shear in the mth mode, induced at Level x; and

w_x = portion of the total gravity load of the structure (W) located or assigned to Level x.

The modal drift in a story (Δ_m) shall be computed as the difference of the deflections (δ_{xm}) at the top and bottom of the story under consideration.

9.5.4.7 Modal Story Shears and Moments

The story shears, story overturning moments, and the shear forces and overturning moments in vertical elements of the structural system at each level due to the seismic forces determined from the appropriate equation in Section 9.5.4.6 shall be computed for each mode by linear static methods.

9.5.4.8 Design Values

The design value for the modal base shear (V_t), each of the story shear, moment and drift quantities, and the deflection at each level shall be determined by combining their modal values as obtained from Sections 9.5.4.6 and 9.5.4.7. The combination shall be carried out by taking the square root of the sum of the squares of each of the modal values or by the complete quadratic combination (CQC) method.

The base shear (V) using the equivalent lateral force procedure in Section 9.5.3 shall be calculated using a fundamental period of the structure (T), in seconds, of 1.2 times the coefficient for upper limit on the calculated period (C_u) times the approximate fundamental period of the structure (T_a). Where the design value for the modal base shear (V_t) is less than the calculated base shear (V) using the equivalent lateral force procedure, the design story shears, moments, drifts, and floor deflections shall be multiplied by the following modification factor:

$$\frac{V}{V_t} \qquad \text{(Eq. 9.5.4.8)}$$

where

V = equivalent lateral force procedure base shear, calculated in accordance with this section and Section 9.5.3; and

V_t = modal base shear, calculated in accordance with this section.

The modal base shear (V_t) is not required to exceed the base shear from the equivalent lateral force procedure in Section 9.5.3.

Exception: For structures in areas with an S_{D1} value of 0.2 and greater with a period of 0.7 s or greater located on Site Class E or F soils, the design base shear shall not be less than that determined using the equivalent lateral force procedure in Section 9.5.3 (see Section 9.5.2.5.3).

9.5.4.9 Horizontal Shear Distribution

The distribution of horizontal shear shall be in accordance with the requirements of Section 9.5.3.5 except that amplification of torsion per Section 9.5.3.5.2 is not required for that portion of the torsion included in the dynamic analysis model.

9.5.4.10 Foundation Overturning

The foundation overturning moment at the foundation-soil interface may be reduced by 10%.

9.5.4.11 P-Delta Effects

The *P*-delta effects shall be determined in accordance with Section 9.5.3.7.2. The story drifts and story shears shall be determined in accordance with Section 9.5.3.7.1.

9.5.5 Soil Structure Interaction

9.5.5.1 General

If the option to incorporate the effects of soil-structure interaction is exercised, the requirements of this section shall be used in the determination of the design earthquake forces and the corresponding displacements of the structure. The use of these provisions will decrease the design values of the base shear, lateral forces, and overturning moments but may increase the computed values of the lateral displacements and the secondary forces associated with the *P*-delta effects.

The provisions for use with the equivalent lateral force procedure are given in Section 9.5.5.2. and those for use with the modal analysis procedure are given in Section 9.5.5.3.

9.5.5.2 Equivalent Lateral Force Procedure

The following requirements are supplementary to those presented in Section 9.5.3.

9.5.5.2.1 Base shear

To account for the effects of soil-structure interaction, the base shear (V) determined from Eqs. 9.5.5.2.1-1 and 9.5.5.2.1-2 shall be reduced to:

$$\tilde{V} = V - \Delta V \qquad \text{(Eq. 9.5.5.2.1-1)}$$

The reduction (ΔV) shall be computed as follows and shall not exceed $0.3V$:

$$\Delta V = \left[C_s - \tilde{C}_s \left(\frac{0.05}{\tilde{\beta}} \right)^{0.4} \right] \bar{W} \le 0.3V$$

$$\text{(Eq. 9.5.5.2.1-2)}$$

where

C_s = seismic design coefficient computed from Eq. 9.5.3.2.1-1 using the fundamental natural period of the fixed base structure (T or T_a) as specified in Section 9.5.3.3;

$\tilde{C}_s$ = value of C_s computed from Eq. 9.5.3.2.1-1 using the fundamental natural period of the flexibly supported structure (T) defined in Section 9.5.5.2.1.1;

$\tilde{\beta}$ = fraction of critical damping for the structure-foundation system determined in Section 9.5.5.2.1.2; and

$\bar{W}$ = effective gravity load of the structure, which shall be taken as $0.7W$, except that for structures where the gravity load is concentrated at a single level, it shall be taken equal to W.

9.5.5.2.1.1 *Effective building period*: The effective period ($\tilde{T}$) shall be determined as follows:

$$\tilde{T} = T \sqrt{1 + \frac{\bar{k}}{K_y}\left(1 + \frac{K_y\bar{h}^2}{K_\theta}\right)}$$

(Eq. 9.5.5.2.1.1-1)

where

T = fundamental period of the structure as determined in Section 9.5.3.3;

$\bar{k}$ = stiffness of the structure when fixed at the base, defined by the following:

$$\bar{k} = 4\pi^2\left(\frac{\bar{W}}{gT^2}\right)$$

(Eq. 9.5.5.2.1.1-2)

$\bar{h}$ = effective height of the structure which shall be taken as 0.7 times the total height (h_n) except that for structure where the gravity load is effectively concentrated at a single level, it shall be taken as the height to that level;

K_y = lateral stiffness of the foundation defined as the static horizontal force at the level of the foundation necessary to produce a unit deflection at that level, the force and the deflection being measured in the direction in which the structure is analyzed;

K_θ = rocking stiffness of the foundation defined as the static moment necessary to produce a unit average rotation of the foundation, the moment and rotation being measured in the direction in which the structure is analyzed; and

g = acceleration of gravity.

The foundation stiffnesses (K_y and K_θ) shall be computed by established principles of foundation mechanics using soil properties that are compatible with the soil strain levels associated with the design earthquake motion. The average shear modulus (G) for the soils beneath the foundation at large strain levels and the associated shear wave velocity (v_s) needed in these computations shall be determined from Table 9.5.5.2.1.1.

TABLE 9.5.5.2.1.1. Values of G/G_o and v_s/v_{so}

	Spectral Response Acceleration, S_{D1}			
Value	≤ 0.10	≤ 0.15	≤ 0.20	≥ 0.30
G/G_o	0.81	0.64	0.49	0.42
v_s/v_{so}	0.9	0.8	0.7	0.65

where

v_{so} = average shear wave velocity for the soils beneath the foundation at small strain levels ($10^{-3}\%$ or less);

$G_o = \gamma v_{so}^2/g$ = average shear modulus for the soils beneath the foundation at small strain levels; and

γ = average unit weight of the soils.

Alternatively, for structures supported on mat foundations that rest at or near the ground surface or are embedded in such a way that the side wall contact with the soil are not considered to remain effective during the design ground motion, the effective period of the structure shall be determined as follows:

$$\tilde{T} = T \sqrt{1 + \frac{25\alpha r_a\bar{h}}{v_s^2 T^2}\left(1 + \frac{1.12 r_a\bar{h}^2}{r_m^3}\right)}$$

(Eq. 9.5.5.2.1.1-3)

where α = relative weight density of the structure and the soil defined by:

$$\alpha = \frac{\bar{W}}{\gamma A_0\bar{h}}$$

(Eq. 9.5.5.2.1.1-4)

where r_a and r_m = characteristic foundation lengths defined by:

$$r_a = \sqrt{\frac{A_0}{\pi}}$$

(Eq. 9.5.5.2.1.1-5)

and

$$r_m = \sqrt[4]{\frac{4I_0}{\pi}}$$

(Eq. 9.5.5.2.1.1-6)

where

A_0 = area of the load carrying foundation; and

I_0 = static moment of inertia of the load carrying foundation about a horizontal centroidal axis normal to the direction in which the structure is analyzed.

9.5.5.2.1.2 Effective damping: The effective damping factor for the structure-foundation system ($\tilde{\beta}$) shall be computed as follows:

$$\tilde{\beta} = \beta_0 + \frac{0.05}{(\tilde{T}/T)^3} \quad \text{(Eq. 9.5.5.2.1.2-1)}$$

where β_0 = foundation damping factor as specified in Fig. 9.5.5.2.1.2.

The values of β_0 corresponding to $0.4S_{DS} = 0.15$ in Fig. 9.5.5.2.1.2 shall be determined by averaging the results obtained from the solid lines and the dashed lines.

The quantity r in Fig. 9.5.5.2.1.2 is a characteristic foundation length that shall be determined as follows:

For $\bar{h}/L_0 \leq 0.5$

$$r = r_a = \sqrt{\frac{A_0}{\pi}} \quad \text{(Eq. 9.5.5.2.1.2-2)}$$

For $\bar{h}/L_0 \geq 1$

$$r = r_m = \sqrt[4]{\frac{4I_0}{\pi}} \quad \text{(Eq. 9.5.5.2.1.2-3)}$$

where

L_0 = overall length of the side of the foundation in the direction being analyzed;

A_0 = area of the load-carrying foundation; and

I_0 = static moment of inertia of the load-carrying foundation about a horizontal centroidal axis normal to the direction in which the structure is analyzed.

For intermediate values of $\bar{h}/L_0$, the value of r shall be determined by linear interpolation.

EXCEPTION: For structures supported on point bearing piles and in all other cases where the foundation soil consists of a soft stratum of reasonably uniform properties underlain by a much stiffer, rock-like deposit with an abrupt increase in stiffness, the factor β_0 in Eq. 9.5.5.2.1.2-1 shall be replaced by:

$$\beta_0' = \left(\frac{4D_s}{V_s\tilde{T}}\right)^2 \beta_0 \quad \text{(Eq. 9.5.5.2.1.2-4)}$$

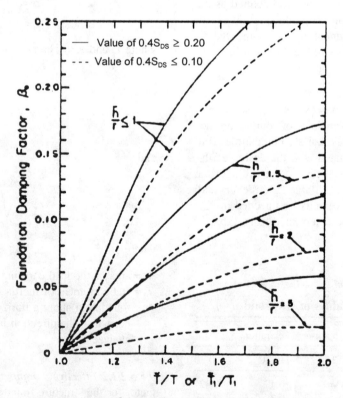

FIGURE 9.5.5.2.1.2. Foundation Damping Factor

if $4D_s/v_s\tilde{T} < 1$ where D_s is the total depth of the stratum.

The value of $\tilde{\beta}$ computed from Eq. 9.5.5.2.1.2-1, both with or without the adjustment represented by Eq. 9.5.5.2.1.2-4, shall in no case be taken as less than $\tilde{\beta} = 0.05$.

9.5.5.2.2 Vertical distribution of seismic forces

The distribution over the height of the structure of the reduced total seismic force ($\tilde{V}$) shall be considered to be the same as for the structure without interaction.

9.5.5.2.3 Other effects

The modified story shears, overturning moments, and torsional effects about a vertical axis shall be determined as for structures without interaction using the reduced lateral forces.

The modified deflections ($\tilde{\delta}_x$) shall be determined as follows:

$$\tilde{\delta}_x = \frac{\tilde{V}}{V}\left[\frac{M_0 h_x}{K_\theta} + \delta_x\right] \qquad \text{(Eq. 9.5.5.2.3-1)}$$

where

M_0 = overturning moment at the base determined in accordance with Section 9.5.3.6 using the unmodified seismic forces and not including the reduction permitted in the design of the foundation;

h_x = height above the base to the level under consideration; and

δ_x = deflections of the fixed base structure as determined in Section 9.5.3.7.1 using the unmodified seismic forces.

The modified story drifts and P-delta effects shall be evaluated in accordance with the provisions of Section 9.5.3.7 using the modified story shears and deflections determined in this section.

9.5.5.3 Modal Analysis Procedure

The following provisions are supplementary to those presented in Section 9.5.4.

9.5.5.3.1 Modal base shears

To account for the effects of soil-structure interaction, the base shear corresponding to the fundamental mode of vibration (V_1) shall be reduced to:

$$\tilde{V}_1 = V_1 - \Delta V_1 \qquad \text{(Eq. 9.5.5.3.1-1)}$$

The reduction (ΔV_1) shall be computed in accordance

with Eq. 9.5.5.2.1-2 with $\bar{W}$ taken as equal to the gravity load $\bar{W}_1$ defined by Eq. 9.5.4.5-2, C_s computed from Eq. 9.5.4.5-3 using the fundamental period of the fixed base structure (T_1), and $\tilde{C}_s$ computed from Eq. 9.5.4.5-3 using the fundamental period of the elastically supported structure ($\tilde{T}_1$).

The period $\tilde{T}_1$ shall be determined from Eq. 9.5.5.2.1.1-1, or from Eq. 9.5.5.2.1.1-3 when applicable, taking $T = \tilde{T}_1$, evaluating $\bar{k}$ from Eq. 9.5.5.2.1.1-2 with $\bar{W} = \bar{W}_1$, and computing $\bar{h}$ as follows:

$$\bar{h} = \frac{\displaystyle\sum_{i=1}^{n} w_i \varphi_{i1} h_i}{\displaystyle\sum_{i=1}^{n} w_i \varphi_{i1}} \qquad \text{(Eq. 9.5.5.3.1-2)}$$

The above designated values of $\bar{W}$, $\bar{h}$, T, and $\tilde{T}$ also shall be used to evaluate the factor α from Eq. 9.5.5.2.1.1-4 and factor β_0 from Fig. 9.5.5.2.1.2. No reduction shall be made in the shear components contributed by the higher modes of vibration. The reduced base shear ($\tilde{V}_1$) shall in no case be taken less than $0.7V_1$.

9.5.5.3.2 Other modal effects

The modified modal seismic forces, story shears, and overturning moments shall be determined as for structures without interaction using the modified base shear ($\tilde{V}_1$) instead of V_1. The modified modal deflections ($\tilde{\delta}_{xm}$) shall be determined as follows:

$$\tilde{\delta}_{xl} = \frac{\tilde{V}_1}{V_1}\left[\frac{M_{01} h_x}{K_\theta} + \delta_{x1}\right] \qquad \text{(Eq. 9.5.5.3.2-1)}$$

and

$$\tilde{\delta}_{xm} = \delta_{xm} \quad \text{for } m = 2, 3, \ldots \qquad \text{(Eq. 9.5.5.3.2-2)}$$

where

M_{01} = overturning base moment for the fundamental mode of the fixed-base structure, as determined in Section 9.5.4.7 using the unmodified modal base shear V_1; and

δ_{xm} = modal deflections at Level x of the fixed-base structure as determined in Section 9.5.4.6 using the unmodified modal shears, V_m.

The modified modal drift in a story ($\tilde{\Delta}_m$) shall be computed as the difference of the deflections ($\tilde{\delta}_{xm}$) at the top and bottom of the story under consideration.

9.5.5.3.3 Design values

The design values of the modified shears, moments, deflections, and story drifts shall be determined as for structures without interaction by taking the square root of the sum of the squares of the respective modal contributions. In the design of the foundation, it shall be permitted to reduce the overturning moment at the foundation-soil interface determined in this manner by 10% as for structures without interaction.

The effects of torsion about a vertical axis shall be evaluated in accordance with the provisions of Section 9.5.3.5 and the *P*-delta effects shall be evaluated in accordance with the provisions of Section 9.5.3.7.2, using the story shears and drifts determined in Section 9.5.5.3.2.

9.6 ARCHITECTURAL, MECHANICAL, AND ELECTRICAL COMPONENTS AND SYSTEMS

9.6.1 General

Section 9.6 establishes minimum design criteria for architectural, mechanical, electrical, and nonstructural systems, components and elements permanently attached to structures, including supporting structures and attachments (hereinafter referred to as "components"). The design criteria establish minimum equivalent static force levels and relative displacement demands for the design of components and their attachments to the structure, recognizing ground motion and structural amplification, component toughness and weight, and performance expectations. Seismic Design Categories for structures are defined in Section 9.4.2. For the purposes of this section, components shall be considered to have the same Seismic Design Category as that of the structure that they occupy or to which they are attached unless otherwise noted.

This section also establishes minimum seismic design force requirements for nonbuilding structures that are supported by other structures. Seismic design requirements for nonbuilding structures that are supported at grade are prescribed in Section 9.14; however, the minimum seismic design forces for nonbuilding structures that are supported by another structure shall be determined in accordance with the requirements of Section 9.6.1.3 with R_p equal to the value of R specified in Section 9.14 and $a_p = 2.5$ for nonbuilding structures with flexible dynamic characteristics and $a_p = 1.0$ for nonbuilding structures with rigid dynamic characteristics. The distribution of lateral forces for the supported nonbuilding structure and all nonforce requirements specified in Section 9.14 shall apply to supported nonbuilding structures.

Exception: For structures in Seismic Design Categories D, E and F if the combined weight of the supported components and nonbuilding structures with flexible dynamic characteristics exceeds 25% of the weight of the structure, the structure shall be designed considering interaction effects between the structure and the supported items.

In addition, all components are assigned a component importance factor (I_p) in this chapter. The default value for I_p is 1.00 for typical components in normal service. Higher values for I_p are assigned for components which contain hazardous substances, must have a higher level of assurance of function, or otherwise require additional attention because of their life-safety characteristics. Component importance factors are prescribed in Section 9.6.1.5.

All architectural, mechanical, electrical and other nonstructural components in structures shall be designed and constructed to resist the equivalent static forces and displacements determined in accordance with this Section. The design and evaluation of support structures and architectural components and equipment shall consider their flexibility as well as their strength.

Exceptions:

1. All components in Seismic Design Category A;
2. Architectural components in Seismic Design Category B other than parapets supported by bearing walls or shear walls provided that the importance factor (I_p) is equal to 1.0;
3. Mechanical and electrical components in Seismic Design Category B.
4. Mechanical and electrical components in structures assigned to Seismic Design Category C provided that the importance factor (I_p) is equal to 1.0;
5. Mechanical and electrical components in Seismic Design Categories D, E, and F that are mounted at 4 ft (1.2 m) or less above a floor level and weigh 400 lb (180 kg) or less, are not critical to the continued operation of the structure and will not create a hazard by its failure or the failure of distribution components connected to it.
6. Mechanical and electrical components in Seismic Design Categories C, D, E, and F that weigh 20 lb (9 kg) or less or, for distribution systems not containing hazardous substances, weigh 5 lb/ft (7.4 kg/m) or less.

The functional and physical interrelationship of components and their effect on each other shall be designed so that the failure of an essential or nonessential architectural, mechanical, or electrical component shall not cause the failure of a nearby essential architectural, mechanical, or electrical component.

9.6.1.1 Reference Standards

9.6.1.1.1 Consensus standards

The following references are consensus standards and are to be considered part of these provisions to the extent referred to in this chapter:

Ref. 9.6-1 American Society of Mechanical Engineers (ASME), ASME A17.1, *Safety Code for Elevators and Escalators*, 1996.

Ref. 9.6-2 American Society of Mechanical Engineers (ASME), *Boiler and Pressure Vessel Code*, including addendums through 1997.

Ref. 9.6-3 American Society For Testing and Materials (ASTM), ASTM C635, *Standard Specification for the Manufacture, Performance, and Testing of Metal Suspension Systems for Acoustical Tile and Lay-in Panel Ceilings*, 1995.

Ref. 9.6-4 American Society For Testing And Materials (ASTM), ASTM C636, *Standard Practice for Installation of Metal Ceiling Suspension Systems for Acoustical Tile and Lay-in Panels*, 1992.

Ref. 9.6-5 American National Standards Institute/American Society of Mechanical Engineers, ANSI/ASME B31.1-95, Power Piping.

Ref. 9.6-6 American National Standards Institute/American Society of Mechanical Engineers, ANSI/ASME B31.3-96, Process Piping.

Ref. 9.6-7 American National Standards Institute/American Society of Mechanical Engineers, ANSI/ASME B31.4-92, Liquid Transportation Systems for Hydrocarbons, Liquid Petroleum Gas, Anhydrous Ammonia, and Alcohols.

Ref. 9.6-8 American National Standards Institute/American Society of Mechanical

Engineers, ANSI/ASME B31.5-92, Refrigeration Piping.

Ref. 9.6-9 American National Standards Institute/American Society of Mechanical Engineers, ANSI/ASME B31.9-95, Building Services Piping.

Ref. 9.6-10 American National Standards Institute/American Society of Mechanical Engineers, ANSI/ASME B31.11-86, Slurry Transportation Piping Systems.

Ref. 9.6-11 American National Standards Institute/American Society of Mechanical Engineers, ANSI/ASME B31.8-95, Gas Transmission and Distribution Piping Systems.

Ref. 9.6-12 Institute of Electrical and Electronic Engineers (IEEE), Standard 344, Recommended Practice for Seismic Qualification of Class 1E Equipment for Nuclear Power Generating Stations, 1987.

Ref. 9.6-13 National Fire Protection Association (NFPA), NFPA-13, Standard for the Installation of Sprinkler Systems, 1996.

9.6.1.1.2 Accepted standards

The following references are standards developed within the industry and represent acceptable procedures for design and construction:

Ref. 9.6-14 American Society of Heating, Ventilating, and Air Conditioning (ASHRAE), Handbook, Chapter 50, "Seismic Restraint Design," 1995.

Ref. 9.6-15 Ceilings and Interior Systems Construction Association (CISCA), Recommendations for Direct-Hung Acoustical Tile and Lay-in Panel Ceilings, Seismic Zones 0–2, 1991.

Ref. 9.6-16 Ceilings and Interior Systems Construction Association (CISCA), Recommendations for Direct-Hung Acoustical Tile and Lay-in Panel Ceilings, Seismic Zones 3–4, 1990.

Ref. 9.6-17 Sheet Metal and Air Conditioning Contractors National Association (SMACNA), HVAC Duct Construction Standards, Metal and Flexible, 1985.

Ref. 9.6-18 Sheet Metal and Air Conditioning Contractors National Association (SMACNA), Rectangular Industrial Duct Construction Standards, 1980.

Ref. 9.6-19 Sheet Metal and Air Conditioning Contractors National Association (SMACNA), Seismic Restraint Manual Guidelines for Mechanical Systems, 1991, including Appendix E, 1993 addendum.

9.6.1.2 Component Force Transfer

Components shall be attached such that the component forces are transferred to the structure of the building. Component seismic attachments shall be bolted, welded, adhered or otherwise positively fastened without consideration of frictional resistance produced by the effects of gravity.

The design documents shall include sufficient information relating to the attachments to verify compliance with the requirements of Section 9.6.

9.6.1.3 Seismic Forces

Seismic forces (F_p) shall be determined in accordance with Eq. 9.6.1.3-1:

$$F_p = \frac{0.4a_p S_{DS} W_p}{R_p/I_p} \left(1 + 2\frac{z}{h} \right) \quad \text{(Eq. 9.6.1.3-1)}$$

F_p is not required to be taken as greater than:

$$F_p = 1.6 S_{DS} I_p W_p \quad \text{(Eq. 9.6.1.3-2)}$$

and F_p shall not be taken as less than:

$$F_p = 0.3 S_{DS} I_p W_p \quad \text{(Eq. 9.6.1.3-3)}$$

where:

F_p = seismic design force centered at the component's center of gravity and distributed relative to component's mass distribution;

S_{DS} = spectral acceleration, short period, as determined from Section 9.4.1.2.5;

a_p = component amplification factor that varies from 1.00 to 2.50 (select appropriate value from Table 9.6.2.2 or 9.6.3.2);

I_p = component importance factor that varies from 1.00 to 1.50 (see Section 9.6.1.5);

W_p = component operating weight;

R_p = component response modification factor that varies from 1.50 to 5.00 (select appropriate value from Table 9.6.2.2 or 9.6.3.2);

z = height in structure at point of attachment of component. For items at or below the base, z shall be taken as 0; and

h = average roof height of structure relative to grade elevation.

The force (F_p) shall be applied independently longitudinally, and laterally in combination with service loads associated with the component. Combine horizontal and vertical load effects as indicated in Section 9.5.2.7 substituting F_p for the term Q_E. The reliability/redundancy factor, ρ, is permitted to be taken equal to 1.

When positive and negative wind loads exceed F_p for nonbearing exterior wall, these wind loads shall govern the design. Similarly when the building code horizontal loads exceed F_p for interior partitions, these building code loads shall govern the design.

9.6.1.4 Seismic Relative Displacements

Seismic relative displacements (D_p) shall be determined in accordance with the following equations:

For two connection points on the same Structure A or the same structural system, one at Level x and the other at Level y, D_p shall be determined as:

$$D_p = \delta_{xA} - \delta_{yA} \quad \text{(Eq. 9.6.1.4-1)}$$

D_p is not required to be taken as greater than:

$$D_p = (X - Y)\Delta_{aA}/h_{sx} \quad \text{(Eq. 9.6.1.4-2)}$$

For two connection points on separate Structures A or B or separate structural systems, one at Level X and the other at Level Y, D_p shall be determined as:

$$D_p = |\delta_{xA}| + |\delta_{yB}| \quad \text{(Eq. 9.6.1.4-3)}$$

D_p is not required to be taken as greater than:

$$D_p = X\Delta_{aA}/h_{sx} + Y\Delta_{aB}/h_{sx} \quad \text{(Eq. 9.6.1.4-4)}$$

where

D_p = relative seismic displacement that the component must be designed to accommodate;

δ_{xA} = deflection at building level x of Structure A, determined by an elastic analysis as defined in Section 9.5.3.7.1;

δ_{yA} = deflection at building Level y of Structure A, determined by an elastic analysis as defined in Section 9.5.3.7.1;

δ_{yB} = deflection at building Level y of Structure B, determined by an elastic analysis as defined in Section 9.5.3.7.1;

X = height of upper support attachment at Level x as measured from the base;

Y = height of lower support attachment at Level y as measured from the base;

Δ_{aA} = allowable story drift for Structure A as defined in Table 9.5.2.8;

Δ_{aB} = allowable story drift for Structure B as defined in Table 9.5.2.8; and

h_{sx} = story height used in the definition of the allowable drift Δ_a in Table 9.5.2.8. Note that Δ_a/h_{sx} = the drift index.

The effects of seismic relative displacements shall be considered in combination with displacements caused by other loads as appropriate.

9.6.1.5 Component Importance Factor

The component importance factor (I_p) shall be selected as follows:

I_p = 1.5 life-safety component required to function after an earthquake (e.g., fire protection sprinkler system);

I_p = 1.5 component which contains hazardous content;

I_p = 1.5 storage racks in structures open to the public (e.g. warehouse retails stores); and

I_p = 1.0 all other components.

In addition, for structures in Seismic Use Group III:

I_p = 1.5 all components needed for continued operation of the facility or whose failure could impair the continued operation of the facility.

9.6.1.6 Component Anchorage

Components shall be anchored in accordance with the following provisions.

9.6.1.6.1

The force in the connected part shall be determined based on the prescribed forces for the component specified in Section 9.6.1.3.

9.6.1.6.2

Anchors embedded in concrete or masonry shall be proportioned to carry the least of the following:

a. The design strength of the connected part,

b. 2 times the force in the connected part due to the prescribed forces, or

c. The maximum force that can be transferred to the connected part by the component structural system.

9.6.1.6.3

Determination of forces in anchors shall take into account the expected conditions of installation including eccentricities and prying effects.

9.6.1.6.4

Determination of force distribution of multiple anchors at one location shall take into account the stiffness of the connected system and its ability to redistribute loads to other anchors in the group beyond yield.

9.6.1.6.5

Powder driven fasteners shall not be used for tension load applications in Seismic Design Categories D, E, and F unless approved for such loading.

9.6.1.6.6

The design strength of anchors in concrete shall be determined in accordance with the provisions of Section 9.9.

9.6.1.7 Construction Documents

Construction documents shall be prepared to comply with the requirements of this standard, as indicated in Table 9.6.1.7.

9.6.2 Architectural Component Design

9.6.2.1 General

Architectural systems, components, or elements (hereinafter referred to as "components") listed in Table 9.6.2.2 and their attachments shall meet the requirements of Sections 9.6.2.2 through 9.6.2.9.

9.6.2.2 Architectural Component Forces and Displacements

Architectural components shall meet the force requirements of Section 9.6.1.3 and Table 9.6.2.2.

Components supported by chains or otherwise suspended from the structural system above are not required to meet the lateral seismic force requirements and seismic relative displacement requirements of this section provided that they cannot be damaged to become a hazard or cannot damage any other component when subject to seismic motion and they have ductile or articulating connections to the structure at the point of attachment. The gravity design load for these items shall be three times their operating load.

9.6.2.3 Architectural Component Deformation

Architectural components that could pose a life-safety hazard shall be designed for the seismic relative displacement requirements of Section 9.6.1.4. Architectural components shall be designed for vertical deflection due to joint rotation of cantilever structural members.

9.6.2.4 Exterior Nonstructural Wall Elements and Connections

Exterior nonstructural wall panels or elements that are attached to or enclose the structure shall be designed to resist the forces in accordance with Eqs. 9.6.1.3-1 or 9.6.1.3-2, and shall accommodate movements of the structure resulting from response to the design basis ground motion, D_p, or temperature changes. Such elements shall be supported by means of positive and direct structural supports or by mechanical connections and fasteners. The support system shall be designed in accordance with the following:

a. Connections and panel joints shall allow for the story drift caused by relative seismic displacements (D_p) determined in Section 9.6.1.4, or 1/2 in. (13 mm), whichever is greatest.

b. Connections to permit movement in the plane of the panel for story drift shall be sliding connections using slotted or oversized holes, connections that permit movement by bending of steel, or other connections that provide equivalent sliding or ductile capacity.

c. The connecting member itself shall have sufficient ductility and rotation capacity to preclude fracture of the concrete or brittle failures at or near welds.

d. All fasteners in the connecting system such as bolts, inserts, welds, and dowels and the body of the connectors shall be designed for the force (F_p) determined by Eq. 9.6.1.3-2 with values of R_p and a_p taken from Table 9.6.2.2 applied at the center of mass of the panel.

e. Anchorage using flat straps embedded in concrete or masonry shall be attached to or hooked around reinforcing steel or otherwise terminated so as to effectively transfer forces to the reinforcing steel *or to assure that pullout of anchorage is not the initial failure mechanism.*

TABLE 9.6.1.7. Construction Documents

Component Description	Provisions References		Required Seismic Design Categories
	Quality Assurance	Design	
Exterior wall panels, including anchorage	A.9.3.3.9 No. 1	9.6.2.4	D, E, F
Suspended ceiling system, including anchorage	A.9.3.3.9 No. 2	9.6.2.6	D, E, F
Access floors, including anchorage	A.9.3.3.9 No. 2	9.6.2.7	D, E, F
Steel storage racks, including anchorage	A.9.3.3.9 No. 2	9.6.2.9	D, E, F
HVAC ductwork containing hazardous materials, including anchorage	A.9.3.3.10 No. 4	9.6.3.10	C, D, E
Piping systems and mechanical units containing flammable, combustible, or highly toxic materials	A.9.3.3.10 No. 3	9.6.3.11 9.6.3.12 9.6.3.13	C, D, E, F
Anchorage of electrical equipment for emergency or standby power systems	A.9.3.3.10 No. 1	9.6.3.14	C, D, E, F
Anchorage of all other electrical equipment	A.9.3.3.10 No. 2	9.6.3.14	E, F
Project-specific requirements for mechanical and electrical components and their anchorage	A.9.3.4.5	9.6.3	C, D, E, F

TABLE 9.6.2.2. Architectural Component Coefficients

Architectural Component or Element	a_p[a]	R_p[b]
Interior Nonstructural Walls and Partitions		
Plain (unreinforced) masonry walls	1	1.25
All other walls and partitions	1	2.5
Cantilever Elements (Unbraced or braced to structural frame below its center of mass)		
Parapets and cantilever interior nonstructural walls	2.5	2.5
Chimneys and stacks when laterally braced or supported by the structural frame	2.5	2.5
Cantilever Elements (Branched to structural frame above its center of mass)		
Parapets	1.0	2.5
Chimneys and Stacks	1.0	2.5
Exterior Nonstructural Walls	1.0[c]	2.5
Exterior Nonstructural Wall Elements and Connections		
Wall Element	1	2.5
Body of wall panel connections	1	2.5
Fasteners of the connecting system	1.25	1
Veneer		
Limited deformability elements and attachments	1	2.5
Low deformability elements and attachments	1	1.25
Penthouses (except when framed by an extension of the building frame)	2.5	3.5
Ceilings		
All	1	2.5
Cabinets		
Storage cabinets and laboratory equipment	1	2.5
Access Floors		
Special access floors (designed in accordance with Section 9.6.2.7.2)	1	2.5
All other	1	1.25
Appendages and Ornamentations	2.5	2.5
Signs and Billboards	2.5	2.5
Other Rigid Components		
High deformability elements and attachments	1	3.5
Limited deformability elements and attachments	1	2.5
Low deformability materials and attachments	1	1.25
Other Flexible Components		
High deformability elements and attachments	2.5	3.5
Limited deformability elements and attachments	2.5	2.5
Low deformability materials and attachments	2.5	1.25

[a]A lower value for a_p shall not be used unless justified by detailed dynamic analysis. The value for a_p shall not be less than 1.00. The value of $a_p = 1$ is for equipment generally regarded rigid and rigidly attached. The value of $a_p = 2.5$ is for equipment generally regarded as flexible or flexibly attached. See Section 9.2.1 for definitions of rigid and flexible.

[b]$R_p = 1.25$ for anchorage design when component anchorage is provided by expansion anchor bolts, shallow chemical anchors, or shallow (nonductile) cast-in-place anchors or when the component is constructed of nonductile materials. Powder-actuated fasteners (shot pins) shall not be used for component anchorage in tension applications in Seismic Design Categories D, E, or F. Shallow anchors are those with an embedment length-to-bold diameter ratio of less than 8.

[c]Where flexible diaphragms provide lateral support for walls and partitions, the design forces for anchorage to the diaphragm shall be as specified in Section 9.5.2.5.4.

9.6.2.5 Out-of-Plane Bending

Transverse or out-of-plane bending or deformation of a component or system that is subjected to forces as determined in Section 9.6.2.2 shall not exceed the deflection capability of the component or system.

9.6.2.6 Suspended Ceilings

Suspended ceilings shall be designed to meet the seismic force requirements of Section 9.6.2.6.1. In addition, suspended ceilings shall meet the requirements of either Industry Standard Construction as modified in Section 9.6.2.6.2 or Integral Construction as specified in Section 9.6.2.6.3.

9.6.2.6.1 Seismic forces

Suspended ceilings shall be designed to meet the force requirements of Section 9.6.1.3.

The weight of the ceiling, W_p, shall include the ceiling grid and panels; light fixtures if attached to, clipped to, or laterally supported by the ceiling grid; and other components which are laterally supported by the ceiling. W_p shall be taken as not less than 4 lb/ft^2 (19 N/m^2).

The seismic force, F_p, shall be transmitted through the ceiling attachments to the building structural elements or the ceiling-structure boundary.

Design of anchorage and connections shall be in accordance with these Provisions.

9.6.2.6.2 Industry standard construction

Unless designed in accordance with Section 9.6.2.6.3, suspended ceilings shall be designed and constructed in accordance with this section.

9.6.2.6.2.1 Seismic Design Category C: Suspended ceilings in Seismic Design Category C shall be designed and installed in accordance with the Ceilings and Interior Systems Construction Association (CISCA) recommendations for seismic zones 0–2 (Ref. 9.6-15), except that seismic forces shall be determined in accordance with Section 9.6.1.3 and 9.6.2.6.1.

Sprinkler heads and other penetrations in Seismic Design Category C shall have a minimum of 1/4 in. (6 mm) clearance on all sides.

9.6.2.6.2.2 Seismic Design Categories D, E, and F: Suspended ceilings in Seismic Design Categories D, E, and F shall be designed and installed in accordance with the Ceilings and Interior Systems Construction Association (CISCA) recommendations for seismic zones 3–4 (Ref. 9.6-16) and the additional requirements listed in this subsection.

a. A heavy duty T-bar grid system shall be used.

b. The width of the perimeter supporting closure angle shall be not less than 2.0 in. (50 mm). In each orthogonal horizontal direction, one end of the ceiling grid shall be attached to the closure angle. The other end in each horizontal direction shall have a 3/4 in. (19 mm) clearance from the wall and shall rest upon and be free to slide on a closure angle.

c. For ceiling areas exceeding 1,000 ft^2 (92.9 m^2), horizontal restraint of the ceiling to the structural system shall be provided. The tributary areas of the horizontal restraints shall be approximately equal.

   Exception: Rigid braces are permitted to be used instead of diagonal splay wires. Braces and attachments to the structural system above shall be adequate to limit relative lateral deflections at point of attachment of ceiling grid to less than 1/4 in. (6 mm) for the loads prescribed in Section 9.6.1.3.

d. For ceiling areas exceeding 2,500 ft^2 (232 m^2), a seismic separation joint or full height partition that breaks the ceiling up into areas not exceeding 2,500 ft^2 shall be provided unless structural analyses are performed of the ceiling bracing system for the prescribed seismic forces which demonstrate ceiling system penetrations and closure angles provide sufficient clearance to accommodate the additional movement. Each area shall be provided with closure angles in accordance with Item b and horizontal restraints or bracing in accordance with Item c.

e. Except where rigid braces are used to limit lateral deflections, sprinkler heads and other penetrations shall have a 2 in. (50 mm) oversize ring, sleeve, or adapter through the ceiling tile to allow for free movement of at least 1 in. (25 mm) in all horizontal directions. Alternatively, a swing joint that can accommodate 1 in. (25 mm) of ceiling movement in all horizontal directions are permitted to be provided at the top of the sprinkler head extension.

f. Changes in ceiling plan elevation shall be provided with positive bracing.

g. Cable trays and electrical conduits shall be supported independently of the ceiling.

h. Suspended ceilings shall be subject to the special inspection requirements of Sec. A.9.3.3.9 of this Standard.

9.6.2.6.3 Integral Ceiling/Sprinkler Construction

As an alternate to providing large clearances around sprinkler system penetrations through ceiling systems, the sprinkler system and ceiling grid are permitted to be designed and tied together as an integral unit. Such a design shall consider the mass and flexibility of all elements involved, including: ceiling system, sprinkler system, light fixtures, and mechanical (HVAC) appurtenances. The design shall be performed by a registered design professional.

9.6.2.7 Access Floors

9.6.2.7.1 General

Access floors shall be designed to meet the force provisions of Section 9.6.1.3 and the additional provisions of this section. The weight of the access floor, W_p, shall include the weight of the floor system, 100% of the weight of all equipment fastened to the floor, and 25% of the weight of all equipment supported by, but not fastened to the floor. The seismic force, F_p, shall be transmitted from the top surface of the access floor to the supporting structure.

Overturning effects of equipment fastened to the access floor panels also shall be considered. The ability of "slip on" heads for pedestals shall be evaluated for suitability to transfer overturning effects of equipment.

When checking individual pedestals for overturning effects, the maximum concurrent axial load shall not exceed the portion of W_p assigned to the pedestal under consideration.

9.6.2.7.2 Special Access Floors

Access floors shall be considered to be "special access floors" if they are designed to comply with the following considerations:

1. Connections transmitting seismic loads consist of mechanical fasteners, concrete anchors, welding, or bearing. Design load capacities comply with recognized design codes and/or certified test results.
2. Seismic loads are not transmitted by friction, produced solely by the effects of gravity, powder-actuated fasteners (shot pins), or adhesives.
3. The design analysis of the bracing system includes the destabilizing effects of individual members buckling in compression.
4. Bracing and pedestals are of structural or mechanical shape produced to ASTM specifications that specify minimum mechanical properties. Electrical tubing shall not be used.
5. Floor stringers that are designed to carry axial seismic loads and that are mechanically fastened to the supporting pedestals are used.

9.6.2.8 Partitions

Partitions that are tied to the ceiling and all partitions greater than 6 ft (1.8 m) in height shall be laterally braced to the building structure. Such bracing shall be independent of any ceiling splay bracing. Bracing shall be spaced to limit horizontal deflection at the partition head to be compatible with ceiling deflection requirements as determined in Section 9.6.2.6 for suspended ceilings and Section 9.6.2.2 for other systems.

9.6.2.9 Steel Storage Racks

Steel storage racks supported at the base of the structure shall be designed to meet the force requirements of Section 9.14. Steel storage racks supported above the base of the structure shall be designed to meet the force requirements of Sections 9.6.1 and 9.6.2.

9.6.3 Mechanical and Electrical Component Design

9.6.3.1 General

Attachments and equipment supports for the mechanical and electrical systems, components, or elements (hereinafter referred to as "components") shall meet the requirements of Sections 9.6.3.2 through 9.6.3.16.

9.6.3.2 Mechanical and Electrical Component Forces and Displacements

Mechanical and electrical components shall meet the force and seismic relative displacement requirements of Section 9.6.1.3, Section 9.6.1.4, and Table 9.6.3.2.

When complex equipment such as valves and valve operators, turbines and generators, and pumps and motors are functionally connected by mechanical links not capable of transferring the seismic loads or accommodating seismic relative displacements, the design shall protect such links by alternative methods.

Components supported by chains or otherwise suspended from the structural system above are not required to meet the lateral seismic force requirements and seismic relative displacement requirements of this section provided that they cannot be damaged to become a hazard or cannot damage any other

TABLE 9.6.3.2. Mechanical and Electrical Components Seismic Coefficients

Mechanical and Electrical Component or Element[c]	a_p[a]	R_p[b]
General Mechanical Equipment		
Boilers and furnaces	1.0	2.5
Pressure vessels on skirts and free-standing	2.5	2.5
Stacks	2.5	2.5
Cantilevered chimneys	2.5	2.5
Other	1.0	2.5
Manufacturing and Process Machinery		
General	1.0	2.5
Conveyors (nonpersonnel)	2.5	2.5
Piping Systems		
High deformability elements and attachments	1.0	3.5
Limited deformability elements and attachments	1.0	2.5
Low deformability elements and attachments	1.0	1.25
HVAC Systems		
Vibration isolated	2.5	2.5
Nonvibration isolated	1.0	2.5
Mounted-in-line with ductwork	1.0	2.5
Other	1.0	2.5
Elevator Components	1.0	2.5
Escalator Components	1.0	2.5
Trussed Towers (free-standing or guyed)	2.5	2.5
General Electrical		
Distributed systems (bus ducts, conduit, cable tray)	2.5	5.0
Equipment	1.0	2.5
Lighting Fixtures	1.0	1.25

[a]A lower value for a_p shall not be used unless justified by detailed dynamic analyses. The value for a_p shall not be less than 1.00. The value of $a_p = 1$ is for equipment generally regarded as rigid or rigidly attached. The value of $a_p = 2.5$ is for equipment generally regarded as flexible or flexibly attached. See Section 9.2.2 for definitions of rigid and flexible.

[b]$R_p = 1.25$ for anchorage design when component anchorage is provided by expansion anchor bolts, shallow chemical anchors, or shallow low deformability cast-in-place anchors or when the component is constructed of nonductile materials. Powder-actuated fasteners (shot pins) shall not be used for component anchorage in Seismic Design categories D, E, or F. Shallow anchors are those with an embedment length-to-bold diameter ratio of less than 8.

[c]Components mounted on vibration isolation systems shall have a bumper restraint or snubber in each horizontal direction. The design force shall be taken as $2F_p$.

component when subject to seismic motion and they have ductile or articulating connections to the structure at the point of attachment. The gravity design load for these items shall be three times their operating load.

9.6.3.3 Mechanical and Electrical Component Period

The fundamental period of the mechanical and electrical component (and its attachment to the building), T_p, shall be determined by the following equation provided that the component and attachment can be reasonably represented and analytically by a simple spring and mass single-degree-of-freedom system:

$$T_p = 2\pi \sqrt{\frac{W_p}{K_p g}} \qquad \text{(Eq. 9.6.3.3)}$$

where

T_p = component fundamental period;
W_p = component operating weight;
g = gravitational acceleration; and
K_p = stiffness of resilient support system of the component and attachment, determined in terms of load per unit deflection at the center of gravity of the component.

Note that consistent units must be used.

Otherwise, determine the fundamental period of the component in seconds (T_p) from experimental test data or by a properly substantiated analysis.

9.6.3.4 Mechanical and Electrical Component Attachments

The stiffness of mechanical and electrical component attachments shall be designed such that the load path for the component performs its intended function.

9.6.3.5 Component Supports

Mechanical and electrical component supports and the means by which they are attached to the component shall be designed for the forces determined in Sections 9.6.1.3 and in conformance with Sections 9.8 through 9.12, as appropriate, for the material comprising the means of attachment. Such supports include structural members, braces, frames, skirts, legs, saddles, pedestals, cables, guys, stays, snubbers, and tethers, as well as element forged or cast as a part of the mechanical or electrical component. If standard or proprietary supports are used, they shall be designed by either load rating (i.e., testing) or for the calculated seismic forces. In addition, the stiffness of the support, when appropriate, shall be designed such that the seismic load path for the component performs its intended function.

Component supports shall be designed to accommodate the seismic relative displacement between points of support determined in accordance with Section 9.6.1.4.

In addition, the means by which supports are attached to the component, except when integral (i.e., cast or forged), shall be designed to accommodate both the forces and displacements determined in accordance with Section 9.6.1.3 and 9.6.1.4. If the value of $I_p = 1.5$ for the component, the local region of the support attachment point to the component shall be evaluated for the effect of the load transfer on the component wall.

9.6.3.6 Component Certification

The manufacturer's certificate of compliance with the force requirements of the Section shall be submitted to the regulatory agency when required by the contract documents or when required by the regulatory agency.

9.6.3.7 Utility and Service Lines at Structure Interfaces

At the interface of adjacent structures or portions of the same structure that may move independently, utility lines shall be provided with adequate flexibility to accommodate the anticipated differential movement between the portions that move independently. Differential displacement calculations shall be determined in accordance with Section 9.6.1.4.

9.6.3.8 Site-Specific Considerations

The possible interruption of utility service shall be considered in relation to designated seismic systems in Seismic Use Group III as defined in Section 9.1.3.4. Specific attention shall be given to the vulnerability of underground utilities and utility interfaces between the structure and the ground where Site Class E or F soil is present, and where the seismic coefficient C_a at the underground utility or at the base of the structure is equal to or greater than 0.15.

9.6.3.9 Storage Tanks Mounted in Structures

Storage tanks, including their attachments and supports, shall be designed to meet the force requirements of Section 9.14.

9.6.3.10 HVAC Ductwork

Attachments and supports for HVAC ductwork systems shall be designed to meet the force and displacement provisions of Section 9.6.1.3 and 9.6.1.4 and the additional provisions of this section. In addition to their attachments and supports, ductwork systems designated as having an $I_p = 1.5$ themselves shall be designed to meet the force and displacement provisions of Section 9.6.1.3 and 9.6.1.4 and the additional provisions of this section.

Seismic restraints are not required for HVAC ducts with $I_p = 1.0$ if either of the following conditions are met:

a. HVAC ducts are suspended from hangers 12 in. (305 mm) or less in length from the top of the duct to the supporting structure. The hangers shall be detailed to avoid significant bending of the hangers and their attachements, or
b. HVAC ducts have a cross-sectional area of less than 6 ft^2 (0.557 m^2).

HVAC duct systems fabricated and installed in accordance with standards approved by the authority having jurisdiction shall be deemed to meet the lateral bracing requirements of this section.

Equipment items installed in-line with the duct system (e.g., fans, heat exchangers, and humidifiers) weighing more than 75 lb (344 N) shall be supported and laterally braced independent of the duct system and shall meet the force requirements of Section 9.6.1.3.

9.6.3.11 Piping Systems

Attachments and supports for piping systems shall be designed to meet the force and displacement provisions of Sections 9.6.1.3 and 9.6.1.4 and the additional provisions of this section. In addition to their attachments and supports, piping systems designated as having $I_p = 1.5$ themselves shall be designed to meet the force and displacement provisions of Section 9.6.1.3 and 9.6.1.4 and the additional provisions of this section.

Seismic effects that shall be considered in the design of a piping system include the dynamic effects of the piping system, its contents, and, when appropriate, its supports. The interaction between the piping system and the supporting structures, including other mechanical and electrical equipment shall also be considered

9.6.3.11.1 Pressure piping systems

Pressure piping systems designed and constructed in accordance with ASME B31, *Code for Pressure Piping* (Ref. 9.6-3) shall be deemed to meet the force, displacement, and other provisions of this section. In lieu of specific force and displacement provisions provided in the ASME B31, the force and displacement provisions of Sections 9.6.1.3 and 9.6.1.4 shall be used.

9.6.3.11.2 Fire protection sprinkler systems

Fire protection sprinkler systems designed and constructed in accordance with NFPA 13, *Standard for the Installation of Sprinkler Systems* (Ref. 9.6-12) shall be deemed to meet the other requirements of this section, except the force and displacement requirements of Sections 9.6.1.3 and 9.6.1.4 shall be satisfied.

9.6.3.11.3 Other Piping Systems

Piping designated as having an $I_p = 1.5$ but not designed and constructed in accordance with ASME B31 (Ref. 9.6-3) or NFPA 13 (Ref. 9.6-12) shall meet the following:

a. The design strength for seismic loads in combination with other service loads and appropriate environmental effects shall not exceed the following:
(1) For piping systems constructed with ductile materials (e.g., steel, aluminum or copper), 90% of the piping material yield strength.

(2) For threaded connections with ductile materials, 70% of the piping material yield strength.
(3) For piping constructed with nonductile materials (e.g., plastic, cast iron, or ceramics), 25% of the piping material minimum specified tensile strength.
(4) For threaded connections in piping constructed with nonductile materials, 20% of the piping material minimum specified tensile strength.
b. Provisions shall be made to mitigate seismic impact for piping components constructed of nonductile materials or in cases where material ductility is reduced (e.g., low temperature applications).
c. Piping shall be investigated to ensure that the piping has adequate flexibility between support attachment points to the structure, ground, other mechanical and electrical equipment, or other piping.
d. Piping shall be investigated to ensure that the interaction effects between it and other piping or constructions are acceptable.

9.6.3.11.4 Supports and attachments for other piping

Attachments and supports for piping not designed and constructed in accordance with ASME B31 [Ref 9.6-3] or NFPA 13 [Ref 9.6-12] shall meet the following provisions:

a. Attachments and supports transferring seismic loads shall be constructed of materials suitable for the application and designed and constructed in accordance with a nationally recognized structural code such as, when constructed of steel, the AISC *Manual of Steel Construction* (Ref. 9.8-1 or 9.8-2) or MSS SP-58, *Pipe Hangers and Supports—Materials, Design, and Manufacture* (Ref. 9.6-11).
b. Attachments embedded in concrete shall be suitable for cyclic loads.
c. Rod hangers shall not be used as seismic supports unless the length of the hanger from the supporting structure is 12 in. (305 mm) or less. Rod hangers shall not be constructed in a manner that subjects the rod to bending moments.
d. Seismic supports are not required for:
(1) Ductile piping in Seismic Design Category D, E, or F designated as having an $I_p = 1.5$ and a nominal pipe size of 1 in. (25 mm) or less when provisions are made to protect the piping from impact or to avoid the impact of larger piping or other mechanical equipment.
(2) Ductile piping in Seismic Design Category C designated as having an $I_p = 1.5$ and a nominal pipe size of 2 in. (50 mm) or less when

provisions are made to protect the piping from impact or to avoid the impact of larger piping or other mechanical equipment.

(3) Ductile piping in Seismic Design Category D, E, or F designated as having an $I_p = 1.0$ and a nominal pipe size of 3 in. (75 mm) or less.

(4) Ductile piping in Seismic Design Category A, B, or C designated as having an $I_p = 1.0$ and a nominal pipe size of 6 in. (150 mm) or less.

e. Seismic supports shall be constructed so that support engagement is maintained.

9.6.3.12 Boilers and Pressure Vessels

Attachments and supports for boilers and pressure vessels shall be designed to meet the force and displacement provisions of Sections 9.6.1.3 and 9.6.1.4 and the additional provisions of this section. In addition to their attachments and supports, boilers and pressure vessels designated as having an $I_p = 1.5$ themselves shall be designed to meet the force and displacement provisions of Sections 9.6.1.3 and 9.6.1.4.

The seismic design of a boiler or pressure vessel shall include analysis of the following: the dynamic effects of the boiler or pressure vessel, its contents, and its supports; sloshing of liquid contents; loads from attached components such as piping; and the interaction between the boiler or pressure vessel and its support.

9.6.3.12.1 ASME boilers and pressure vessels

Boilers or pressure vessels designed in accordance with the ASME *Boiler and Pressure Vessel Code* (Ref. 9.6-4) shall be deemed to meet the force, displacement, and other requirements of this section. In lieu of the specific force and displacement provisions provided in the ASME code, the force and displacement provisions of Sections 9.6.1.3 and 9.6.1.4 shall be used.

9.6.3.12.2 Other boilers and pressure vessels

Boilers and pressure vessels designated as having an $I_p = 1.5$ but not constructed in accordance with the provisions of the ASME code (Ref. 9.6-4) shall meet the following provisions:

a. The design strength for seismic loads in combination with other service loads and appropriate environmental effects shall not exceed the following:

(1) For boilers and pressure vessels constructed with ductile materials (e.g., steel, aluminum or copper), 90% of the material minimum specified yield strength.

(2) For threaded connections in boilers or pressure

vessels or their supports constructed with ductile materials, 70% of the material minimum specified yield strength.

(3) For boilers and pressure vessels constructed with nonductile materials (e.g., plastic, cast iron, or ceramics), 25% of the material minimum specified tensile strength.

(4) For threaded connections in boilers or pressure vessels or their supports constructed with nonductile materials, 20% of the material minimum specified tensile strength.

b. Provisions shall be made to mitigate seismic impact for boiler and pressure vessel components constructed of nonductile materials or in cases where material ductility is reduced (e.g., low temperature applications).

c. Boilers and pressure vessels shall be investigated to ensure that the interaction effects between them and other constructions are acceptable.

9.6.3.12.3 Supports and attachments for other boilers and pressure vessels

Attachments and supports for boilers and pressure vessels shall meet the following provisions:

a. Attachments and supports transferring seismic loads shall be constructed of materials suitable for the application and designed and constructed in accordance with nationally recognized structural code such as, when constructed of steel, the AISC *Manual of Steel Construction* (Ref 9.8-1 or 9.8-2).

b. Attachments embedded in concrete shall be suitable for cyclic loads.

c. Seismic supports shall be constructed so that support engagement is maintained.

9.6.3.13 Mechanical Equipment, Attachments and Supports

Attachments and supports for mechanical equipment not covered in Sections 9.6.3.8 through 9.6.3.12 or 9.6.3.16 shall be designed to meet the force and displacement provisions of Sections 9.6.1.3 and 9.6.1.4 and the additional provisions of this section. In addition to their attachments and supports, such mechanical equipment designated as having an $I_p = 1.5$, itself, shall be designated to meet the force and displacement provisions of Sections 9.6.1.3 and 9.6.1.4 and the additional provisions of this section.

The seismic design of mechanical equipment, attachments and their supports shall include analysis of the following: the dynamic effects of the equipment, its contents, and when appropriate its supports. The interaction between the equipment and the supporting structures, including other mechanical and electrical equipment, shall also be considered.

9.6.3.13.1 Mechanical equipment

Mechanical equipment designated as having an $I_p = 1.5$ shall meet the following provisions.

a. The design strength for seismic loads in combination with other service loads and appropriate environmental effects shall not exceed the following:

(1) For mechanical equipment constructed with ductile materials (e.g., steel, aluminum, or copper), 90% of the equipment material minimum specified yield strength.

(2) For threaded connections in equipment constructed with ductile materials, 70% of the material minimum specified yield strength.

(3) For mechanical equipment constructed with nonductile materials (e.g., plastic, cast iron, or ceramics), 25% of the equipment material minimum tensile strength.

(4) For threaded connections in equipment constructed with nonductile materials, 20% of the material minimum specified yield strength.

b. Provisions shall be made to mitigate seismic impact for equipment components constructed of nonductile materials or in cases where material ductility is reduced (e.g., low temperature applications).

c. The possibility for loadings imposed on the equipment by attached utility or service lines due to differential motions of points of support from separate structures shall be evaluated.

9.6.3.13.2 Attachments and supports for mechanical equipment

Attachments and supports for mechanical equipment shall meet the following provisions:

a. Attachments and supports transferring seismic loads shall be constructed of materials suitable for the application and designed and constructed in accordance with a nationally recognized structural code such as, when constructed of steel, AISC, *Manual of Steel Construction* (Ref. 9.8-1 or 9.8-2).

b. Friction clips shall not be used for anchorage attachment.

c. Expansion anchors shall not be used for mechanical equipment rated over 10 hp (7.45 kW).

   EXCEPTION: Undercut expansion anchors.

d. Drilled and grouted-in-place anchors for tensile load applications shall use either expansive cement or expansive epoxy grout.

e. Supports shall be specifically evaluated if weak-axis bending of light-gage support steel is relied on for the seismic load path.

f. Components mounted on vibration isolation systems shall have a bumper restraint or snubber in each horizontal direction. The design force shall be taken as $2F_p$. The intent is to prevent excessive movement and to avoid fracture of support springs and any non-ductile components of the isolators.

g. Seismic supports shall be constructed so that support engagement is maintained.

9.6.3.14 Electrical Equipment, Attachments and Supports

Attachments and supports for electrical equipment shall be designed to meet the force and displacement provisions of Sections 9.6.1.3 and 9.6.1.4 and the additional provisions of this section. In addition to their attachments and supports, electrical equipment designated as having $I_p = 1.5$, itself, shall be designed to meet the force and displacement provisions of Sections 9.6.1.3 and 9.6.1.4 and the additional provisions of this section.

The seismic design of other electrical equipment shall include analysis of the following: the dynamic effects of the equipment, its contents, and when appropriate its supports. The interaction between the equipment and the supporting structures, including other mechanical and electrical equipment, shall also be considered.

9.6.3.14.1 Electrical equipment

Electrical equipment designated as having an $I_p = 1.5$ shall meet the following provisions:

a. The design strength for seismic loads in combination with other service loads and appropriate environmental effects shall not exceed the following:

(1) For electrical equipment constructed with ductile material (e.g. steel, aluminum, or copper), 90% of the equipment material minimum specified yield strength.

(2) For threaded connections in equipment constructed with ductile materials, 70% of the material minimum specified yield strength.

(3) For electrical equipment constructed with nonductile materials (e.g., plastic, cast iron, or ceramics), 25% of the equipment material minimum tensile strength.

(4) For threaded connections in equipment constructed with nonductile materials, 20% of the material minimum specified yield strength.

b. Provisions shall be made to mitigate seismic impact for equipment components constructed of nonductile materials or in cases where material ductility is reduced (e.g., low temperature applications).

c. The possibility for loadings imposed on the equipment by attached utility or service lines due to differential motion of points of support from separate structures shall be evaluated.

d. Batteries on racks shall have wrap-around restraints to ensure that the batteries will not fall off the rack. Racks shall be evaluated for sufficient lateral and longitudinal load capacity.

e. Internal coils of dry type transformers shall be positively attached to their supporting substructure within the transformer enclosure.

f. Slide out components in electrical control panels shall have a latching mechanism to hold contents in place.

g. Structural design of electrical cabinets shall be in conformance with standards of the industry that are acceptable to the authority having jurisdiction. Large cut-outs in the lower shear panel shall be specifically evaluated if an evaluation is not provided by the manufacturer.

h. The attachment of additional items weighing more than 100 lb (440 N) shall be specifically evaluated if not provided by the manufacturer.

9.6.3.14.2 Attachments and supports for electrical equipment

Attachments and supports for electrical equipment shall meet the following provisions:

a. Attachments and supports transferring seismic loads shall be constructed of materials suitable for the application and designed and constructed in accordance with a nationally recognized structural code such as, when constructed of steel, AISC, *Manual of Steel Construction* (Ref. 9.8-1 or 9.8-2).

b. Friction clips shall not be used for anchorage attachment.

c. Oversized washers shall be used at bolted connections through the base sheet metal if the base is not reinforced with stiffeners.

d. Supports shall be specifically evaluated if weak-axis bending of light gage support steel is relied on for the seismic load path.

e. The supports for linear electrical equipment such as cable trays, conduit, and bus ducts shall be designed to meet the force and displacement provisions of Sections 9.6.1.3 and 9.6.1.4 only if any of the following situations apply:

(1) Supports are cantilevered up from the floor;

(2) Supports include bracing to limit deflection;

(3) Supports are constructed as rigid welded frames;

(4) Attachments into concrete utilize nonexpanding insets, shot pins, or cast iron embedments; or

(5) Attachments utilize spot welds, plug welds, or minimum size welds as defined by AISC. (Ref. 9.8-1 or 9.8-2)

f. Components mounted on vibration isolation systems shall have a bumper restraint or snubber in each horizontal direction, and vertical restraints shall be provided where required to resist overturning. Isolator housings and restraints shall not be constructed of cast iron or other materials with limited ductility. (See additional design force requirements in Table 9.6.3.2.) A viscoelastic pad or similar material of appropriate thickness shall be used between the bumper and equipment item to limit the impact load.

9.6.3.15 Alternate Seismic Qualification Methods

As an alternative to the analysis methods implicit in the design methodology described above, equipment testing is an acceptable method to determine seismic capacity. Thus, adaptation of a nationally recognized standard for qualification by testing that is acceptable to the authority having jurisdiction is an acceptable alternative, so long as the equipment seismic capacity equals or exceeds the demand expressed in Sections 9.6.1.3 and 9.6.1.4.

9.6.3.16 Elevator Design Requirements

9.6.3.16.1 Reference document

Elevators shall meet the force and displacement provisions of Section 9.6.3.2 unless exempted by either Section 9.1.2.1 or 9.6.1 Elevators designed in accordance with the seismic provisions of the ASME *Safety Code for Elevators and Escalators* (Ref. 9.6-2) shall be deemed to meet the seismic force requirements of this section, except as modified below.

9.6.3.16.2 Elevators and hoistway structural system

Elevators and hoistway structural systems shall be designed to meet the force and displacement provisions of Sections 9.6.1.3 and 9.6.1.4.

9.6.3.16.3 Elevator machinery and controller supports and attachments

Elevator machinery and controller supports and attachments shall be designed to meet with the force and displacement provisions of Sections 9.6.1.3 and 9.6.1.4.

9.6.3.16.4 Seismic controls

Seismic switches shall be provided for all elevators addressed by Section 9.6.3.16.1 including those meeting the requirements of the ASME reference, provided they operate with a speed of 150 ft/min (46 m/min) or greater.

Seismic switches shall provide an electrical signal indicating that structural motions are of such a magnitude that the operation of elevators may be impaired. Upon activation of the seismic switch, elevator operations shall conform to provisions in the ASME *Safety Code for Elevators and Escalators* (Ref. 9.6-2 except as noted below). The seismic switch shall be located at or above the highest floor serviced by the elevators. The seismic switch shall have two horizontal perpendicular axes of sensitivity. Its trigger level shall be set to 30% of the acceleration of gravity.

In facilities where the loss of the use of an elevator is a life-safety issue, the elevator shall only be used after the seismic switch has triggered provided that:

1. The elevator shall operate no faster than the service speed,
2. Before the elevator is occupied, it is operated from top to bottom and back to top to verify that it is operable, and
3. The individual putting the elevator back in service shall ride the elevator from top to bottom and back to top to verify acceptable performance.

9.6.3.16.5 Retainer plates

Retainer plates are required at the top and bottom of the car and counterweight.

9.7 FOUNDATION DESIGN REQUIREMENTS

9.7.1 General

This section includes only those foundation requirements that are specifically related to seismic resistant construction. It assumes compliance with all other basic requirements. These requirements include, but are not limited to, requirements for the extent of the foundation investigation, fills to be present or to be placed in the area of the structure, slope stability, subsurface drainage, and settlement control. Also included are pile requirements and capacities and bearing and lateral soil pressure requirements.

9.7.2 Seismic Design Category A

There are no special requirements for the foundations of structures assigned to Category A.

9.7.3 Seismic Design Category B

The determination of the site coefficient (Section 9.4.1.2.4) shall be documented and the resisting capacities of the foundations, subjected to the prescribed seismic forces of Sections 9.1 through 9.9, shall meet the following requirements:

9.7.3.1 Structural Components

The design strength of foundation components subjected to seismic forces alone or in combination with other prescribed loads and their detailing requirements shall conform to the requirements of Sections 9.8 through 9.12. The strength of foundation components shall not be less than that required for forces acting without seismic forces.

9.7.3.2 Soil Capacities

The capacity of the foundation soil in bearing or the capacity of the soil interface between pile or pier and the soil shall be sufficient to support the structure with all prescribed loads, without seismic forces, taking due account of the settlement that the structure can withstand. For the load combination including earthquake as specified in Section 9.5.2.7 the soil capacities must be sufficient to resist loads at acceptable strains considering both the short duration of loading and the dynamic properties of the soil.

9.7.4 Seismic Design Category C

Foundations for structures assigned to Category C shall conform to all of the requirements for Categories A and B and to the additional requirements of this section.

9.7.4.1 Investigation

When required by the authority having jurisdiction, a written report shall be submitted. The report shall include, in addition to the requirements of Section 9.7.1 and the evaluations required in Section 9.7.3, the results of an investigation to determine the potential hazards due to slope instability, liquefaction, and surface rupture due to faulting or lateral spreading, all as a result of earthquake motions.

9.7.4.2 Pole-Type Structures

When construction employing posts or poles as columns embedded in earth or embedded in concrete footings in the earth is used to resist lateral loads, the depth of embedment required for posts or poles to resist seismic forces shall be determined by means of the design criteria established in the foundation investigation report.

9.7.4.3 Foundation Ties

Individual pile caps, drilled piers, or caissons shall be interconnected by ties. All ties shall have a design strength in tension or compression, greater than a force equal to 10% of S_{DS} times the larger pile cap or column factored dead plus factored live load unless it is demonstrated that equivalent restraint will be provided by reinforced concrete beams within slabs on grade or reinforced concrete slabs on grade or confinement by competent rock, hard cohesive soils, very dense granular soils, or other approved means.

9.7.4.4 Special Pile Requirements

Concrete piles, concrete filled steel pipe piles, drilled piers, or caissons require minimum bending, shear, tension, and elastic strain capacities. Refer to Section A.9.7.4.4 for supplementary provisions.

9.7.5 Foundation Requirements for Seismic Design Categories D, E, and F

Foundations for structures assigned to Categories D, E, and F shall conform to all of the requirements for Category C construction and to the additional requirements of this section.

9.7.5.1 Investigation

The owner shall submit to the authority having jurisdiction a written report that includes an evaluation of the items in Section 9.7.4.1 and the determination of lateral pressures on basement and retaining walls due to earthquake motions.

9.7.5.2 Foundation Ties

Individual spread footings founded on soil defined in Section 9.4.1.2.1 as Site Class E or F shall be interconnected by ties. Ties shall conform to Section 9.7.4.3.

9.7.5.3 Liquefaction Potential and Soil Strength Loss

The geotechnical report shall assess potential consequences of any liquefaction and soil strength loss, including estimation of differential settlement, lateral movement or reduction in foundation soil-bearing capacity, and shall discuss mitigation measures. Such measures shall be given consideration in the design of the structure and can include, but are not limited to, ground stabilization, selection of appropriate foundation type and depths, selection of appropriate structural systems to accommodate anticipated displacements, or any combination of these measures. Where deemed appropriate by the authority having jurisdiction, a site specific geotechnical report is not required when prior evaluations of nearby sites with similar soil conditions provide sufficient direction relative to the proposed construction.

The potential for liquefaction and soil strength loss shall be evaluated for site peak ground accelerations, magnitudes, and source characteristics consistent with the design earthquake ground motions. Peak ground acceleration may be determined based on a site-specific study taking into account soil amplification effects or, in the absence of such a study, peak ground accelerations shall be assumed equal to $S_{DS}/2.5$.

9.7.5.4 Special Pile Requirements

Piling shall be designed and constructed to withstand maximum imposed curvatures from earthquake ground motions and structure response. Curvatures shall include free-field soil strains (without the structure) modified for soil-pile-structure interaction coupled with pile deformations induced by lateral pile resistance to structure seismic forces. Concrete piles in Site Class E or F shall be designed and detailed in accordance with special moment frames within seven pile diameters of the pile cap and the interfaces of soft to medium stiff clay or liquefiable strata. Refer to Section A.9.7.5 for supplementary provisions in addition to those given in Section A.9.7.4.4.

9.8 STEEL

9.8.1 Reference Documents

The design, construction and quality of steel components that resist seismic forces shall conform to the requirements of the references listed in this section except that modifications are necessary to make the references compatible with the provisions of this document. Section A.9.8 provides supplementary provisions for this compatibility.

Ref. 9.8-1 *Load and Resistance Factor Design Specification for Structural Steel Buildings (LRFD)*, American Institute of Steel Construction (AISC),1993.

Ref. 9.8-2 *Allowable Stress Design and Plastic Design Specification for Structural Steel Buildings (ASD)*, American Institute of Steel Construction, June 1, 1989.

Ref. 9.8-3 *Seismic Provisions for Structural Steel Buildings*, American Institute of Steel Construction, 1997, including Supplement 1, 1999, Part I and III.

Ref. 9.8-4 *Specification for the Design of Cold-Formed Steel Structural Members*, American Iron and Steel Institute (AISI), 1996.

Ref. 9.8-5 ASCE 8-90, *Specification for the Design of Cold-Formed Stainless Steel Structural Members*, American Society of Civil Engineers, 1990.

Ref. 9.8-6 *Standard Specification, Load Tables and Weight Tables for Steel Joists and Joist Girders*, Steel Joist Institute, 1994 Edition.

Ref. 9.8-7 *Structural Applications for Steel Cables for Buildings*, ASCE 19, 1995 Edition.

9.9 STRUCTURAL CONCRETE

9.9.1 Reference Documents

The quality and testing of materials and the design and construction of structural concrete components that resist seismic forces shall conform to the requirements of the references listed in this section except that modifications are necessary to make the reference compatible with the provisions of this document. Section A.9.9 provides the supplementary provisions for this compatibility. The load combinations of Section 2.4.1 are not applicable for design of reinforced concrete to resist earthquake loads.

Ref 9.9-1 *Building Code Requirements for Structural Concrete, American Concrete Institute*, ACI 318-99, excluding Appendix A.

9.10 RESERVED FOR COMPOSITE STRUCTURES

9.11 MASONRY

9.11.1 Reference Documents

The design, construction, and quality assurance of masonry components that resist seismic forces shall conform to the requirements of the reference listed in this section except that modifications are necessary to make the reference compatible with the provisions of this document. Section A.9.11 provides the supplementary provisions for this compatibility.

Ref. 9.11-1 *Building Code Requirements for Masonry Structures*, ACI 530-99/ASCE 5-99/TMS 402-99, and *Specifications for Masonry Structures*, ACI 530.1-99/ASCE 6-99/TMS 602-99.

9.12 WOOD

9.12.1 Reference Documents

The quality, testing, design, and construction of members and their fastenings in wood systems that resist seismic forces shall conform to the requirements of the reference documents listed in this section except that modifications are necessary to make the references compatible with the provisions of this document. Section A.9.12 provides the details of such modifications to the application of these references, for both conventional and engineered wood construction.

Ref. 9.12-1 *National Design Specification for Wood Construction* including *Design Values for Wood Construction (NDS Supplement)*, ANSI/NFoPA NDS-1991 (1997).

Ref. 9.12-2 *American Softwood Lumber Standard, Voluntary Product Standard PS 20-94*, National Institute of Standards and Technology (1994).

Ref. 9.12-3 *Softwood Plywood—Construction and Industrial*, PS 1-95 (1995).

Ref. 9.12-4 *Wood Particle Board*, ANSI A208.1 (1993).

Ref. 9.12-5 *Performance Standard for Wood Based Structural Use Panels*, PS 2-92 (1992).

Ref. 9.12-6 *American National Standard for Wood Products—Structural Glued Laminated Timber*, ANSI/AITC A190.1-1992 (1992).

Ref. 9.12-7 *Wood Poles–Specifications and Dimensions*, ANSI 05.1 (1992).

Ref. 9.12-8 *One- and Two-Family Dwelling Code*, Council of American Building Officials (CABO) 1995.

Ref. 9.12-9 *Standard Specification for Establishing and Monitoring Structural Capacities of Prefabricated Wood I-Joists*, ASTM D 5055-95A (1995).

Ref. 9.12-10 *Load and Resistance Factor Design (LRFD) Standard for Engineered Wood Construction*, including supplements, ASCE 16-95 (1995).

9.13 PROVISIONS FOR SEISMICALLY ISOLATED STRUCTURES

9.13.1 General

Every seismically isolated structure and every portion thereof shall be designed and constructed in accordance with the requirements of this section and the applicable requirements of Section 9.1.

The lateral-force-resisting system and the isolation system shall be designed to resist the deformations and stresses produced by the effects of seismic ground motions as provided in this section.

9.13.2 Criteria Selection

9.13.2.1 Basis for Design

The procedures and limitations for the design of seismically isolated structures shall be determined considering zoning, site characteristics, vertical acceleration, cracked section properties of concrete and masonry members, Seismic Use Group, configuration, structural system, and height in accordance with Section 9.5.2 except as noted below.

9.13.2.2 Stability of the Isolation System

The stability of the vertical load-carrying elements of the isolation system shall be verified by analysis and test, as required, for lateral seismic displacement equal to the total maximum displacement.

9.13.2.3 Seismic Use Group

All portions of the structure, including the structure above the isolation system, shall be assigned a Seismic Use Group in accordance with the requirements of Section 9.1.3.

9.13.2.4 Configuration Requirements

Each structure shall be designated as being regular or irregular on the basis of the structural configuration above the isolation system.

9.13.2.5 Selection of Lateral Response Procedure

9.13.2.5.1 General

Seismically isolated structures except those defined in Section 9.13.2.5.2 shall be designed using the dynamic lateral response procedure of Section 9.13.4.

9.13.2.5.2 Equivalent-lateral-force procedure

The equivalent-lateral-response procedure of Section 9.13.3 is permitted to be used for design of a seismically isolated structure provided that:

1. The structure is located at a site with S_1 less than or equal to $0.60g$;
2. The structure is located on a Class A, B, C, or D site;
3. The structure above the isolation interface is less than or equal to four stories or 65 ft (19.8 m) in height;
4. The effective period of the isolated structure at maximum displacement, T_M, is less than or equal to 3.0 s.
5. The effective period of the isolated structure at the design displacement, T_D, is greater than three times the elastic, fixed-base period of the structure above the isolation system as determined by Eq. 9.5.3.3-1 or 9.5.3.3-2;
6. The structure above the isolation system is of regular configuration; and
7. The isolation system meets all of the following criteria:
    a. The effective stiffness of the isolation system at the design displacement is greater than one third of the effective stiffness at 20% of the design displacement,
    b. The isolation system is capable of producing a restoring force as specified in Section 9.13.6.2.4,
    c. The isolation system has force-deflection properties that are independent of the rate of loading,
    d. The isolation system has force-deflection properties that are independent of vertical load and bilateral load, and
    e. The isolation system does not limit maximum considered earthquake displacement to less than S_{M1}/S_{D1} times the total design displacement.

9.13.2.5.3 Dynamic analysis

The dynamic lateral response procedure of Section 9.13.4 shall be used as specified below.

9.13.2.5.3.1 Response-spectrum analysis: Response-spectrum analysis shall not be used for design of a seismically isolated structure unless:

1. The structure is located on a Class A, B, C, or D Site; and
2. The isolation system meets the criteria of Item 7 of Section 9.13.2.5.2.

9.13.2.5.3.2 Time-history analysis: Time-history analysis shall be permitted for design of any seismically isolated structure and shall be used for design of all

seismically isolated structures not meeting the criteria of Section 9.13.2.5.3.1.

9.13.2.5.3.3 Site-specific design spectra: Site-specific ground-motion spectra of the design earthquake and the maximum considered earthquake developed in accordance with Section 9.13.4.4.1 shall be used for design and analysis of all seismically isolated structures if any one of the following conditions apply:

1. The structure is located on a Class E or F Site; or
2. The structure is located at a site with S_1 greater than $0.60g$, as determined in Section 9.4.1.

9.13.3 Equivalent-Lateral Force Procedure

9.13.3.1 General

Except as provided in Section 9.13.4, every seismically isolated structure or portion thereof shall be designed and constructed to resist minimum earthquake displacements and forces as specified by this section and the applicable requirements of Section 9.5.3.

9.13.3.2 Deformation Characteristics of the Isolation System

Minimum lateral earthquake design displacements and forces on seismically isolated structures shall be based on the deformation characteristics of the isolation system.

The deformation characteristics of the isolation system shall explicitly include the effects of the wind-restraint system if such a system is used to meet the design requirements of this document.

The deformation characteristics of the isolation system shall be based on properly substantiated tests performed in accordance with Section 9.13.9.

9.13.3.3 Minimum Lateral Displacements

9.13.3.3.1 Design displacement

The isolation system shall be designed and constructed to withstand minimum lateral earthquake displacements, D_D, that act in the direction of each of the main horizontal axes of the structure in accordance with the following:

$$D_D = \frac{gS_{D1}T_D}{4\pi^2 B_D} \quad \text{(Eq. 9.13.3.3.1)}$$

where

g = acceleration of gravity. The units of the acceleration of gravity, g, are in./s^2 (mm/s^2) if the units of the design displacement, D, are inches (mm);

S_{D1} = design 5% damped spectral acceleration at 1 s period, in units of g-s, as determined in Section 9.4.1.1;

T_D = effective period of seismically isolated structure in seconds (s), at the design displacement in the direction under consideration, as prescribed by Eq. 9.13.3.3.2; and

B_D = numerical coefficient related to the effective damping of the isolation system at the design displacement, β_D, as set forth in Table 9.13.3.3.1.

9.13.3.3.2 Effective period at design displacement

The effective period of the isolated structure at design displacement, T_D, shall be determined using the deformational characteristics of the isolation system in accordance with the following equation:

$$T_D = 2\pi \sqrt{\frac{W}{k_{D\min}g}} \quad \text{(Eq. 9.13.3.3.2)}$$

where

W = total seismic dead load weight of the structure above the isolation interface as defined in Sections 9.5.3.2 and 9.5.5.3 (kip or kN);

$k_{D\min}$ = minimum effective stiffness in kips/in. (kN/mm) of the isolation system at the design displacement in the horizontal direction under

TABLE 9.13.3.3.1. Damping Coefficient, B_I

Effective Damping, β_D or β_M (Percentage of Critical)[a,b]	B_D or B_M Factor
≤2%	0.8
5%	1.0
10%	1.2
20%	1.5
30%	1.7
40%	1.9
≥50%	2.0

[a]The damping coefficient shall be based on the effective damping of the isolation system determined in accordance with the requirements of Section 9.13.9.5.2.
[b]The damping coefficient shall be based on linear interpolation for effective damping values other than those given.

consideration as prescribed by Eq. (9.13.9.5.1-2); and

g = acceleration due to gravity.

9.13.3.3.3 Maximum lateral displacement

The maximum displacement of the isolation system, D_M, in the most critical direction of horizontal response shall be calculated in accordance with the formula:

$$D_M = \frac{g S_{M1} T_M}{4\pi^2 B_M} \quad \text{(Eq. 9.13.3.3.3)}$$

where

g = acceleration of gravity;

S_{M1} = maximum considered 5% damped spectral acceleration at 1 s period, in units of g-s, as determined in Section 9.4.1.2;

T_M = effective period, in seconds (s), of seismic-isolated structure at the maximum displacement in the direction under consideration as prescribed by Eq. 9.13.3.3.4; and

B_M = numerical coefficient related to the effective damping of the isolation system at the maximum displacement, β_D, as set forth in Table 9.13.3.3.1.

9.13.3.3.4 Effective period at maximum displacement

The effective period of the isolated structure at maximum displacement, T_M, shall be determined using the deformational characteristics of the isolation system in accordance with the equation:

$$T_M = 2\pi \sqrt{\frac{W}{k_{M\min} g}} \quad \text{(Eq. 9.13.3.3.4)}$$

where

W = total seismic dead load weight of the structure above the isolation interface as defined in Section 9.5.3.2 and 9.5.5.3 (kip or kN);

$k_{M\min}$ = minimum effective stiffness, in kips/in. (kN/mm), of the isolation system at the maximum displacement in the horizontal direction under consideration as prescribed by Eq. 9.13.9.5.1-4; and

g = acceleration of gravity.

9.13.3.3.5 Total lateral displacement

The total design displacement, D_{TD}, and the total maximum displacement, D_{TM}, of elements of the iso-

lation system shall include additional displacement due to actual and accidental torsion calculated from the spatial distribution of the lateral stiffness of the isolation system and the most disadvantageous location of mass eccentricity.

The total design displacement, D_{TD}, and the total maximum displacement, D_{TM}, of elements of an isolation system with uniform spatial distribution of lateral stiffness shall not be taken as less than that prescribed by the following equations:

$$D_{TD} = D_D \left[1 + y \frac{12e}{b^2 + d^2} \right] \quad \text{(Eq. 9.13.3.3.5-1)}$$

$$D_{TM} = D_M \left[1 + y \frac{12e}{b^2 + d^2} \right] \quad \text{(Eq. 9.13.3.3.5-2)}$$

Exception: The total design displacement, D_{TD}, and the total maximum displacement, D_{TM}, are permitted to be taken as less than the value prescribed by Eqs. 9.13.3.3.5-1 and 9.13.3.3.5-2, respectively, but not less than 1.1 times D_D and D_M, respectively, provided the isolation system is shown by calculation to be configured to resist torsion accordingly.

where

D_D = design displacement, in inches (mm), at the center of rigidity of the isolation system in the direction under consideration as prescribed by Eq. 9.13.3.3.1;

D_M = maximum displacement, in inches (mm), at the center of rigidity of the isolation system in the direction under consideration as prescribed by Eq. 9.13.3.3.3;

y = distance, in feet (mm), between the centers of rigidity of the isolation system and the element of interest measured perpendicular to the direction of seismic loading under consideration;

e = actual eccentricity, in feet (mm), measured in plan between the center of mass of the structure above the isolation interface and the center of rigidity of the isolation system, plus accidental eccentricity, in feet (mm), taken as 5% of the longest plan dimension of the structure perpendicular to the direction of force under consideration;

b = shortest plan dimension of the structure, in feet (mm), measured perpendicular to d; and

d = longest plan dimension of the structure, in feet (mm).

9.13.3.4 Minimum Lateral Forces

9.13.3.4.1 Isolation system and structural elements at or below the isolation system

The isolation system, the foundation, and all structural elements below the isolation system shall be designed and constructed to withstand a minimum lateral seismic force, V_b, using all of the appropriate provisions for a nonisolated structure where:

$$V_b = k_{D\,max} D_D \qquad \text{(Eq. 9.13.3.4.1)}$$

where

V_b = minimum lateral seismic design force or shear on elements of the isolation system or elements below the isolation system as prescribed by Eq. 9.13.3.4.1;

$k_{D\,max}$ = maximum effective stiffness, in kips/in. (kN/mm), of the isolation system at the design displacement in the horizontal direction under consideration; and

D_D = design displacement, in inches (mm), at the center of rigidity of the isolation system in the direction under consideration as prescribed by Eq. 9.13.3.3.1.

V_b shall not be taken as less than the maximum force in the isolation system at any displacement up to and including the design displacement.

9.13.3.4.2 Structural elements above the isolation system

The structure above the isolation system shall be designed and constructed to withstand a minimum shear force, V_s, using all of the appropriate provisions for a nonisolated structure where:

$$V_s = \frac{k_{D\,max} D_D}{R_I} \qquad \text{(Eq. 9.13.3.4.2)}$$

where

$k_{D\,max}$ = maximum effective stiffness, in kips/in. (kN/mm), of the isolation system at the design displacement in the horizontal direction under consideration;

D_D = design displacement, in inches (mm), at the center of rigidity of the isolation system in the direction under consideration as prescribed by Eq. 9.13.3.3.1; and

R_I = numerical coefficient related to the type of lateral-force-resisting system above the isolation system.

The R_I factor shall be based on the type of lateral-force-resisting system used for the structure above the isolation system and shall be 3/8 of the R value given in Table 9.5.2.2 with an upper bound value not to exceed 2.0 and a lower bound value not to be less than 1.0.

9.13.3.4.3 Limits on V_s

The value of V_s shall not be taken as less than the following:

1. The lateral seismic force required by Section 9.5.3 for a fixed-base structure of the same weight, W, and a period equal to the isolated period, T_D;
2. The base shear corresponding to the factored design wind load; and
3. The lateral seismic force required to fully activate the isolation system (e.g., the yield level of a softening system, the ultimate capacity of a sacrificial wind-restraint system, or the break-away friction level of a sliding system) factored by 1.5.

9.13.3.5 Vertical Distribution of Force

The total force shall be distributed over the height of the structure above the isolation interface in accordance with the following equation:

$$F_x = \frac{V_s w_x h_x}{\displaystyle\sum_{i=1}^{n} w_i h_i} \qquad \text{(Eq. 9.13.3.5)}$$

where

V_s = total lateral seismic design force or shear on elements above the isolation system as prescribed by Eq. 9.13.3.4.2;

W_x = portion of W that is located at or assigned to Level i, n, or x, respectively;

h_x = height above the base Level i, n, or x, respectively;

w_i = portion of W that is located at or assigned to Level i, n, or x, respectively; and

h_i = height above the base Level i, n, or x, respectively.

At each level designated as x, the force, F_x, shall be applied over the area of the structure in accordance with the mass distribution at the level. Stresses in each structural element shall be calculated as the effect of force, F_x, applied at the appropriate levels above the base.

9.13.3.6 Drift Limits

The maximum interstory drift of the structure above the isolation system shall not exceed $0.015h_{sx}$. The drift shall be calculated by Eq. 9.5.3.7.1 with the C_d factor of the isolated structure equal to the R_I factor defined in Section 9.13.3.4.2.

9.13.4 Dynamic Lateral Response Procedure

9.13.4.1 General
As required by Section 9.13.2, every seismically isolated structure or portion thereof shall be designed and constructed to resist earthquake displacements and forces as specified in this section and the applicable requirements of Section 9.5.4.

9.13.4.2 Isolation System and Structural Elements Below the Isolation System

The total design displacement of the isolation system shall not be taken as less than 90% of D_{TD} as specified by Section 9.13.3.3.5.

The total maximum displacement of the isolation system shall not be taken as less than 80% of D_{TM} as prescribed by Section 9.13.3.3.5.

The design lateral shear force on the isolation system and structural elements below the isolation system shall not be taken as less than 90% of V_b as prescribed by Eq. 9.13.3.4.1.

The limits on displacements specified by this section shall be evaluated using values of D_{TD} and D_{TM} determined in accordance with Section 9.13.3.5 except that D'_D is permitted to be used in lieu of D_D and D'_M is permitted to be used in lieu of D_M where:

$$D'_D = \frac{D_D}{\sqrt{1 + (T/T_D)^2}} \quad \text{(Eq. 9.13.4.2-1)}$$

$$D'_M = \frac{D_M}{\sqrt{1 + (T/T_M)^2}} \quad \text{(Eq. 9.13.4.2-2)}$$

where

D_D = design displacement, in inches (mm), at the center of rigidity of the isolation system in the direction under consideration as prescribed by Eq. 9.13.3.3.1;

D_M = maximum displacement in inches (mm), at the center of rigidity of the isolation system in the direction under consideration as prescribed by Eq. 9.13.3.3.3;

T = elastic, fixed-base period of the structure above the isolation system as determined by Section 9.5.3.3;

T_D = effective period of seismically isolated structure in seconds (s), at the design displacement in the direction under consideration as prescribed by Eq. 9.13.3.3.2; and

T_M = effective period, in seconds (s), of the seismically isolated structure, at the maximum displacement in the direction under consideration as prescribed by Eq. 9.13.3.3.4.

9.13.4.3 Structural Elements Above the Isolation System

The design lateral shear force on the structure above the isolation system, if regular in configuration, shall not be taken as less than 80% of V_s, or less than the limits specified by Section 9.13.3.4.3.

EXCEPTION: The design lateral shear force on the structure above the isolation system, if regular in configuration, is permitted to be taken as less than 80%, but shall not be less than 60% of V_s, when time-history analysis is used for design of the structure.

The design lateral shear force on the structure above the isolation system, if irregular in configuration, shall not be taken as less than V_s, or less than the limits specified by Section 9.13.3.4.3.

EXCEPTION: The design lateral shear force on the structure above the isolation system, if irregular in configuration, is permitted to be taken as less than 100%, but shall not be less than 80% of V_s, when time-history analysis is used for design of the structure.

9.13.4.4 Ground Motion

9.13.4.4.1 Design spectra
Properly substantiated site-specific spectra are required for design of all structures located on Site Class E or F or located at a site with S_1 greater than $0.60g$. Structures that do not require site-specific spectra and for which site-specific spectra have not been calculated shall be designed using the response spectrum shape given in Fig. 9.4.1.2.6.

A design spectrum shall be constructed for the design earthquake. This design spectrum shall not be taken as less than the design earthquake response spectrum given in Fig. 9.4.1.2.6.

EXCEPTION: If a site-specific spectrum is calculated for the design earthquake, the design spectrum is permitted to be taken as less than 100% but shall not be less than 80% of the design earthquake response spectrum given in Fig. 9.4.1.2.6.

A design spectrum shall be constructed for the maximum considered earthquake. This design spectrum shall not be taken as less than 1.5 times the design earthquake response spectrum given in Fig. 9.4.1.2.6. This design spectrum shall be used to determine the total maximum displacement and overturning forces for design and testing of the isolation system.

EXCEPTION: If a site-specific spectrum is calculated for the maximum considered earthquake, the design spectrum is permitted to be taken as less than 100% but shall not be less than 80% of 1.5 times the design earthquake response spectrum given in Fig. 9.4.1.2.6.

9.13.4.4.2 Time histories

Pairs of horizontal ground-motion time-history components shall be selected and scaled from not less than three recorded events. For each pair of horizontal ground-motion components, the square root sum of the squares (SRSS) of the 5% damped spectrum of the scaled, horizontal components shall be constructed. The motions shall be scaled such that the average value of the SRSS spectra does not fall below 1.3 times the 5% damped spectrum of the design earthquake (or maximum considered earthquake) by more than 10% for periods from $0.5T_D$ s to $1.25T_M$ s, where T_D and T_M are determined in accordance with Sections 9.13.3.3.2 and 9.13.3.3.4, respectively.

9.13.4.5 Mathematical Model

9.13.4.5.1 General

The mathematical models of the isolated structure including the isolation system, the lateral-force-resisting system, and other structural elements shall conform to Section 9.5.4.2 and to the requirements of Sections 9.13.4.5.2 and 9.13.4.5.3, below.

9.13.4.5.2 Isolation system

The isolation system shall be modeled using deformational characteristics developed and verified by test in accordance with the requirements of Section 9.13.3.2. The isolation system shall be modeled with sufficient detail to:

1. Account for the spatial distribution of isolator units;
2. Calculate translation, in both horizontal directions, and torsion of the structure above the isolation interface considering the most disadvantageous location of mass eccentricity;

3. Assess overturning/uplift forces on individual isolator units; and
4. Account for the effects of vertical load, bilateral load, and/or the rate of loading if the force deflection properties of the isolation system are dependent on one or more of these attributes.

9.13.4.5.3 Isolated building

9.13.4.5.3.1 Displacement:
The maximum displacement of each floor and the total design displacement and total maximum displacement across the isolation system shall be calculated using a model of the isolated structure that incorporates the force-deflection characteristics of nonlinear elements of the isolation system and the lateral-force-resisting system.

Isolation systems with nonlinear elements include, but are not limited to, systems that do not meet the criteria of Item 7 of Section 9.13.2.5.2.

Lateral-force-resisting systems with nonlinear elements include, but are not limited to, irregular structural systems designed for a lateral force less than V_s and regular structural systems designed for a lateral force less than 80% of V_s.

9.13.4.5.3.2 Forces and displacements in key elements:
Design forces and displacements in key elements of the lateral force-resisting system shall not be calculated using a linear elastic model of the isolated structure unless:

1. Pseudo-elastic properties assumed for nonlinear isolation-system components are based on the maximum effective stiffness of the isolation system; and
2. All key elements of the lateral-force-resisting system are linear.

9.13.4.6 Description of Analysis Procedures

9.13.4.6.1 General

Response-spectrum and time-history analyses shall be performed in accordance with Section 9.5.3 and the requirements of this section.

9.13.4.6.2 Input earthquake

The design earthquake shall be used to calculate the total design displacement of the isolation system and the lateral forces and displacements of the isolated structure. The maximum considered earthquake shall be used to calculate the total maximum displacement of the isolation system.

9.13.4.6.3 Response-spectrum analysis

Response-spectrum analysis shall be performed using a damping value equal to the effective damping of the isolation system or 30% of critical, whichever is less.

Response-spectrum analysis used to determine the total design displacement and the total maximum displacement shall include simultaneous excitation of the model by 100% of the most critical direction of ground motion and 30% of the ground motion on the orthogonal axis. The maximum displacement of the isolation system shall be calculated as the vectorial sum of the two orthogonal displacements.

The design shear at any story shall not be less than the story shear obtained using Eq. 9.13.3.5 and a value of V_s taken as that equal to the base shear obtained from the response-spectrum analysis in the direction of interest.

9.13.4.6.4 Time-history analysis

Time-history analysis shall be performed with at least three appropriate pairs of horizontal time-history components as defined in Section 9.13.4.4.2.

Each pair of time histories shall be applied simultaneously to the model considering the most disadvantageous location of mass eccentricity. The maximum displacement of the isolation system shall be calculated from the vectorial sum of the two orthogonal components at each time step.

The parameter of interest shall be calculated for each time-history analysis. If three time-history analyses are performed, the maximum response of the parameter of interest shall be used for design. If seven or more time-history analyses are performed, the average value of the response parameter of interest shall be used for design.

9.13.4.7 Design Lateral Force

9.13.4.7.1 Isolation system and structural elements at or below the isolation system

The isolation system, foundation, and all structural elements below the isolation system shall be designed using all of the appropriate provisions for a nonisolated structure and the forces obtained from the dynamic analysis without reduction.

9.13.4.7.2 Structural elements above the isolation system

Structural elements above the isolation system shall be designed using the appropriate provisions for a nonisolated structure and the forces obtained from the dynamic analysis reduced by a factor of R_I. The R_I factor shall be based on the type of lateral-force-resisting system used for the structure above the isolation system.

9.13.4.7.3 Scaling of results

When the factored lateral shear force on structural elements, determined using either response spectrum or time-history analysis, is less than the minimum level prescribed by Sections 9.13.4.2 and 9.13.4.3, all response parameters, including member forces and moments, shall be adjusted upward proportionally.

9.13.4.7.4 Drift limits

Maximum interstory drift corresponding to the design lateral force including displacement due to vertical deformation of the isolation system shall not exceed the following limits:

1. The maximum interstory drift of the structure above the isolation system calculated by response spectrum analysis shall not exceed $0.015h_{sx}$, and
2. The maximum interstory drift of the structure above the isolation system calculated by time-history analysis based on the force-deflection characteristics of nonlinear elements of the lateral force-resisting system shall not exceed $0.020h_{sx}$.

Drift shall be calculated using Eq. 9.5.3.7.1 with the C_d factor of the isolated structure equal to the R_I factor defined in Section 9.13.3.4.2.

The secondary effects of the maximum considered earthquake lateral displacement Δ of the structure above the isolation system combined with gravity forces shall be investigated if the interstory drift ratio exceeds $0.010/R_I$.

9.13.5 Lateral Load on Elements of Structures and Nonstructural Components Supported by Buildings

9.13.5.1 General

Parts or portions of an isolated structure, permanent nonstructural components and the attachments to them, and the attachments for permanent equipment supported by a structure shall be designed to resist seismic forces and displacements as prescribed by this section and the applicable requirements of Section 9.6.

9.13.5.2 Forces and Displacements

9.13.5.2.1 Components at or above the isolation interface

Elements of seismically isolated structures and nonstructural components, or portions thereof, that are at or above the isolation interface shall be designed to resist a total lateral seismic force equal to the maximum dynamic response of the element or component under consideration.

EXCEPTION: Elements of seismically isolated structures and nonstructural components or portions designed to resist total lateral seismic force as prescribed by Eq. 9.5.2.7-1 or 9.5.2.7-2 as appropriate.

9.13.5.2.2 Components crossing the isolation interface

Elements of seismically isolated structures and nonstructural components, or portions thereof, that cross the isolation interface shall be designed to withstand the total maximum displacement.

9.13.5.2.3 Components below the isolation interface

Elements of seismically isolated structure and nonstructural components, or portions thereof, that are below the isolation interface shall be designed and constructed in accordance with the requirements of Section 9.5.2.

9.13.6 Detailed System Requirements

9.13.6.1 General

The isolation system and the structural system shall comply with the material requirements of Sections 9.8 through 9.12. In addition, the isolation system shall comply with the detailed system requirements of this section and the structural system shall comply with the detailed system requirements of this section and the applicable portions of Section 9.5.2.

9.13.6.2 Isolation System

9.13.6.2.1 Environmental conditions

In addition to the requirements for vertical and lateral loads induced by wind and earthquake, the isolation system shall provide for other environmental conditions including aging effects, creep, fatigue, operating temperature, and exposure to moisture or damaging substances.

9.13.6.2.2 Wind forces

Isolated structures shall resist design wind loads at all levels above the isolation interface. At the isolation interface, a wind restraint system shall be provided to limit lateral displacement in the isolation system to a value equal to that required between floors of the structure above the isolation interface.

9.13.6.2.3 Fire resistance

Fire resistance for the isolation system shall meet that required for the structure columns, walls, or other structural elements.

9.13.6.2.4 Lateral-restoring force

The isolation system shall be configured to produce a restoring force such that the lateral force at the total design displacement is at least $0.025W$ greater than the lateral force at 50% of the total design displacement.

EXCEPTION: The isolation system need not be configured to produce a restoring force, as required above, provided the isolation system is capable of remaining stable under full vertical load and accommodating a total maximum displacement equal to the greater of either 3.0 times the total design displacement or $36S_{M1}$ in. (or $915S_{M1}$ mm).

9.13.6.2.5 Displacement restraint

The isolation system shall not be configured to include a displacement restraint that limits lateral displacement due to the maximum considered earthquake to less than M_M times the total design displacement unless the seismically isolated structure is designed in accordance with the following criteria when more stringent than the requirements of Section 9.13.2:

1. Maximum considered earthquake response is calculated in accordance with the dynamic analysis requirements of Section 9.13.4 explicitly considering the nonlinear characteristics of the isolation system and the structure above the isolation system.
2. The ultimate capacity of the isolation system and structural elements below the isolation system shall exceed the strength and displacement demands of the maximum considered earthquake.
3. The structure above the isolation system is checked for stability and ductility demand of the maximum considered earthquake.
4. The displacement restraint does not become effective at a displacement less than 0.75 times the total design displacement unless it is demonstrated by analysis that earlier engagement does not result in unsatisfactory performance.

9.13.6.2.6 Vertical-load stability

Each element of the isolation system shall be designed to be stable under the design vertical load at a horizontal displacement equal to the total maximum displacement. The design vertical load shall be computed using load combination 5 of Section 2.3.2 for the maximum vertical load and load combination 7 of Section 2.3.2 for the minimum vertical load. The seismic load E is given by Eqs. 9.5.2.7-1 and 9.5.2.7-2 where S_{DS} in these equations is replaced by S_{MS}. The vertical load due to earthquake, Q_E, shall be based on peak response due to the maximum considered earthquake.

9.13.6.2.7 Overturning

The factor of safety against global structural overturning at the isolation interface shall not be less than 1.0 for required load combinations. All gravity and seismic loading conditions shall be investigated. Seismic forces for overturning calculations shall be based on the maximum considered earthquake and W shall be used for the vertical restoring force.

Local uplift of individual elements shall not be allowed unless the resulting deflections do not cause overstress or instability of the isolator units or other structure elements.

9.13.6.2.8 Inspection and replacement

1. Access for inspection and replacement of all components of the isolation system shall be provided.
2. A registered design professional shall complete a final series of inspections or observations of structure separation areas and components that cross the isolation interface prior to the issuance of the certificate of occupancy for the seismically isolated structure. Such inspections and observations shall indicate that the conditions allow free and unhindered displacement of the structure to maximum design levels and that all components that cross the isolation interface as installed are able to accommodate the stipulated displacements.
3. Seismically isolated structures shall have a periodic monitoring, inspection and maintenance program for the isolation system established by the registered design professional responsible for the design of the system.
4. Remodeling, repair or retrofitting at the isolation system interface, including that of components that cross the isolation interface, shall be performed under the direction of a registered design professional.

9.13.6.2.9 Quality control

A quality control testing program for isolator units shall be established by the engineer responsible for the structural design.

9.13.6.3 Structural System

9.13.6.3.1 Horizontal distribution of force

A horizontal diaphragm or other structural elements shall provide continuity above the isolation interface and shall have adequate strength and ductility to transmit forces (due to nonuniform ground motion) from one part of the structure to another.

9.13.6.3.2 Building separations

Minimum separations between the isolated structure and surrounding retaining walls or other fixed obstructions shall not be less than the total maximum displacement.

9.13.6.3.3 Nonbuilding structures

These shall be designed and constructed in accordance with the requirements of Section 9.14 using design displacements and forces calculated in accordance with Section 9.13.3 or 9.13.4.

9.13.7 Foundations

Foundations shall be designed and constructed in accordance with the requirements of Section 4 using design forces calculated in accordance with Section 9.13.3 or 9.13.4, as appropriate.

9.13.8 Design and Construction Review

9.13.8.1 General

A design review of the isolation system and related test programs shall be performed by an independent engineering team including persons licensed in the appropriate disciplines and experienced in seismic analysis methods and the theory and application of seismic isolation.

9.13.8.2 Isolation System

Isolation system design review shall include, but not be limited to, the following:

1. Review of site-specific seismic criteria including the development of site-specific spectra and ground motion time histories and all other design criteria developed specifically for the project;
2. Review of the preliminary design including the determination of the total design displacement of

the isolation system design displacement and the lateral force design level;

3. Overview and observation of prototype testing (Section 9.13.9);

4. Review of the final design of the entire structural system and all supporting analyses; and

5. Review of the isolation system quality control testing program (Section 9.13.6.2.9).

9.13.9 Required Tests of the Isolation System

9.13.9.1 General

The deformation characteristics and damping values of the isolation system used in the design and analysis of seismically isolated structures shall be based on tests of a selected sample of the components prior to construction as described in this section.

The isolation system components to be tested shall include the wind-restraint system if such a system is used in the design.

The tests specified in this section are for establishing and validating the design properties of the isolation system and shall not be considered as satisfying the manufacturing quality control tests of Section 9.13.6.2.9.

9.13.9.2 Prototype Tests

9.13.9.2.1 General

Prototype tests shall be performed separately on two full-size specimens of each predominant type and size of isolator unit of the isolation system. The test specimens shall include the wind restraint system as well as individual isolator units if such systems are used in the design. Specimens tested shall not be used for construction.

9.13.9.2.2 Record

For each cycle of tests, the force-deflection and hysteretic behavior of the test specimen shall be recorded.

9.13.9.2.3 Sequence and cycles

The following sequence of tests shall be performed for the prescribed number of cycles at a vertical load equal to the average dead load plus one-half the effects due to live load on all isolator units of a common type and size:

1. Twenty fully reversed cycles of loading at a lateral force corresponding to the wind design force;

2. Three fully reversed cycles of loading at each of the following increments of the total design displacement—$0.25D_D$, $0.5D_D$, $1.0D_D$, and $1.0D_M$, where D_D and D_M are as determined in Sections 9.13.3.3.1 and 9.13.3.3.3, respectively or Section 9.13.4 as appropriate.

3. Three fully reversed cycles of loading at the total maximum displacement, $1.0D_{TM}$; and

4. $15S_{D1}/B_D S_{DS}$, but not less than ten, fully reversed cycles of loading at 1.0 times the total design displacement, $1.0D_{TD}$.

If an isolator unit is also a vertical-load-carrying element, then Item 2 of the sequence of cyclic tests specified above shall be performed for two additional vertical load cases specified in Section 9.13.6.2.6. The load increment due to earthquake overturning, Q_E, shall be equal to or greater than the peak earthquake vertical force response corresponding to the test displacement being evaluated. In these tests, the combined vertical load shall be taken as the typical or average downward force on all isolator units of a common type and size.

9.13.9.2.4 Units dependent on loading rates

If the force-deflection properties of the isolator units are dependent on the rate of loading, each set of tests specified in Section 9.13.9.2.3 shall be performed dynamically at a frequency, f, equal to the inverse of the effective period, T_D.

If reduced-scale prototype specimens are used to quantify rate-dependent properties of isolators, the reduced-scale prototype specimens shall be of the same type and material and be manufactured with the same processes and quality as full-scale prototypes and shall be tested at a frequency that represents full-scale prototype loading rates.

The force-deflection properties of an isolator unit shall be considered to be dependent on the rate of loading if there is greater than a plus or minus 15% difference in the effective stiffness and the effective damping at the design displacement when tested at a frequency equal to the inverse of the effective period of the isolated structure and when tested at any frequency in the range of 0.1 to 2.0 times the inverse of the effective period of the isolated structure.

9.13.9.2.5 Units dependent on bilateral load

If the force-deflection properties of the isolator units are dependent on bilateral load, the tests specified in Sections 9.13.9.2.3 and 9.13.9.2.4 shall be augmented to include bilateral load at the following increments of the total design displacement: 0.25 and 1.0, 0.50 and 1.0, 0.75 and 1.0, and 1.0 and 1.0.

If reduced-scale prototype specimens are used to quantify bilateral-load-dependent properties, the reduced scale specimens shall be of the same type and material and manufactured with the same processes and quality as full-scale prototypes.

The force-deflection properties of an isolator unit shall be considered to be dependent on bilateral load if the bilateral and unilateral force-deflection properties have greater than a 15% difference in effective stiffness at the design displacement.

9.13.9.2.6 Maximum and minimum vertical load

Isolator units that carry vertical load shall be statically tested for maximum and minimum downward vertical load at the total maximum displacement. In these tests, the combined vertical loads shall be taken as specified in Section 9.13.6.2.6 on any one isolator of a common type and size. The dead load, D, and live load, L, are specified in Section 9.5.2.7. The seismic load E is given by Eqs. 9.5.2.7-1 and 9.5.2.7-2 where S_{DS} in these equations is replaced by S_{MS} and the vertical load, Q_E, is based on the peak earthquake vertical force response corresponding to the maximum considered earthquake.

9.13.9.2.7 Sacrificial-wind-restraint systems

If a sacrificial-wind-restraint system is to be utilized, the ultimate capacity shall be established by test.

9.13.9.2.8 Testing similar units

The prototype tests are not required if an isolator unit is of similar size and of the same type and material as a prototype isolator unit that has been previously tested using the specified sequence of tests.

9.13.9.3 Determination of Force-Deflection Characteristics

The force-deflection characteristics of the isolation system shall be based on the cyclic load tests of isolator prototypes specified in Section 9.13.9.2.

The effective stiffness of an isolator unit, k_{eff}, shall be calculated for each cycle of loading as follows:

$$k_{eff} = \frac{|F^+| + |F^-|}{|\Delta^+| + |\Delta^-|} \quad \text{(Eq. 9.13.9.3-1)}$$

where F^+ and F^- are the positive and negative forces, at Δ^+ and Δ^-, respectively.

As required, the effective damping, β_{eff}, of an isolator unit shall be calculated for each cycle of loading by the equation:

$$\beta_{eff} = \frac{2}{\pi} \frac{E_{loop}}{k_{eff}(|\Delta^+| + |\Delta^-|)^2} \quad \text{(Eq. 9.13.9.3-2)}$$

where the energy dissipated per cycle of loading, E_{loop}, and the effective stiffness, k_{eff}, shall be based on peak test displacements of Δ^+ and Δ^-.

9.13.9.4 System Adequacy

The performance of the test specimens shall be assessed as adequate if the following conditions are satisfied:

1. For each increment of test displacement specified in Item 2 of Section 9.13.9.2.3 and for each vertical load case specified in Section 9.13.9.2.3:

    there is no greater than a 15% difference between the effective stiffness at each of the three cycles of test and the average value of effective stiffness for each test specimen;

2. For each increment of test displacement specified in Item 2 of Section 9.13.9.2.3 and for each vertical load case specified in Section 9.13.9.2.3;

    there is no greater than a 15% difference in the average value of effective stiffness of the two test specimens of a common type and size of the isolator unit over the required three cycles of test;

3. For each specimen there is no greater than a plus or minus 20% change in the initial effective stiffness of each test specimen over the $30S_{DI}B_D/S_{DS}$, but not less than 10, cycles of test specified in Item 3 of Section 9.13.9.2.3;

4. For each specimen there is no greater than a 20% decrease in the initial effective damping over for the $30S_{DI}B_D/S_{DS}$, but not less than 10, cycles of test specified in Section 9.13.9.2.3; and

5. All specimens of vertical-load-carrying elements of the isolation system remain stable up to the total maximum displacement for static load as prescribed in Section 9.13.9.2.6 and shall have a positive incremental force-carrying capacity.

9.13.9.5 Design Properties of the Isolation System

9.13.9.5.1 Maximum and minimum effective stiffness

At the design displacement, the maximum and minimum effectiveness stiffness of the isolated system, k_{Dmax} and k_{Dmin}, shall be based on the cyclic tests of Section 9.13.9.2.3 and calculated by the equations:

$$k_{Dmax} = \frac{\sum |F_D^+|_{max} + \sum |F_D^-|_{max}}{2D_D}$$

(Eq. 9.13.9.5.1-1)

$$k_{Dmin} = \frac{\sum |F_D^+|_{min} + \sum |F_D^-|_{min}}{2D_D}$$ (Eq. 9.13.9.5.1-2)

At the maximum displacement, the maximum and minimum effective stiffness of the isolation system, k_{Mmax} and k_{Mmin}, shall be based on the cyclic tests of Item 3 of Section 9.13.9.2.3 and calculated by the equations:

$$k_{Mmax} = \frac{\sum |F_M^+|_{max} + \sum |F_M^-|_{max}}{2D_M}$$

(Eq. 9.13.9.5.1-3)

$$k_{Mmin} = \frac{\sum |F_M^+|_{min} + \sum |F_M^-|_{min}}{2D_M}$$

(Eq. 9.13.9.5.1-4)

The maximum effective stiffness of the isolation system, k_{Dmax} (or k_{Mmax}), shall be based on forces from the cycle of prototype testing at a test displacement equal to D_D (or D_M) that produces the largest value of effective stiffness. Minimum effective stiffness of the isolation system, k_{Dmin} (or k_{Mmin}), shall be based on forces from the cycle of prototype testing at a test displacement equal to D_D (or D_M) that produces the smallest value of effective stiffness.

For isolator units that are found by the tests of Sections 9.13.9.3, 9.13.9.4 and 9.13.9.5 to have force-deflection characteristics that vary with vertical load, rate of loading or bilateral load, respectively, the values of k_{Dmax} and k_{Mmax} shall be increased and the values of k_{Dmin} and k_{Mmin} shall be decreased, as necessary, to bound the effects of measured variation in effective stiffness.

9.13.9.5.2 Effective damping

At the design displacement, the effective damping of the isolation system, β_D, shall be based on the cyclic tests of Item 2 of Section 9.13.9.3 and calculated by the equation:

$$\beta_D = \frac{\sum E_D}{2\pi k_{Dmax}D_D^2}$$ (Eq. 9.13.9.5.2-1)

In Eq. 9.13.9.5.2-1, the total energy dissipated per cycle of design displacement response, ΣE_D, shall be taken as the sum of the energy dissipated per cycle in all isolator units measured at a test displacement equal to D_D. The total energy dissipated per cycle of design displacement response, ΣE_D, shall be based on forces and deflections from the cycle of prototype testing at test displacement D_D that produces the smallest values of effective damping.

At the maximum displacement, the effective damping of the isolation system, β_M, shall be based on the cyclic tests of Item 2 of Section 9.13.9.3 and calculated by the equation:

$$\beta_M = \frac{\sum E_M}{2\pi k_{Mmax}D_M^2}$$ (Eq. 9.13.9.5.2-2)

In Eq. 9.13.9.5.2-2, the total energy dissipated per cycle of design displacement response, ΣE_M, shall be taken as the sum of the energy dissipated per cycle in all isolator units measured at a test displacement equal to D_M. The total energy dissipated per cycle of maximum displacement response, ΣE_M, shall be based on forces and deflections from the cycle of prototype testing at test displacement D_M that produces the smallest value of effective damping.

9.14 NONBUILDING STRUCTURES

9.14.1 General

9.14.1.1

Nonbuilding structures include all self-supporting structures that are supported by the earth, that carry gravity loads, and that may be required to resist the effects of earthquake, with the exception of: buildings, vehicular bridges, dams, and other structures excluded in Section 9.1.2.1. Nonbuilding structures shall be designed to resist the minimum lateral forces specified in this section. Design shall conform to the applicable provisions of other sections as modified by this section.

9.14.1.2

The design of nonbuilding structures shall provide sufficient stiffness, strength and ductility, consistent with the requirements specified herein for buildings, to resist the effects of seismic ground motions as represented by these design forces:

a. Applicable strength and other design criteria shall be obtained from other portions of Section 9 or its referenced codes and standards.
b. When applicable strength and other design criteria are not contained in or referenced by Section 9, such criteria shall be obtained from approved national standards. Where approved national standards define acceptance criteria in terms of allowable stresses as opposed to strength, the design seismic forces shall be obtained from this Section and used in combination with other loads as specified in Section 2.4 of this standard and used directly with allowable stresses specified in the national standards. Detailing shall be in accordance with the approved national standards.

9.14.1.3

Architectural, mechanical, and electrical components supported by nonbuilding structures shall be designed in accordance with Section 9.6 of this Standard.

9.14.1.4

The weight W for nonbuilding structures shall include all dead load as defined for buildings in Section 9.5.3.2. For purposes of calculating design seismic forces in nonbuilding structures, W also shall include all normal operating contents for items such as tanks, vessels, bins, and piping. W shall include snow and ice loads when these loads constitute 25% or more of W.

9.14.1.5

The fundamental period of the nonbuilding structure shall be determined by rational methods as prescribed in Section 9.5.3.3.

9.14.1.6

The drift limitations of Section 9.5.2.8 need not apply to nonbuilding structures if a rational analysis indicates they can be exceeded without adversely affecting structural stability. Drift limitations shall be established for structural and nonstructural elements whose failure causes life-safety hazards. *P*-delta effects shall be considered as specified in Section 9.5.3.7.2 for stability of the structure.

9.14.1.7

For nonbuilding structures supporting flexible nonstructural elements whose combined gravity weight exceeds 25% of the structure at sites where the seismic coefficient S_{DS} is greater than or equal to 0.50, the interaction effect between the structure and the supported element shall be analyzed.

9.14.1.8 Reference Standard

Ref. 9.14-1 Rack Manufacturers Institute (RMI), Specification for the Design, Testing, and of Industrial Steel Storage Racks, 1990.

9.14.2 Nonbuilding Structures Similar to Buildings

The lateral force procedure for nonbuilding structures with structural systems similar to buildings (those with structural systems listed in Table 9.5.2.2) shall be selected in accordance with the force and detailing requirements of Section 9.5.2.1.

EXCEPTION: Intermediate moment frames of reinforced concrete shall not be used at sites where the seismic coefficient S_{DS} is greater than or equal to 0.50 unless:

1. The nonbuilding structure is less than 50 ft (15.2 m) in height; and
2. $R = 3.0$ is used for design.

9.14.2.1.2

The importance factor (I) and seismic use group for nonbuilding structures are based on the relative hazard of the contents and the function. The value of I shall be the largest value determined by the approved standards, or the largest value as selected from Table 9.14.2.1.2.

9.14.2.2 Rigid Nonbuilding Structures

Nonbuilding structures that have a fundamental period, T, less than 0.06 s, including their anchorages, shall be designed for the lateral force obtained from the following:

$$V = 0.3S_{DS}WI \qquad \text{(Eq. 9.14.2.2)}$$

where

V = total design lateral seismic base shear force applied to a nonbuilding structure;
S_{DS} = site design response acceleration as determined from Section 9.4.1.2.5;
W = nonbuilding structure operating weight; and
I = importance factor as determined from Table 9.14.2.1.

Importance Factor	$I = 1.0$	$I = 1.25$	$I = 1.5$
Seismic Use Group	I	II	III
Hazard	H-I	H-II	H-III
Function	F-I	F-II	F-III

Notes:

H-I—The stored product is biologically or environmentally benign; low fire or low physical hazard.

H-II—The stored product is rated low explosion, moderate fire, or moderate physical hazard as determined by the authority having jurisdiction.

H-III—The stored product is rated high or moderate explosion hazard, high fire hazard, or high physical hazard as determined by the authority having jurisdiction.

F-I—Nonbuilding structures not classified as F-III.

F-II—Not applicable.

F-III—Seismic use group III nonbuilding structures or designated ancillary nonbuilding structures (such as communication towers, fuel storage tanks, cooling towers, or electrical substation structures) required for operation of Seismic Use Group III structures.

The force shall be distributed with height in accordance with Section 9.5.3.4.

9.14.2.3 Deflection Limits and Structure Separation

Deflection limits and structure separation shall be determined in accordance with this Standard unless specifically amended in this chapter.

9.14.2.1 Design Basis

Nonbuilding structures that are not covered by Sections 9.14.2 through 9.14.2.3 shall be designed to resist minimum seismic lateral forces not less than those determined in accordance with the requirements of Section 9.5.3.2 with the following additions and exceptions:

1. The factor R shall be the lesser of the values given in Table 9.14.2.1 or the values in Table 9.5.2.2.
2. The overstrength factor, W_0, shall be taken from the same table as the R factor.
3. The importance factor, I, shall be as given in Table 9.14.2.1.2.
4. The height limitations shall be as given in Table 9.14.2.1 or the values in Table 9.5.2.2.
5. The vertical distribution of the lateral seismic forces in nonbuilding structures covered by this section shall be determined:
   a. Using the requirements of Section 9.5.3.4 or
   b. Using the procedures of Section 9.5.4.
   c. In accordance with an approved Standard applicable to the specific nonbuilding structure.
6. Irregular structures at sites where the S_{DS} is greater than or equal to 0.50 and that cannot be modeled as a single mass shall use the procedures of Section 9.5.4.
7. Where an approved national standard provides a basis for the earthquake resistant design of a particular type of nonbuilding structure covered by Section 9.14, such a standard shall not be used unless the following limitations are met:
   a. The seismic ground acceleration, and the seismic coefficient, shall be in conformance with the requirements of Sections 9.4.1 and 9.1.2.5, respectively.
   b. The values for total lateral force and total base overturning moment used in design shall not be less than 80% of the base shear value and overturning moment, each adjusted for the effects of soil-structure interaction that is obtained using this Standard.
8. The base shear is permitted to be reduced in accordance with Section 9.5.5.2.1 to account for the effects of soil-structure interaction. In no case shall the reduced base shear, V, be less than $0.7V$.

9.14.3 Nonbuilding Structures Similar to Buildings

9.14.3.1

Nonbuilding structures that have structural systems that are designed and constructed in a manner similar to buildings and have a dynamic response similar to building structures shall be designed similar to building structures and in compliance with this Standard with exceptions as contained in this section.

This general category of nonbuilding structures shall be designed in accordance with Sections 9.5.2.1 and 9.14.2 through 9.14.2.1.

The lateral force design procedure for nonbuilding structures with structural systems similar to building structures (those with structural systems listed in Table 9.5.2.2) shall be selected in accordance with the force and detailing requirements of Section 9.5.2.1.

The combination of load effects, E, shall be determined in accordance with Section 9.5.2.7.

TABLE 9.14.2.1. Seismic Coefficients for Nonbuilding Structures

Nonbuilding Structure Type	R	Ω_0	C_d	Structural System and Height Limits (ft)[c]			
				Seismic Design Category			
				A & B	C	D	E & F
Nonbuilding frame systems:	See Table 9.5.2.2						
Concentric Braced Frames of Steel				NL	NL	NL	NL
Special Concentric Braced Frames of Steel				NL	NL	NL	NL
Moment Resisting Frame Systems:	See Table 9.5.2.2						
Special Moment Frames of Steel				NL	NL	NL	NL
Ordinary Moment Frames of Steel				NL	NL	50	50
Special Moment Frames of Concrete				NL	NL	NL	NL
Intermediate Moment Frames of Concrete				NL	NL	50	50
Ordinary Moment Frames of Concrete				NL	50	NP	NP
Steel Storage Racks	4	2	3-1/2	NL	NL	NL	NL
Elevated tanks, vessels, bins, or hoppers[a]:							
On braced legs	3	2	2-1/2	NL	NL	NL	NL
On unbraced legs	3	2	2-1/2	NL	NL	NL	NL
Irregular braced legs single pedestal or skirt supported	2	2	2	NL	NL	NL	NL
Welded steel	2	2	2	NL	NL	NL	NL
Concrete	2	2	2	NL	NL	NL	NL
Horizontal, saddle supported welded steel vessels	3	2	2-1/2	NL	NL	NL	NL
Tanks or vessels supported on structural towers similar to buildings	3	2	2	NL	NL	NL	NL
Flat bottom, ground supported tanks, or vessels:							
Anchored (welded or bolted steel)	3	2	2-1/2	NL	NL	NL	NL
Unanchored (welded or bolted steel)	2-1/2	2	2	NL	NL	NL	NL
Reinforced or prestressed concrete:							
Tanks with reinforced nonsliding base	2	2	2	NL	NL	NL	NL
Tanks with anchored flexible base	3	2	2	NL	NL	NL	NL
Tanks with unanchored and unconstrained:							
Flexible base	1-1/2	1-1/2	1-1/2	NL	NL	NL	NL
Other material	1-1/2	1-1/2	1-1/2	NL	NL	NL	NL
Cast-in-place concrete silos, stacks, and chimneys having walls continuous to the foundation	3	1-3/4	3	NL	NL	NL	NL
All other reinforced masonry structures	3	2	2-1/2	NL	NL	50	50
All other nonreinforced masonry structures	1-1/4	2	1-1/2	NL	50	50	50
All other steel and reinforced concrete distributed mass cantilever structures not covered herein including stacks, chimneys, silos, and skirt-supported vertical vessels	3	2	2-1/2	NL	NL	NL	NL
Trussed towers (freestanding or guyed), guyed stacks and chimneys	3	2	2-1/2	NL	NL	NL	NL
Cooling towers:							
Concrete or steel	3-1/2	1-3/4	3	NL	NL	NL	NL
Wood frame	3-1/2	3	3	NL	NL	50	50
Electrical transmission towers, substation wire support structures, distribution structures							
Truss: Steel and aluminum	3	1-1/2	3	NL	NL	NL	NL
Pole: Steel	1-1/2	1-1/2	1-1/2	NL	NL	NL	NL
Wood	1-1/2	1-1/2	1-1/2	NL	NL	NL	NL
Concrete	1-1/2	1-1/2	1-1/2	NL	NL	NL	NL
Frame: Steel	3	1-1/2	1-1/2	NL	NL	NL	NL
Wood	2-1/2	1-1/2	1-1/2	NL	NL	NL	NL
Concrete	2	1-1/2	1-1/2	NL	NL	NL	NL

TABLE 9.14.2.1. Seismic Coefficients for Nonbuilding Structures (*Continued*)

Nonbuilding Structure Type	R	Ω_0	C_d	Structural System and Height Limits (ft)[c]			
				Seismic Design Category			
				A & B	C	D	E & F
Telecommunication towers							
Truss: Steel	3	1-1/2	3	NL	NL	NL	NL
Pole: Steel	1-1/2	1-1/2	1-1/2	NL	NL	NL	NL
Wood	1-1/2	1-1/2	1-1/2	NL	NL	NL	NL
Concrete	1-1/2	1-1/2	1-1/2	NL	NL	NL	NL
Frame: Steel	3	1-1/2	1-1/2	NL	NL	NL	NL
Wood	2-1/2	1-1/2	1-1/2	NL	NL	NL	NL
Concrete	2	1-1/2	1-1/2	NL	NL	NL	NL
Amusement structures and monuments	2	2	2	NL	NL	NL	NL
Inverted pendulum type structures (not elevated tank)[b]	2	2	2	NL	NL	NL	NL
Signs and billboards	3-1/2	1-3/4	3	NL	NL	NL	NL
All other self-supporting structures, tanks or vessels not covered above or by approved standards	1-1/4	2	2-1/2	NL	50	50	50

Note: NL = No limit. NP = Not permitted.

[a]Support towers similar to building type structures, including those with irregularities (see Section 9.5.2.3 of this Standard for definition of irregular structures) shall comply with the requirements of Section 9.5.2.6.

[b]Light posts, stoplight, etc.

[c]Height shall be measured from the base.

9.14.3.2 Pipe Racks

9.14.3.2.1 Design basis

Pipe racks supported at the base of the structure shall be designed to meet the force requirements of Section 9.5.3 or 9.5.4.

Displacements of the pipe rack and potential for interaction effects (pounding of the piping system) shall be considered using the amplified deflections obtained from the following formula:

$$\delta_x = \frac{C_d \delta_{xe}}{I} \qquad \text{(Eq. 9.14.3.2.1)}$$

where

C_d = deflection amplification factor in Table 9.14.2.1;
δ_{xe} = deflections determined using the prescribed seismic design forces of the Standard; and
I = importance factor determined from Table 9.14.2.1.

Exception: The importance factor, I, shall be determined from Table 9.14.2.1 for the calculation of δ_{xe}.

See Section 9.6.3.11 for the design of piping systems and their attachments. Friction resulting from gravity loads shall not be considered to provide resistance to seismic forces.

9.14.3.3 Steel Storage Racks

This section applies to steel storage racks supported at the base of the structure. Storage racks shall be designed, fabricated, and installed in accordance with Ref. 9.14-1 and the requirements of this section. Steel storage racks supported above the base of the structure shall be designed in accordance with Sections 9.6.1 and 9.6.2.

9.14.3.3.1 General requirements

Steel storage racks shall satisfy the force requirements of this section.

Exception: Steel storage racks supported at the base are permitted to be designed as structures with an R of 4, provided that the requirements of Section 9.5 are met. Higher values of R are permitted to be used when the detailing requirements of reference documents of Section 9.8 as modified in Section A.9.8 are met. The importance factor I shall be taken equal to the I_p values in accordance with Section 9.6.1.5.

9.14.3.3.2 Operating weight

Steel storage racks shall be designed for each of the following conditions of operating weight, W or W_p.

a. Weight of the rack plus every storage level loaded to 67% of its rated load capacity.
b. Weight of the rack plus the highest storage level only loaded to 100% of its rated load capacity.

The design shall consider the actual height of the center of mass of each storage load component.

9.14.3.3.3 Vertical distribution of seismic forces

For all steel storage racks, the vertical distribution of seismic forces shall be as specified in Section 9.5.3.4 and in accordance with the following:

a. The base shear, V, of the typical structure shall be the base shear of the steel storage rack when loaded in accordance with Section 9.14.3.3.2.
b. The base of the structure shall be the floor supporting the steel storage rack. Each steel storage level of the rack shall be treated as a level of the structure, with heights h_i, and h_x measured from the base of the structure.
c. The factor k may be taken as 1.0.
d. The factor I shall be in accordance with Section 9.6.1.5.

9.14.3.3.4 Seismic displacements

Steel storage rack installations shall accommodate the seismic displacement of the storage racks and their contents relative to all adjacent or attached components and elements. The assumed total relative displacement for storage racks shall be not less than 5% of the height above the base.

9.14.3.4 Electrical Power Generating Facilities

9.14.3.4.1 General

Electrical power generating facilities are power plants that generate electricity by steam turbines, combustion turbines, diesel generators or similar turbo machinery.

9.14.3.4.2 Design basis

Electrical power generating facilities shall be de-

signed using this Standard and the appropriate factors contained in Section 9.14.2.

9.14.3.5 Structural Towers for Tanks and Vessels

9.14.3.5.1 General

Structural towers which support tanks and vessels shall be designed to meet the provisions of Section 9.14.1.2. In addition, the following special considerations shall be included:

a. The distribution of the lateral base shear from the tank or vessel onto the supporting structure shall consider the relative stiffness of the tank and resisting structural elements.
b. The distribution of the vertical reactions from the tank or vessel onto the supporting structure shall consider the relative stiffness of the tank and resisting structural elements. When the tank or vessel is supported on grillage beams, the calculated vertical reaction due to weight and overturning shall be increased at least 20% to account for non-uniform support. The grillage beam and vessel attachment shall be designed for this increased design value.
c. Seismic displacements of the tank and vessel shall consider the deformation of the support structure when determining P-delta effects or evaluating required clearances to prevent pounding of the tank on the structure.

9.14.3.6 Piers and Wharves

9.14.3.6.1 General

Piers and wharves are structures located in waterfront areas that project into a body of water or parallel the shore line.

9.14.3.6.2 Design basis

Piers and wharves shall be designed to comply with this Standard and approved standards. Seismic forces on elements below the water level shall include the inertial force of the mass of the displaced water. The additional seismic mass equal to the mass of the displaced water shall be included as a lumped mass on the submerged element, and shall be added to the calculated seismic forces of the pier or wharf structure. Seismic dynamic forces from the soil shall be determined by the registered design professional.

The design shall account for the effects of liquefaction on piers and wharfs as required.

9.14.4 Nonbuilding Structures Not Similar to Buildings

9.14.4.1 General

Nonbuilding structures that have structural systems that are designed and constructed in a manner such that the dynamic response is not similar to buildings shall be designed in compliance with this Standard with exceptions as contained in this section.

This general category of nonbuilding structures shall be designed in accordance with this Standard and the specific applicable approved standards. Loads and load distributions shall not be less than those determined in this Standard.

The combination of load effects, E, shall be determined in accordance with Section 9.5.2.7.

Exception: The redundancy/reliability factor, ρ, per Section 9.5.2.4 shall be taken as 1.

9.14.4.2 Earth Retaining Structures

9.14.4.2.1 General

This section applies to all earth retaining walls. The applied seismic forces shall be determined in accordance with Section 9.7.5.1 with a geotechnical analysis prepared by a registered design professional.

9.14.4.3 Tanks and Vessels

9.14.4.3.1 General

This section applies to all tanks and vessels storing liquids, gases, and granular solids supported at the base. Tanks and vessels covered herein include reinforced concrete, prestressed concrete, steel, and fiber-reinforced plastic materials. Tanks supported on elevated levels in buildings shall be designed in accordance with Section 9.6.3.9.

9.14.4.3.2 Design basis

Tanks and vessels shall be designed in accordance with this Standard and shall be designed to resist seismic lateral forces determined from a substantiated analysis using approved standards.

9.14.4.3.3 Additional requirements

In addition, for sites where S_{DS} is greater than 0.60, flat-bottom tanks designated with an I_p greater than 1.0 or tanks greater than 20 ft (6.2 m) in diameter or tanks that have a height-to-diameter ratio greater than 1.0 shall also be designed to meet the following additional requirements:

1. Sloshing effects shall be calculated and provided for in the design, fabrication, and installation.
2. Piping connections to steel flat-bottom storage tanks shall consider the potential uplift of the tank wall during earthquakes. Unless otherwise calculated, the following displacements shall be assumed for all side-wall connections and bottom penetrations:
   a. Vertical displacement of 2 in. (50 mm) for anchored tanks.
   b. Vertical displacement of 12 in. (300 mm) for unanchored tanks, and
   c. Horizontal displacement of 8 in. (200 mm) for unanchored tanks with a diameter of 40 ft (12.2 m) or less.

9.14.4.4 Electrical Transmission, Substation, and Distribution Structures

9.14.4.4.1 General

This section applies to electrical transmission, substation, and distribution structures.

9.14.4.4.2 Design basis

Electrical transmission, substation wire support and distribution structures shall be designed to resist seismic lateral forces determined from a substantiated analysis using approved standards.

9.14.4.5 Telecommunication Towers

9.14.4.5.1 General

This section applies to telecommunication towers.

9.14.4.5.2 Design basis

Self-supporting and guyed telecommunication towers shall be designed to resist seismic lateral forces determined from a substantiated analysis using approved standards.

9.14.4.6 Stacks and Chimneys

9.14.4.6.1 General

Stacks and chimneys are permitted to be either lined or unlined, and shall be constructed from concrete, steel, or masonry.

9.14.4.6.2 Design basis

Steel stacks, concrete stacks, steel chimneys, concrete chimneys, and liners shall be designed to resist seismic lateral forces determined from a substantiated analysis using approved standards. Interaction of the stack or chimney with the liners shall be considered. A minimum separation shall be provided between the liner and chimney equal to C_d times the calculated differential lateral drift.

9.14.4.7 Amusement Structures

9.14.4.7.1 General

Amusement structures are permanently fixed structures constructed primarily for the conveyance and entertainment of people.

9.14.4.7.2 Design basis

Amusement structures shall be designed to resist seismic lateral forces determined from a substantiated analysis using approved standards.

9.14.4.8 Special Hydraulic Structures

9.14.4.8.1 General

Special hydraulic structures are structures that are contained inside liquid containing structures. These structures are exposed to liquids on both wall surfaces at the same head elevation under normal operating conditions. Special hydraulic structures are subjected to out of plane forces only during an earthquake when the structure is subjected to differential hydrodynamic fluid forces. Examples of special hydraulic structures include: separation walls, baffle walls, weirs, and other similar structures.

9.14.4.8.2 Design basis

Special hydraulic structures shall be designed for out-of-phase movement of the fluid. Unbalanced forces from the motion of the liquid must be applied simultaneously "in front of" and "behind" these elements.

Structures subject to hydrodynamic pressures induced by earthquakes shall be designed for rigid body and sloshing liquid forces and their own inertia force. The height of sloshing shall be determined and compared to the freeboard height of the structure.

Interior elements, such as baffles or roof supports, also shall be designed for the effects of unbalanced forces and sloshing.

9.14.4.9 Buried Structures

9.14.4.9.1 General

Buried structures are subgrade structures such as tanks, tunnels, and pipes. Buried structures that are designated as Seismic Use Group II or III, or are of such a size or length to warrant special seismic design as determined by the registered design professional shall be identified in the geotechnical report.

9.14.4.9.2 Design basis

Buried structures shall be designed to resist minimum seismic lateral forces determined from a substantiated analysis using approved standards. Flexible couplings shall be provided for buried structures requiring special seismic considerations where changes in the support system, configuration, or soil condition occur.

9.14.4.10 Inverted Pendulums

These structures are a special category of structures which support an elevated lumped mass, and exclude water tanks.

10.0 ICE LOADS—ATMOSPHERIC ICING

10.1 DEFINITIONS

Freezing rain or drizzle: Rain or drizzle falling into a shallow layer of subfreezing air at the earth's surface. The water freezes on contact with the ground or a structure to form glaze ice.

Snow: Considered to be an atmospheric ice accretion when it adheres to a structure by capillary forces, freezing and/or sintering. Roof snow loads are covered in Section 7.

In-cloud icing: Occurs on structures in supercooled clouds and fogs. The droplets colliding with the structure freeze to form lower density rime ice or glaze ice.

Hoarfrost: An accumulation of ice crystals formed by direct deposition of water vapor from the air onto a structure.

Ice-sensitive structures: Structures including, but not limited to, lattice structures, guyed masts, overhead lines, suspension and cable-stayed bridges, aerial cable systems (e.g., for ski lifts and logging operations), amusement rides, open catwalks, ladders, railings, flagpoles and signs.

Components and appurtenances: Elements attached to the structure will be exposed to atmospheric icing. Examples are antennas and handrails.

10.2 GENERAL

Atmospheric ice loads due to ice accretions formed by freezing rain and drizzle, snow, and in-cloud icing shall be considered in the design of ice-sensitive structures throughout the United States. Hoarfrost need not be considered in design.

10.3 DESIGN FOR ICE LOADS

The radial thickness of an ice accretion is assumed to be uniform over the exposed surface of all structural members, components, and appurtenances. The application of a uniform equivalent radial ice thickness to a variety of cross-sectional shapes is shown in Figure 10-1. When historical ice accretion thicknesses and associated densities are known for the region of interest, the design ice thickness and accretion density shall be determined from this data for the appropriate mean recurrence interval. When ice accretion data are not available, the design values shall be estimated by the analysis of meteorological information for the appropriate mean recurrence interval. The quality of the measured or estimated ice load information and the length of the record must be taken into account in determining the design ice load. Alternatively, the local jurisdiction is not prohibited from establishing the design ice thickness and density. Mean recurrence intervals for each structure category in Table 1-1 are given in Table 10-1.

Dynamic loads such as galloping and aeolian vibration that are caused or enhanced by an ice accretion on a flexible structural member, component or appurtenance are not covered in this section.

10.3.1 Weight of Ice

The weight of ice per unit of length is the product of the accretion density and the cross-sectional area of the accretion.

10.3.2 Wind on Ice Covered Structures

Ice accreted on structural members, components and appurtenances increases the projected area of the structure exposed to wind. Wind loads on this increased projected area shall be used in the design of ice-sensitive structures.

10.3.3 Partial Loading

The effects of a partial ice load shall be considered. Partial loading of ice-sensitive structures occurs, for example, when the exposure of an overhead line varies either vertically or horizontally or when accreted ice falls off only part of the overhead line.

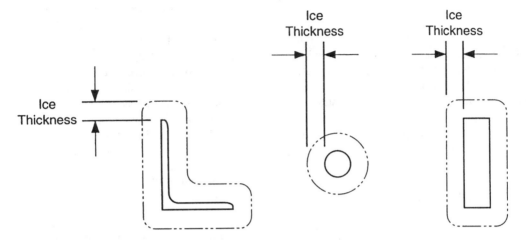

FIGURE 10-1. Application of Uniform Equivalent Radial Ice Thicknesses to a Variety of Cross-Sectional Shapes

TABLE 10-1. Mean Recurrence Intervals for Structure Categories

Structure Category	Mean Recurrence Interval (years)
I	25
II	50
III	50
IV	100

APPENDICES

A.9.0 SUPPLEMENTAL PROVISIONS

A.9.1 PURPOSE

These provisions are not directly related to computation of earthquake loads, but they are deemed essential for satisfactory performance in an earthquake when designing with the loads determined from Section 9, due to the substantial cyclic inelastic strain capacity assumed to exist by the load procedures in Section 9. These supplemental provisions form an integral part of Section 9.

A.9.3 QUALITY ASSURANCE

This section provides minimum requirements for quality assurance for seismic force resisting systems and other designated seismic systems. These requirements supplement the testing and inspection requirements contained in the reference standards given in Sections 9.6 through 9.12.

A.9.3.1 Scope
As a minimum, the quality assurance provisions apply to the following:

1. The seismic force-resisting systems in structures assigned to Seismic Design Categories C, D, E, and F.
2. Other designated seismic systems in structures assigned to Seismic Design Categories D, E, and F that are required in Table 9.6.1.7.

   Exception: Structures that comply with the following criteria are exempt from the preparation of a quality assurance plan but those structures are not exempt from special inspection(s) or testing requirements:

   i. The structure is constructed of light wood framing or light gauge cold-formed steel framing, S_{DS} does not exceed 0.50g, the height of the structure does not exceed 35 ft above grade, and the structure meets the requirements in Items iii and iv below; or
   ii. The structure is constructed using a reinforced masonry structural system or reinforced concrete structural system, S_{DS} does not exceed 0.50g, the height of the structure does not exceed 25 ft above grade, and the structure meets the requirements in Items iii and iv below;
   iii. The structure is classified as Seismic Use Group I;
   iv. The structure does not have any of the following plan irregularities as defined in Table 9.5.2.3.2 or any of the following vertical irregularities as defined in Table 9.5.2.3.3:
       (1) Torsional irregularity,
       (2) Extreme torsional irregularity,
       (3) Nonparallel systems,
       (4) Stiffness irregularity—soft story,
       (5) Stiffness irregularity—extreme soft story,
       (6) Discontinuity in capacity-weak story.

The following standards are referenced in the provisions for inspection and testing:

Ref. 9.1.6-1 ANSI/AWS D1.1-98, Structural Welding Code-Steel.

Ref. 9.1.6-2 ASTM A435-90, Specification for Straight Beam Ultrasound Examination of Steel Plates, 1990.

Ref. 9.1.6-3 ASTM A898-91, Specification for Straight Beam Ultrasound Examination for Rolled Steel Shapes, 1991.

A.9.3.2 Quality Assurance Plan
A quality assurance plan shall be submitted to the authority having jurisdiction.

A.9.3.2.1 Details of Quality Assurance Plan
The quality assurance plan shall specify the designated seismic systems or seismic force-resisting system in accordance with Section A.9.3 that are subject to quality assurance. The registered design professional in responsible charge of the design of a seismic force resisting system and a designated seismic system shall be responsible for the portion of the quality assurance plan applicable to that system. The special inspections and special tests needed to establish that the construction is in conformance with these provisions shall be included in the portion of the quality assurance plan applicable to the designated seismic system. The quality assurance plan shall include:

a. The seismic-force-resisting systems and designated seismic systems in accordance with this chapter that are subject to quality assurance.
b. The special inspections and testing to be provided as required by these Provisions and the reference standards in Section 9.
c. The type and frequency of testing.
d. The type and frequency of special inspections.
e. The frequency and distribution of testing and special inspection reports.
f. The structural observations to be performed.
g. The frequency and distribution of structural observation reports.

A.9.3.2.2 Contractor Responsibility

Each contractor responsible for the construction of a seismic-force-resisting system, designated seismic system, or component listed in the quality assurance plan shall submit a written contractor's statement of responsibility to the regulatory authority having jurisdiction and to the owner prior to the commencement of the work on the system or component. The contractor's statement of responsibility shall contain the following:

1. Acknowledgment of awareness of the special requirements contained in the quality assurance plan.
2. Acknowledgment that control will be exercised to obtain conformance with the design documents approved by the authority having jurisdiction.
3. Procedures for exercising control within the contractor's organization, the method and frequency of reporting, and the distribution of the reports.
4. Identification and qualifications of the person(s) exercising such control and their position(s) in the organization.

A.9.3.3 Special Inspection

The building owner shall employ a special inspector(s) to observe the construction of all designated seismic systems in accordance with the quality assurance plan for the following construction work:

A.9.3.3.1 Foundations

Continuous special inspection is required during driving of piles, and placement of concrete in piers or piles. Periodic special inspection is required during construction of drilled piles, piers, and caisson work, the placement of concrete in shallow foundations, and the placement of reinforcing steel.

A.9.3.3.2 Reinforcing Steel

A.9.3.3.2.1

Periodic special inspection during and upon completion of the placement of reinforcing steel in intermediate and special moment frames of concrete and concrete shear walls.

A.9.3.3.2.2

Continuous special inspection during the welding of reinforcing steel resisting flexural and axial forces in intermediate and special moment frames of concrete, in boundary members of concrete shear walls, and welding of shear reinforcement.

A.9.3.3.3 Structural Concrete

Periodic special inspection during and on completion of the placement of concrete in intermediate and special moment frames, and in boundary members of concrete shear walls.

A.9.3.3.4 Prestressed Concrete

Periodic special inspection during the placement and after the completion of placement of prestressing steel and continuous special inspection is required during all stressing and grouting operations and during the placement of concrete.

A.9.3.3.5 Structural Masonry

A.9.3.3.5.1

Periodic special inspection during the preparation of mortar, the laying of masonry units, and placement of reinforcement; and prior to placement of grout.

A.9.3.3.5.2

Continuous special inspection during welding of reinforcement, grouting, consolidation, and reconsolidation, and placement of bent-bar anchors as required by Section A.9.11.

A.9.3.3.6 Structural Steel

A.9.3.3.6.1

Continuous special inspection is required for all structural welding.

Exception: Periodic special inspection for single-pass fillet or resistance welds and welds loaded to less than 50% of their design strength shall be the minimum requirement, provided the qualifications of the welder and the welding electrodes are inspected at the beginning of the work and all welds are inspected for compliance with the approved construction documents at the completion of welding.

A.9.3.3.6.2

Periodic special inspection is required in accordance with Ref. 9.8-1 or 9.8-2 for installation and tightening of fully tensioned high-strength bolts in slip-critical connections and in connections subject to direct tension. Bolts in connections identified as not being slip-critical or subject to direction tension need not be inspected for bolt tension other than to ensure that the plies of the connected elements have been brought into snug contact.

A.9.3.3.7 Structural Wood

A.9.3.3.7.1

Continuous special inspection during all field gluing operations of elements of the seismic-force-resisting system.

A.9.3.3.7.2

Periodic special inspection is required for nailing, bolting, anchoring and other fastening of components within the seismic-force-resisting system including drag struts, braces, and hold downs.

A.9.3.3.8 Cold-Formed Steel Framing

A.9.3.3.8.1

Periodic special inspection is required during all welding operations of elements of the seismic-force-resisting system.

A.9.3.3.8.2

Periodic special inspection is required for screw attachment, bolting, anchoring, and other fastening of components within the seismic-force-resisting system, including struts, braces, and hold-downs.

A.9.3.3.9 Architectural Components

Special inspection for architectural components shall be as follows:

1. Periodic special inspection during the erection and fastening of exterior cladding, interior and exterior nonloadbearing walls, and veneer in Seismic Design Categories D, E, and F; and

    Exceptions: a) Structures 30 ft (9.2 m) or less in height; and b) Cladding and veneer weighing 5 lb/ft^2 (25 kg/m^2) or less.
2. Periodic special inspection during the anchorage of access floors, suspended ceilings, and storage racks 8 ft (2.5 m) or greater in height in Seismic Design Categories D, E, and F.

A.9.3.3.10 Mechanical and Electrical Components

Special inspection for mechanical and electrical components shall be as follows:

1. Periodic special inspection during the anchorage of electrical equipment for emergency or standby power systems in Seismic Design Categories C, D, E, and F;
2. Periodic special inspection during the installation of anchorage of all other electrical equipment in Seismic Design Categories E and F;
3. Periodic special inspection during the installation for flammable, combustible, or highly toxic piping systems and their associated mechanical units in Seismic Design Categories C, D, E, and F; and
4. Periodic special inspection during the installation of HVAC ductwork that will contain hazardous materials in Seismic Design Categories C, D, E, and F.

A.9.3.3.11 Seismic Isolation System

Periodic special inspection is required during the fabrication and installation of isolator units and energy dissipation devices if used as part of the seismic isolation system.

A.9.3.4 Testing

The special inspector(s) shall be responsible for verifying that the special test requirements are performed by an approved testing agency for the types of work in designated seismic systems listed below.

A.9.3.4.1 Reinforcing and Prestressing Steel

Special testing of reinforcing and prestressing steel shall be as follows:

A.9.3.4.1.1

Examine certified mill test reports for each shipment of reinforcing steel used to resist flexural and axial forces in reinforced concrete intermediate and special moment frames and boundary members of reinforced concrete shear walls or reinforced masonry shear walls and determine conformance with construction documents.

A.9.3.4.1.2

Where ASTM A615 reinforcing steel is used to resist earthquake-induced flexural and axial forces in special moment frames and in wall boundary elements of shear walls in structures of Seismic Design Categories D, E, and F, verify that the requirements of Section 21.2.5.1 of Ref. 9.9-1 have been satisfied.

A.9.3.4.1.3

Where ASTM A615 reinforcing steel is to be welded, verify that chemical tests have been performed to determine weldability in accordance with Section 3.5.2 of Ref. 9.9-1.

A.9.3.4.2 Structural Concrete

Samples of structural concrete shall be obtained at the project site and tested in accordance with requirements of Ref. 9.9-1.

A.9.3.4.3 Structural Masonry

Quality assurance testing of structural masonry shall be in accordance with the requirements of Ref. 9.11-1.

A.9.3.4.4 Structural Steel

The testing needed to establish that the construction is in conformance with these Provisions shall be included in a quality assurance plan. The minimum testing contained in the quality assurance plan shall be as required in Ref. 9.8-3 and the following requirements:

A.9.3.4.4.1 Base Metal Testing

Base metal thicker than 1.5 in. (38 mm), when subject to through-thickness weld shrinkage strains, shall be ultrasonically tested for discontinuities behind and adjacent to such welds after joint completion. Any material discontinuities shall be accepted or rejected on the basis of ASTM A435, *Specification for Straight Beam Ultrasound Examination of Steel Plates*, or ASTM A898, *Specification for Straight Beam Ultrasound Examination for Rolled Steel Shapes (Level 1 Criteria)*, and criteria as established by the registered design professional(s) in responsible charge and the construction documents.

A.9.3.4.5 Mechanical and Electrical Equipment

As required to ensure compliance with the seismic design provisions herein, the registered design professional in responsible charge shall clearly state the applicable requirements on the construction documents. Each manufacturer of these designated seismic system components shall test or analyze the component and its mounting system or anchorage as required and shall submit a certificate of compliance for review and acceptance by the registered design professional in responsible charge of the design of the designated seismic system and for approval by the authority having jurisdiction. The basis of certification shall be by actual test on a shaking table, by

three-dimensional shock tests, by an analytical method using dynamic characteristics and forces, by the use of experience data (i.e., historical data demonstrating acceptable seismic performance), or by more rigorous analysis providing for equivalent safety. The special inspector shall examine the designated seismic system and shall determine whether its anchorages and label conform with the certificate of compliance.

A.9.3.4.6 Seismic-Isolated Structures

For required system tests, see Section 9.13.9.

A.9.3.5 Structural Observations

Structural observations shall be provided for those structures included in Seismic Design Categories D, E, and F when one or more of the following conditions exist:

1. The structure is included in Seismic Use Group II or Seismic Use Group III, or
2. The height of the structure is greater than 75 ft above the base, or
3. The structure is in Seismic Design Category E or F and Seismic Use Group I and is greater than two stories in height.

Observed deficiencies shall be reported in writing to the owner and the authority having jurisdiction.

A.9.3.6 Reporting and Compliance Procedures

Each special inspector shall furnish to the authority having jurisdiction, registered design professional in responsible charge, the owner, the persons preparing the quality assurance plan, and the contractor copies of regular weekly progress reports of his observations, noting therein any uncorrected deficiencies and corrections of previously reported deficiencies. All deficiencies shall be brought to the immediate attention of the contractor for correction. At completion of construction, each special inspector shall submit a final report to the authority having jurisdiction certifying that all inspected work was completed substantially in accordance with approved construction documents. Work not in compliance shall be described in the final report. At completion of construction, the building contractor shall submit a final report to the authority having jurisdiction certifying that all construction work incorporated into the seismic-force-resisting system and other designated seismic systems was constructed substantially in accordance with the approved construction documents and applicable workmanship requirements. Work not in compliance shall be described in the final report. The contractor shall correct all deficiencies as required.

A.9.7 SUPPLEMENTARY FOUNDATION REQUIREMENTS

A.9.7.4.4 Special Pile Requirements for Category C

All concrete piles and concrete filled pipe piles shall be connected to the pile cap by embedding the pile reinforcement in the pile cap for a distance equal to the development length as specified in Ref. 9.9-1 or by the use of field-placed dowels anchored in the concrete pile. For deformed bars, the development length is the full development length for compression without reduction in length for excess area.

Where special reinforcement at the top of the pile is required, alternative measures for laterally confining concrete and maintaining toughness and ductile-like behavior at the top of the pile shall be allowed provided due consideration is given to forcing the hinge to occur in the confined region. Where a minimum length for reinforcement or the extent of closely spaced confinement reinforcement is specified at the top of the pile, provisions shall be made so that those specified lengths or extents are maintained after pile cut-off.

A.9.7.4.4.1 Uncased concrete piles

A minimum reinforcement ratio of 0.0025 shall be provided for uncased cast-in-place concrete drilled piles or drilled piers in the top one-third of the pile length or a minimum length of 10 ft below the ground. There shall be a minimum of four bars with closed ties (or equivalent spirals) of a minimum 1/4 in. diameter provided at 16-longitudinal-bar-diameter maximum spacing with a maximum spacing of 4 in. in the top 2 ft of the pile. Reinforcement detailing requirements shall be in conformance with Section A.9.9.6.2.

A.9.7.4.4.2 Metal-cased concrete piles

Reinforcement requirements are the same as for uncased concrete piles.

Exception: Spiral welded metal-casing of a thickness not less than No. 14 gauge can be considered as providing concrete confinement equivalent to the closed ties or equivalent spirals required in an uncased concrete pile, provided that the metal casing is adequately protected against possible deleterious action due to soil constituents, changing water levels, or other factors indicated by boring records of site conditions.

A.9.7.4.4.3 Concrete-filled pipe

Minimum reinforcement 0.01 times the cross-sectional area of the pile concrete shall be provided in the top of the pile with a length equal to two times the required cap embedment anchorage into the pile cap.

A.9.7.4.4.4 Precast concrete piles

Longitudinal reinforcement shall be provided for precast concrete piles with a minimum steel ratio of 0.01. Ties or equivalent spirals shall be provided at a maximum 16-bar-diameter spacing with a maximum spacing of 4 in. in the top 2 ft. Reinforcement shall be full length.

A.9.7.4.4.5 Precast-prestressed piles

The minimum volumetric ratio of spiral reinforcement in the ductile region shall be equal to 0.007. The spiral reinforcement shall not be less than the amount required by the following formula:

$$\rho_s = 0.12 f_c'/f_{yh}$$

where

ρ_s = volumetric ratio (vol. spiral/vol. core);
$f_c' \leq 6,000$ psi (41.4 MPa); and
f_{yh} = yield strength of spiral reinforcement ≤85 ksi (586 MPa).

Where pile cap connection is made by means of developing pile reinforcing strand, a connection capable of dissipating energy shall be provided.

A.9.7.5 Special Pile Requirements for Category D, E, and F

A.9.7.5.4.1 Uncased concrete piles

A minimum reinforcement ratio of 0.005 shall be provided for uncased cast-in-place concrete piles, drilled piers, or caissons in the top one-half of the pile length or a minimum length of 10 ft below ground. There shall be a minimum of four bars with closed ties or equivalent spirals provided at 8-longitudinal-bar-diameter maximum spacing with a maximum spacing of 3 in. in the top 4 ft of the pile. Ties shall be a minimum of No. 3 bars for up to 20-in.-diameter piles and No. 4 bars for piles of larger diameter.

A.9.7.5.4.2 Metal-cased concrete piles

Reinforcement requirements are the same as for uncased concrete piles.

Exception: Spiral welded metal-casing of a thickness not less than No. 14 gauge can be considered as providing concrete confinement equivalent to the closed ties or equivalent spirals required in an uncased concrete pile, provided that the metal casing is adequately protected against possible deleterious action due to soil constituents, changing water levels, or other factors indicated by boring records of site conditions.

A.9.7.5.4.3 Precast concrete piles

Ties in precast concrete piles shall conform to the requirements of Section A.9.11 for at least the top half of the pile.

A.9.7.5.4.4 Precast-prestressed piles

In addition to the requirements for Seismic Design Category C, the following requirements shall be met:

1. Requirements of Ref. 9.9-1, Chapter 21, need not apply.
2. Where the total pile length in the soil is 35 ft (10,668 mm) or less, the lateral transverse reinforcement in the ductile region shall occur through the length of the pile. Where the pile length exceeds 35 ft, the ductile pile region shall be taken as the greater of 35 ft or the distance from the underside of the pile cap to the point of zero curvature plus three times the least pile dimension.
3. In the ductile region, the center to center spacing of the spirals or hoop reinforcement shall not exceed one-fifth of the least pile dimension, six times the diameter of the longitudinal strand, or 8 in. (203 mm), whichever is smaller.
4. Spiral reinforcement shall be spliced by lapping one full turn, by welding or by the use of a mechanical connector. Where spiral reinforcement is lap spliced, the ends of the spiral shall terminate in a seismic hook in accordance with Ref. 9.9-1, except that the bend shall be not less than 135°. Welded splices and mechanical connectors shall comply with Section 12.14.3 of Ref. 9.9-1.
5. Where the transverse reinforcement consists of circular spirals, the volumetric ratio of spiral transverse reinforcement in the ductile region shall comply with:

$$\rho_s = 0.25(f_c'/f_{yh})(A_g/A_{ch} - 1.0)[0.5 + 1.4P/(f_c'A_g)]$$

but not less than

$$\rho_s = 0.12(f_c'/f_{yh})[0.5 + 1.4P/(f_c'A_g)]$$

and not to exceed $\rho_s = 0.021$
where

ρ_s = volumetric ratio (vol. spiral/vol. core);
$f_c' \leq 6,000$ psi (41.4 MPa);
f_{yh} = yield strength of spiral reinforcement ≤ 85 ksi (586 MPa);
A_g = pile cross-sectional area, in.^{2} (mm^2);
A_{ch} = core area defined by spiral outside diameter, in.^{2} (mm^2); and
P = axial load on pile resulting from the load combination 1.2D + 0.5L + 1.0E, pounds (kN).

This required amount of spiral reinforcement is permitted to be obtained by providing an inner and outer spiral.

6. When transverse reinforcement consists of rectangular hoops and cross ties, the total cross-sectional area of lateral transverse reinforcement in the ductile region with spacings, and perpendicular to dimension, h_c, shall conform to:

$$A_{sh} = 0.3sh_c(f_c'/f_{yh})(A_g/A_{ch} - 1.0)$$
$$\cdot [0.5 + 1.4P/(f_c'A_g)]$$

but not less than

$$A_{sh} = 0.12sh_c(f_c'/f_{yh})[0.5 + 1.4P/(f_c'A_g)]$$

where

s = spacing of transverse reinforcement measured along length of pile, in. (mm);
h_c = cross-sectional dimension of pile core measured center to center of hoop reinforcement, in. (mm); and
$f_{yh} \leq 70$ ksi (483 MPa).

The hoops and cross ties shall be equivalent to deformed bars not less than No. 3 in size. Rectangular hoop ends shall terminate at a corner with seismic hooks.

A.9.7.5.4.5 Steel piles

The connection between the pile cap and steel piles or unfilled steel pipe piles shall be designed for a tensile force equal to 10% of the pile compression capacity.

A.9.8 SUPPLEMENTARY PROVISIONS FOR STEEL

A.9.8.1 General

A.9.8.1.1 Structural Steel by Strength Design

If the allowable stress load combinations of Section 2.4.1 are used, Section A.9.8.1.2 shall be satisfied. Otherwise, the strength load combinations of Section 2.3.2 shall be used to design structural steel for the earthquake loads given by Section 9.

A.9.8.1.2 Allowable Stress Design

When using the load combinations of Section 2.4.1, the allowable strength of members and connections shall be determined from allowable stress set forth in the following:

Structural Steel, Ref. 9.8-2

Cold Formed Steel, Ref. 9.8-4

Cold Formed Stainless Steel, Ref. 9.8-5, Appendix E

Steel Joists and Joist Girders, Ref. 9.8-6

Steel Cables, Ref. 9.8-7

The one-third increase in allowable stress given in Refs. 9.8-2 and 9.8-5, Appendix E, or the 0.75 factor on loads given in Ref. 9.8-4 for use with seismic loads is permitted. The load combination adjustment factors of Section 2.4.3 shall not be used. The increase in allowable stress given in Part 2 of Ref. 9.8-3 for structural steel or Section A.9.8.5 for cold formed steel to approximate strength shall not be used in conjunction with the load combination of Section 2.4.1.

For structural steel members designed using Ref. 9.8-2, Section A.9.8.1.3.1 and the provisions of Ref. 9.8-3 shall also be satisfied, including the detailed proportioning rules that are stated in terms of strength for Seismic Performance Categories C (as limited in Section 2.2 of Ref. 9.8-3) D, and E.

For light framed walls, the provisions of Section A.9.8.6 shall also be satisfied, when required by Section A.9.8.2.3.

A.9.8.1.3 Structural Steel Seismic Requirements

The design of structural steel members and connections to resist seismic forces shall be in accordance with Ref. 9.8-1 or 9.8-2. When using the provisions of Ref. 9.8-2 to compute the capacity of members to resist seismic forces, allowable stresses shall be converted into design strengths using the provisions of Part II, Sections 3.2 and 3.3, of Ref. 9.8-3. When required, structural steel members also shall be designed in accordance with Ref. 9.8-3 as modified by the requirements of this section, except that the definition of E shall be as defined in these provisions and that the term C_a shall be substituted for A_v throughout. Also, Section 8.2C of Ref. 9.8-3 shall be deleted and replaced with the following: 8.2.C Connection Strength: Connection configurations utilizing welds or high strength bolts shall demonstrate, by approved cyclic testing results or calculations, the ability to sustain inelastic rotation and to develop the strength criteria in Section 8.2a considering the expected value of yield strength and strain hardening.

A.9.8.5 Cold-Formed Steel Seismic Requirements

The design of cold-formed carbon or low-alloy steel to resist seismic loads shall be in accordance with the provisions of Ref. 9.8-4 and the design of cold-formed stainless steel structural members to resist seismic loads shall be in accordance with the provisions of Ref. 9.8-5, except as modified by this section. The reference to section and paragraph numbers are to those of the particular specification modified.

A.9.8.5.1

Ref. 9.8-4—Revise Section A5.1.3 of Ref. 9.8-4 by deleting the reference to earthquake or seismic loads in the sentence permitting the 0.75 factor. Seismic load combinations shall be as determined by ASCE 7.

A.9.8.5.2

Ref. 9.8-5—Modify Section 1.5.2 by substituting a load factor of 1.0 in place of 1.5 for nominal earthquake load.

A.9.8.7 Seismic Requirements for Steel Deck Diaphragms

Steel deck diaphragms shall be made from materials conforming to the requirements of Ref. 9.8-4 or 9.8-5. Nominal strengths shall be determined in accordance with approved analytical procedures or with test procedures prepared by a registered design professional experienced in testing of cold-formed steel assemblies and approved by the authority having jurisdiction. Design strengths shall be determined by multiplying the nominal strength by a resistance factor, ϕ, equal to 0.60 for mechanically connected diaphragms and equal to 0.50 for welded diaphragms. The steel deck installation for the building, including fasteners, shall comply with the test assembly arrangement. Quality standards established for the nominal strength test shall be the minimum standards required for the steel deck installation, including fasteners.

A.9.8.8 Steel Cables

The design strength of steel cables shall be determined by the provisions of Ref. 9.8-7 except as modified by this Section. Ref. 9.8-7, Section 5d, shall be modified by substituting $1.5(T_4)$ when T_4 is the net tension in cable due to dead load, prestress, live load, and seismic load. A load factor of 1.1 shall be applied to the prestress force to be added to the load combination of Section 3.1.2 of Ref. 9.5-7.

A.9.8.2 Seismic Provisions for Steel Structural Members

Steel structures and structural elements therein that resist seismic forces shall be designed in accordance with the applicable provisions of Sections A.9.8.1.3 through A.9.8.8. In addition, steel structures also shall be designed in accordance with the requirements of Sections A.9.8.3, A.9.8.2.2, and A.9.8.2.3 for the appropriate Seismic Performance Category.

A.9.8.3 Seismic Design Categories A, and B, and C

Buildings assigned to Seismic Design Category A, B, or C shall be of any construction permitted by the references in Section 9.8.1. An R factor as set forth in Table 9.5.2 for the appropriate steel system is permitted when the structure is designed and detailed in accordance with the provisions of Ref. 9.8-3, Part I, or Section A.9.8.6, for light framed cold-formed steel wall systems. Systems not detailed in accordance with the above shall use the R factor in Table 9.5.2 designated for "steel systems not detailed for seismic."

A.9.8.4 Seismic Design Categories D, E, and F

Buildings assigned to Seismic Design D, E, or F shall be designed in accordance with the additional provisions of Ref. 9.8-3, Part I, for structural steel buildings and Section A.9.8.6 for light framed cold-formed steel wall systems.

A.9.8.6 Light Framed Wall Requirements

Cold-formed steel stud wall systems designed in accordance with Ref. 9.8-4 or 9.8-5 shall, when re-

quired by the provisions of Section A.9.8.3 or A.9.8.4, also comply with the requirements of this section.

A.9.8.6.1 Boundary Members

All boundary members, chords, and collectors shall be designed to transmit the axial force induced by the specified loads of Section 9.

A.9.8.6.2 Connections

Connections of diagonal bracing members, top chord splices, boundary members and collectors shall have a design strength equal to or greater than the nominal tensile strength of the members being connected or Ω_o times the design seismic forces. The pullout resistance of screws shall not be used to resist seismic forces.

A.9.8.6.3 Braced Bay Members

In stud systems where the lateral forces are resisted by braced frames, the vertical and diagonal members of braced bays shall be anchored such that the bottom tracks are not required to resist forces by bending of the track or track web. Both flanges of studs in a bracing bay shall be braced to prevent lateral torsional buckling. In braced shear walls, the vertical boundary members shall be anchored so the bottom track is not required to resist uplift forces by bending of the track web.

A.9.8.6.4 Diagonal Braces

Provision shall be made for pretensioning or other methods of installation of tension-only bracing to prevent loose diagonal straps.

A.9.8.6.5 Shear Walls

Nominal shear values for wall sheathing materials are given in Table A.9.8.6.5. Design shear values shall be determined by multiplying the nominal values therein by a ϕ factor of 0.55. In structures over one story in height the assemblies in Table A.9.8.6.5 shall not be used to resist horizontal loads contributed by forces imposed by masonry or concrete construction.

Panel thicknesses shown in Table A.9.8.6.5 shall be considered to be minimums. No panels less than 24 in. wide shall be used. Plywood or oriented strand board structural panels shall be of a type that is manufactured using exterior glue. Framing members, blocking or strapping shall be provided at the edges of all sheets. Fasteners along the edges in shear panels shall be placed not less than 3/8 in. (9.5 mm) in

TABLE A.9.8.6.5. Nominal Shear Values for Seismic Forces for Shear Walls Framed with Cold-Formed Steel Studs (in lb/ft)[a,b]

Assembly Description	Fastener Spacing at Panel Edges[c] (in.)				Framing Spacing (in. o.c.)
	6	4	3	2	
15/32 rated Structural I sheathing (4-ply) plywood one side[d]	780	990	1,465	1,625	24
7/16 in. oriented strand board one side[d]	700	915	1,275	1,625	24

Note: For fastener and framing spacing, multiply inches by 25.4 to obtain metric mm.

[a]Nominal shear values shall be multiplied by the appropriate strength reduction factor ϕ to determine design strength as set forth in Section A.9.8.6.5.

[b]Studs shall be a minimum 1-5/8 in. by 3-1/2 in. with a 3/8-in. return lip. Track shall be a minimum 1-1/4 in. by 3-1/2 in. Both studs and track shall have a minimum uncoated base metal thickness of 0.033 in. and shall be ASTM A446 Grade A (or ASTM A653 SQ Grade 33 [new designation]). Framing screws shall be No. 8 × 5/8 in. wafer head self-drilling. Plywood and OSB screws shall be a minimum No. 8 × 1 in. bugle head. Where horizontal straps are used to provide blocking they shall be a minimum 1-1/2 in. wide and of the same material and thickness as the stud and track.

[c]Screws in the field of the panel shall be installed 12 in. o.c. unless otherwise shown.

[d]Both flanges of the studs shall be braced in accordance with Section A.9.8.6.3.

from panel edges. Screws shall be of sufficient length to assure penetration into the steel stud by at least two full diameter threads.

The height to length ratio of wall systems listed in Table A.9.8.6.5 shall not exceed 2:1.

Perimeter members at openings shall be provided and shall be detailed to distribute the shearing stresses. Wood sheathing shall not be used to splice these members.

Wall studs and track shall have a minimum uncoated base thickness of not less than 0.033 in. (0.84 mm) and shall not have an uncoated base metal thickness greater than 0.048 in. (1.22 mm). Panel end studs and their uplift anchorage shall have the design strength to resist the forces determined by the seismic loads determined by Eqs. 9.5.2.7-3 and 9.5.2.7-4.

A.9.9 SUPPLEMENTARY PROVISIONS FOR CONCRETE

A.9.9.1 Modifications to Ref. 9.9-1 (ACI 318-99)

The text of Ref. 9.9-1 (ACI 318-99) shall be modified as indicated in Sections A.9.9.1.1 through A.9.9.1.10. Italics are used for text within Sections A.9.9.1.1 through A.9.9.1.10 to indicate provisions which differ from Ref. 9.9-1 (ACI 318-99).

A.9.9.1.1 ACI 318, Section 8.1.2

Modify Section 8.1.2 to read as follows:

8.1.2 Except for load combinations which include earthquake loads, design of nonprestressed rein-

forced concrete members using Appendix A, Alternate Design Method, is permitted.

A.9.9.1.2 ACI 318, Section 9.2.3

Modify Section 9.2.3 to read as follows:

9.2.3 Where resistance to specified earthquake loads or forces E are included in design, the load combinations of Section 2.3 of ASCE 7-98 for strength design and the φ factors of Appendix C of Ref. 9.6-1 shall apply.

A.9.9.1.3 ACI 318, Section 21.0

Add the following notations to Section 21.0:

h = *overall dimension of member in the direction of action considered.*

S_e Connection = *moment, shear, or axial force at connection cross section other than the nonlinear action location corresponding to probable strength at the nonlinear action location, taking gravity load effects into consideration per Section 21.2.8.3.*

S_n Connection = *nominal strength of connection cross section in flexural, shear, or axial action per Section 21.2.8.3.*

$\Delta_m = C_d\Delta_s$

Δ_s = *design level response displacement, which is the total drift or total story drift that occurs when the structure is subjected to the design seismic forces.*

ψ = *dynamic amplification factor from Sections 21.2.8.3 and 21.2.8.4*

A.9.9.1.4 ACI 318, Section 21.1

Modify existing definitions and add definitions to Section 21.1.

Connection. *An element that joins two precast members or a precast member and a cast-in-place member.*

Design Load Combinations. *Combinations of factored loads and forces specified in Section 2.3.2 of ASCE 7.98.*

Dry Connection. *Connection used between precast members which does not qualify as a wet connection.*

Joint. *The geometric volume common to intersecting members.*

Non-linear Action Location. *Center of the region of yielding in flexure, shear, or axial action.*

Non-linear Action Region. *The member length over which nonlinear action takes place. It shall be taken as extending a distance of no less than h/2 on either side of the nonlinear action location.*

Strong Connection. *A connection that remains elastic while the designated nonlinear action regions undergo inelastic response under the design basis ground motion.*

Wall Pier. *A wall segment with a horizontal length-to-thickness ratio of at least 2.5, but not exceeding 6, whose clear height is at least two times its horizontal length.*

Wet Connection. *A connection that uses any of the splicing methods permitted by Sections 21.3.2.3 or 21.3.2.4 to connect precast members and uses cast-in-place concrete or grout to fill the splicing closure.*

A.9.9.1.5 ACI 318, Section 21.2.1

Add new Sections 21.2.1.6 and 21.2.1.7:

21.2.1.6 Precast lateral force resisting systems shall satisfy either of the following criteria:

1. *It emulates the behavior of monolithic reinforced concrete construction and satisfies Section 21.2.2.5, or*
2. *It relies on the unique properties of a structural system composed of interconnected precast elements and it is demonstrated by experimental evidence and analysis to safely sustain the seismic loading requirements of a comparable monolithic reinforced concrete structure satisfying Chapter 21. Substantiating experimental evidence of acceptable perfor-*

mance of those elements required to sustain inelastic deformations shall be based upon cyclic inelastic testing of specimens representing those elements.

21.2.1.7 In structures having precast gravity load carrying systems, the lateral force resisting system shall be one of the systems listed in Table 9.5.2.2 of ASCE 7-98 and shall be well distributed using one of the following methods:

1. *The lateral force resisting system shall be spaced such that the span of the diaphragm or diaphragm segment between lateral force resisting systems shall be no more than three times the width of the diaphragm or diaphragm segment. Where the lateral force resisting system consists of moment resisting frames, at least $(N_b/4) + 1$ of the bays (rounded up to the nearest integer) along any frame line at any story shall be part of the lateral force resisting system where N_b is the total number of bays along that line at that story. This requirement applies to only the lower two thirds of the stories of buildings three stories or taller.*
2. *Beam-to-column connections that are not part of the lateral force resisting system shall be designed in accordance with the following:*

Connection Design Force. *The connection shall be designed to develop strength M. M is the moment developed at the connection when the frame is displaced by Δ_s assuming fixity at the connection and a beam flexural stiffness of no less than one half of the gross section stiffness. M shall be sustained through a deformation of Δ_m.*

Connection Characteristics. *The connection shall be permitted to resist moment in one direction only, positive or negative. The connection at the opposite end of the member shall resist moment with the same positive or negative sign. The connection shall be permitted to have zero flexural stiffness up to a frame displacement of Δ_s.*

In addition, complete calculations for the deformation compatibility of the gravity load carrying system shall be made in accordance with Section 9.5.2.2.4.3 of ASCE 7-98 using cracked section stiffness in the lateral force resisting system and the diaphragm.

Where gravity columns are not provided with lateral support on all sides, a positive connection shall be provided along each unsupported direction parallel to a principal plan axis of the structure. The connection shall be designed for a horizontal force equal to 4% of the axial load strength, P_o, of the column.

The bearing length shall be calculated to include end rotation, sliding, and other movements of precast ends at supports due to earthquake motions in addition to other movements and shall be at least 2 in. (51 mm) more than that required.

A.9.9.1.6 ACI 318, Section 21.2.2

Add new Sections 21.2.2.5, 21.2.2.6 and 21.2.2.7 to Section 21.2.2 to read as follows:

21.2.2.5 Precast structural systems using frames and emulating the behavior of monolithic reinforced concrete construction shall satisfy either Sections 21.2.2.6 or 21.2.2.7.

21.2.2.6 Precast structural systems utilizing wet connections shall comply with the applicable requirements of monolithic concrete construction for resisting seismic forces.

21.2.2.7 Precast structural systems not meeting Section 21.2.2.6 shall utilize strong connections resulting in nonlinear response away from connections. Design shall satisfy the requirements of Section 21.2.8 in addition to the applicable requirements of monolithic concrete construction for resisting seismic forces, except that the provisions of Section 21.3.1.2 shall apply to the segments between nonlinear action locations.

A.9.9.1.7 ACI 318, Section 21.2.5

Modify Section 21.2.5 by renumbering as Section 21.2.5.1 and adding new Sections 21.2.5.2 and 21.2.5.3 to read as follows:

21.2.5 Reinforcements in members resisting earthquake-induced forces

21.2.5.1 Except as permitted in Sections 21.2.5.2 and 21.2.5.3, reinforcement resisting earthquake-induced flexural and axial forces in frame members and in wall boundary elements shall comply with ASTM A706. ASTM A615 Grades 40 and 60 reinforcement shall be permitted in these members if (a) the actual yield strength based on mill tests does not exceed the specified yield strength by more than 18,000 psi (retests shall not exceed this value by more than an additional 3,000 psi), and (b) the ratio of the actual ultimate tensile strength to the actual tensile yield strength is not less than 1.25.

21.2.5.2 Prestressing tendons shall be permitted in flexural members of frames provided the average

prestress, f_{pc}, calculated for an area equal to the member's shortest cross-sectional dimension multiplied by the perpendicular dimension shall be the lesser of 700 psi (4.83 MPa) or $f'_c/6$ at locations of nonlinear action where prestressing tendons are used in members of frames.

21.2.5.3 For members in which prestressing tendons are used together with mild reinforcement to resist earthquake-induced forces, prestressing tendons shall not provide more than one quarter of the strength for both positive moments and negative moments at the joint face and shall extend through exterior joints and be anchored at the exterior face of the joint or beyond. Anchorages for tendons shall be demonstrated to perform satisfactorily for seismic loadings. Anchorage assemblies shall withstand, without failure, a minimum of 50 cycles of loading ranging between 40 and 85 percent of the minimum specified tensile strength of the tendon.

A.9.9.1.8 ACI 318, Section 21.2

Modify Section 21.2 by adding a new Section 21.2.8 to read as follows:

21.2.8 Emulation of monolithic construction using strong connections. Members resisting earthquake-induced forces in precast frames using strong connections shall satisfy the following:

21.2.8.1 Location. Nonlinear action location shall be selected so that there is a strong column/weak beam deformation mechanism under seismic effects. The nonlinear action location shall be no closer to the near face of the strong connection than h/2. For column-to-footing connections where nonlinear action may occur at the column base to complete the mechanism, the nonlinear action location shall be no closer to the near face of the connection than h/2.

21.2.8.2 Anchorage and splices. Reinforcement in the nonlinear action region shall be fully developed outside both the strong connection region and the nonlinear action region. Noncontinuous anchorage reinforcement of the strong connection shall be fully developed between the connection and the beginning of the nonlinear action region. Lap splices are prohibited within connections adjacent to a joint.

21.2.8.3 Design forces. Design strength of strong connections shall be based on:

ϕS_n Connection $> \Psi s_e$ Connection

Dynamic amplification factor, Ψ, shall be taken as 1.0.

21.2.8.4 Column-to-column connection. *The strength of column-to-column connections shall comply with Section 21.2.8.3 with Ψ taken as 1.4. Where column-to-column connections occur, the columns shall be provided with transverse reinforcement as specified in Sections 21.4.4.1 through 21.4.4.3 over their full height if the factored axial compressive force in these members, including seismic effects, exceeds $A_g f_c'/10$.*

> **Exception:** *Where the column-to-column connection is located within the middle third of the column clear height, the following shall apply: (a) the design moment strength, ϕM_n, of the connection shall not be less than 0.4 times the maximum M_{pr} for the column within the story height, and (b) the design shear strength, ϕV_n, of the connection shall not be less than that determined by Section 21.4.5.1.*

21.2.8.5 Column-face connection. *Any strong connection located outside the middle half of a beam span shall be a wet connection unless a dry connection can be substantiated by approved cyclic test results. Any mechanical connector located within such a column-face strong connection shall be a Type 2 mechanical splice as defined in Section 21.2.6.1.*

A.9.9.1.9 ACI 318, Section 21.6

Modify Section 21.6 by adding a new Section 21.6.10 to read as follows:

21.6.10 Wall Piers and Wall Segments.

21.6.10.1 *Wall piers not designed as a part of a special moment-resisting frame shall have transverse reinforcement designed to satisfy the requirements in Section 21.6.10.2.*

> **Exceptions:**
>
> 1. *Wall piers that satisfy Section 21.9.*
> 2. *Wall piers along a wall line within a story where other shear wall segments provide lateral support to the wall piers, and such segments have a total stiffness of at least six times the sum of the stiffnesses of all the wall piers.*

21.6.10.2 *Transverse reinforcement shall be designed to resist the shear forces determined from Sections 21.4.5.1 and 21.3.4.2. Where the axial compressive force, including earthquake effects, is less than $A_g f_c'/20$, transverse reinforcement in wall piers is permitted to have standard hooks at each end in lieu of hoops. Spacing of transverse reinforcement shall not exceed 6 in. (152 mm). Trans-*

verse reinforcement shall be extended beyond the pier clear height for at least the development length of the largest longitudinal reinforcement in the wall pier.

21.6.10.3 *Wall segments with a horizontal length-to-thickness ratio less than 2 1/2, shall be designed as columns.*

A.9.9.1.10 ACI 318, Section 21.9

Modify Section 21.9.2.2 to read as follows:

21.9.2.2 *Members with factored gravity axial forces exceeding $(A_g f_c'/10)$ shall satisfy Sections 21.4.3.1, 21.4.4.1(3), 21.4.4.3 and 21.4.5. The maximum longitudinal spacing of ties shall be s_o for the full column height. The spacing s_o shall not be more than six diameters of the smallest longitudinal bar enclosed or 6 in. (152 mm), whichever is smaller.*

A.9.9.2 Classification of Shear Walls

Structural concrete shear walls which resist seismic forces shall be classified in accordance with Sections A.9.9.2.1 through A.9.9.2.4.

A.9.9.2.1 Ordinary Plain Concrete Shear Walls

Ordinary plain concrete shear walls are walls conforming to the requirements of Ref. 9.9-1 for ordinary structural plain concrete walls.

A.9.9.2.2 Detailed Plain Concrete Shear Walls

Detailed plain concrete shear walls are walls conforming to the requirements for ordinary plain concrete shear walls and shall have reinforcement as follows. Vertical reinforcement of at least 0.20 in.[2] (129 mm[2]) in cross-sectional area shall be provided continuously from support to support at each corner, at each side of each opening, and at the ends of walls. The continuous vertical bar required beside an opening is permitted to substitute for one of the two No. 5 bars required by Section 22.6.6.5 of Ref. 9.9-1. Horizontal reinforcement at least 0.20 in.[2] (120 mm[2]) in cross-sectional area shall be provided.

1. Continuously at structurally connected roof and floor levels and at the top of walls,
2. At the bottom of load-bearing walls or in the top of foundations where doweled to the wall, and
3. At a maximum spacing of 120 in. (3048 mm).

Reinforcement at the top and bottom of openings, where used in determining the maximum spacing specified in Item 3 above, shall be continuous in the wall.

A.9.9.2.3 Ordinary Reinforced Concrete Shear Walls

Ordinary reinforced concrete shear walls are walls conforming to the requirements of Ref. 9.9-1 for ordinary reinforced concrete structural walls.

A.9.9.2.4 Special Reinforced Concrete Shear Walls

Special reinforced concrete shear walls are walls conforming to the requirements of Ref. 9.9-1 for special reinforced concrete structural walls.

A.9.9.3 Seismic Design Category B

Structures assigned to Seismic Design Category B, as determined in Section 9.4.2 shall conform to the requirements for Seismic Design Category A and to the additional requirements for Seismic Design Category B of this Section.

A.9.9.3.1 Ordinary Moment Frames

In flexural members of ordinary moment frames forming part of the seismic-force-resisting system, at least two main flexural reinforcing bars shall be provided continuously top and bottom throughout the beams, through or developed with exterior columns or boundary elements.

Columns of ordinary moment frames having a clear height to maximum plan dimension ratio of five or less shall be designed for shear in accordance with Section 21.10.3 of Ref. 9.9-1.

A.9.9.4 Seismic Design Category C

Structures assigned to Seismic Design Category C, as determined in Section 9.4.2 shall conform to the requirements for Seismic Design Category B and to the additional requirements for Seismic Design Category C of this Section.

A.9.9.4.1 Seismic Force Resisting Systems

Moment frames used to resist seismic forces shall be Intermediate Moment Frames or Special Moment Frames. Shear walls used to resist seismic forces shall be ordinary reinforced concrete shear walls, or special reinforced concrete shear walls.

A.9.9.4.2 Discontinuous Members

Columns supporting reactions from discontinuous stiff members such as walls shall be designed for the special load combinations in Section 9.5.2.7 and shall be provided with transverse reinforcement at the spacing s_o as defined in Section 21.10.5.1 of Ref. 9.9-1 over their full height beneath the level at which the discontinuity occurs. This transverse reinforce-

ment shall be extended above and below the column as required in Section 21.4.4.5 of Ref. 9.9-1.

A.9.9.4.3 Anchor Bolts in the Top of Columns

Anchor bolts set in the top of a column shall be provided with ties which enclose at least four longitudinal column bars. There shall be at least two No. 4, or three No. 3 ties within 5 in. of the top of the column. The ties shall have hooks on each free end that comply with Section 7.1.3(c) of Ref. 9.9-1.

A.9.9.4.4 Plain Concrete

Structural plain concrete members in structures assigned to Seismic Design Category C shall conform to Ref. 9.9-1 and Sections A.9.9.4.4.1 through A.9.9.4.4.3.

A.9.9.4.4.1 Walls

Structural plain concrete walls are not permitted in structures assigned to Seismic Design Category C.

Exception: Structural plain concrete basement, foundation, or other walls below the base are permitted in detached one- and two-family dwellings constructed with stud bearing walls. Such walls shall have reinforcement in accordance with Section 22.6.6.5 of Ref. 9.9-1.

A.9.9.4.4.2 Footings

Isolated footings of plain concrete supporting pedestals or columns are permitted provided the projection of the footing beyond the face of the supported member does not exceed the footing thickness.

Exception: In detached one- and two-family dwellings three stories or less in height, the projection of the footing beyond the face of the supported member is permitted to exceed the footing thickness.

Plain concrete footings supporting walls shall be provided with not less than two continuous longitudinal reinforcing bars. Bars shall not be smaller than No. 4 and shall have a total area of not less than 0.002 times the gross cross-sectional area of the footing. For footings which exceed 8 in. (203 mm) in thickness, a minimum of one bar shall be provided at the top and bottom of the footing. For foundation systems consisting of a plain concrete footing and a plain concrete stemwall, a minimum of one bar shall be provided at the top of the stemwall and at the bottom of the footing. Continuity of reinforcement shall be provided at corners and intersections.

Exceptions:

1. In detached one- and two-family dwellings three stories or less in height and constructed with

stud bearing walls, plain concrete footings supporting walls are permitted without longitudinal reinforcement.

2. Where a slab-on-ground is cast monolithically with the footing, one No. 5 bar is permitted to be located at either the top or bottom of the footing.

A.9.9.4.4.3 Pedestals

Plain concrete pedestals shall not be used to resist lateral seismic forces.

A.9.9.5 Seismic Design Categories D, E, and F

Structures assigned to Seismic Design Category D, E or F, as determined in Section 9.4.2, shall conform to the requirements for Seismic Design Category C and to the additional requirements of this section.

A.9.9.5.1 Seismic Force Resisting Systems

Moment frames used to resist seismic forces shall be special moment frames. Shear walls used to resist seismic forces shall be special reinforced concrete shear walls.

A.9.9.5.2 Frame Members Not Proportioned to Resist Forces Induced by Earthquake Motions

Frame components assumed not to contribute to lateral force resistance shall conform to Section 21.9 of Ref. 9.9-1, as modified by Section A.9.9.1.10.

A.9.9.6 Bolts and Headed Stud Anchors in Concrete[1]

A.9.9.6.1 Scope

The provisions of this section shall govern the strength design of anchors cast in concrete for purposes of transmitting structural loads from one connected element to the other. These provisions apply to headed bolts, headed studs, and hooked (J- or L-) bolts. These provisions do not apply to anchors installed in hardened concrete, or load applications that are predominantly high cycle fatigue or impact. The heads of headed studs and headed bolts shall have a geometry such that the pullout strength of the anchor in uncracked concrete, as demonstrated by approved tests, equals or exceeds $1.4N_p$ (where N_p is given by Eq. (1913-11a)). Hooked bolts shall have a geometry such that the pullout strength of the anchor without the benefit of friction in uncracked concrete, as demonstrated by approved tests, equals of exceeds $1.4N_p$ (where N_p is given by Eq. (1913-11b)). Reinforcement used as part of the embedment shall be designed in accordance with applicable parts of ACI 318.

A.9.9.6.2 Notations and Definitions

The notations and definitions used in this section shall be as set forth in Sections A.9.9.6.2.1 and A.9.9.6.2.2, respectively.

A.9.9.6.2.1 Notations

A_b = bearing area of the head of stud or anchor bolt, in.^{2};

A_{No} = projected concrete failure area of one anchor, for calculation of strength in tension, when not limited by edge distance or spacing, as defined in Section A.9.9.6.5.2.1, in.^{2};

A_N = projected concrete failure area of an anchor or group of anchors, for calculation of strength in tension, as defined in Section A.9.9.6.5.2.1, in.^{2}. A_N shall not be taken greater than nA_{No};

A_{se} = effective cross-sectional area of anchor, in.^{2};

A_{Vo} = projected concrete failure area of one anchor, for calculation of strength in shear, when not limited by corner influences, spacing, or member thickness, as defined in Section A.9.9.6.6.2.1, in.^{2};

A_V = projected concrete failure area of an anchor or group of anchors, for calculation of strength in shear, as defined in Section A.9.9.6.6.2.1, in.^{2}. A_V shall not be taken greater than nA_{Vo};

c = distance from center of an anchor shaft to the edge of concrete, in.;

c_1 = distance from the center of an anchor shaft to the edge of concrete in one direction, in. Where shear force is applied to anchor, c_1 is in the direction of the shear force;

c_2 = distance from center of an anchor shaft to the edge of concrete in the direction orthogonal to c_1, in.;

c_{max} = largest of the edge distances that are less than or equal to $1.5h_{ef}$, in. (used only for the case of 3 or 4 edges);

c_{min} = smallest of the edge distances that are less than or equal to $1.5h_{ef}$, in.;

d_o = shaft diameter of headed stud, headed bolt, or hooked anchor, in. (See also Section A.9.9.6.8.4);

d_u = diameter of head of stud or anchor bolt or equivalent diameter of effective perimeter of an added plate or washer at the head of the anchor, in.;

---

[1]The provisions contained in Section A.9.9.6 are copyrighted by ACI and reprinted with permission.

e_h = distance from the inner surface of the shaft of a J-bolt or L-bolt to the outer tip of the J- or L-bolt, in.;

e_N' = eccentricity of normal force on a group of anchors; the distance between the resultant tension load on a group of anchors in tension and the centroid of the group of anchors loaded in tension, in.;

e_v' = eccentricity of shear force on a group of anchors; the distance between the point of shear force application and the centroid of the group of anchors resisting shear in the direction of the applied shear, in.;

f_c' = specified compressive strength of concrete, psi;

f_{ct} = specified tensile strength of concrete, psi;

f_r = modulus of rupture of concrete, psi. (See Section 9.5.2.3 of ACI 318);

f_t = calculated tensile stress in a region of a member, psi;

f_y = specified yield strength of anchor steel, psi;

f_{ut} = specified tensile strength of anchor steel, psi;

h = thickness of member in which an anchor is embedded measured parallel to anchor axis, in.;

h_{ef} = effective anchor embedment depth, in.;

k = coefficient for basic concrete breakout strength in tension;

k_{cp} = coefficient for pryout strength;

l = load bearing length of anchor for shear, not to exceed $8d_o$, in.;

= h_{ef} for anchors with a constant stiffness over the full length of the embedded section, such as headed studs;

n = number of anchors in a group;

N_b = basic concrete breakout strength in tension of a single anchor in cracked concrete, as defined in Section A.9.9.6.5.2.2, lb;

N_{cb} = nominal concrete breakout strength in tension of a single anchor, as defined in Section A.9.9.6.5.2.1, lb;

N_{cbg} = nominal concrete breakout strength in tension of a group of anchors, as defined in Section A.9.9.6.5.2.1, lb;

N_n = nominal strength in tension, lb;

N_p = pullout strength in tension of a single anchor in cracked concrete, as defined in Section A.9.9.6.5.3.4 or A.9.9.6.5.3.5, lb;

N_{pn} = nominal pullout strength in tension of a single anchor, as defined in Section A.9.9.6.5.3.1, lb;

N_{sb} = side-face blowout strength of a single anchor, lb;

N_{sbg} = side-face blowout strength of a group of anchors, lb;

N_s = nominal strength of a single anchor in tension as governed by the steel strength, as defined in Section A.9.9.6.5.1.2, lb;

N_u = factored tensile load, lb;

s = anchor center-to-center spacing, in.;

s_o = spacing of the outer anchors along the edge in a group, in.;

t = thickness of washer or plate, in.;

V_b = basic concrete breakout strength in shear of a single anchor in cracked concrete, as defined in Section A.9.9.6.6.2.2 or A.9.9.6.6.2.3, lb;

V_{cb} = nominal concrete breakout strength in shear of a single anchor, as defined in Section A.9.9.6.6.2.1, lb;

V_{cbg} = nominal concrete breakout strength in shear of a group of anchors, as defined in Section A.9.9.6.6.2.1, lb;

V_{cp} = nominal concrete pryout strength, as defined in Section A.9.9.6.6.3, lb;

V_n = nominal shear strength, lb;

V_s = nominal strength in shear of a single anchor as governed by the steel strength, as defined in Section A.9.9.6.6.1.1, lb;

V_u = factored shear load, lb;

ϕ = strength reduction factor (see Sections A.9.9.6.4.4 and A.9.9.6.4.5);

y_1 = modification factor, for strength in tension, to account for anchor groups loaded eccentrically, as defined in Section A.9.9.6.5.2.4;

y_2 = modification factor, for strength in tension, to account for edge distances smaller than $1.5h_{ef}$, as defined in Section A.9.9.6.5.2.5;

y_3 = modification factor, for strength in tension, to account for cracking, as defined in Sections A.9.9.6.5.2.6 and A.9.9.6.5.2.7;

y_4 = modification factor, for pullout strength, to account for cracking, as defined in Sections A.9.9.6.5.3.1 and A.9.9.6.5.3.6;

y_5 = modification factor, for strength in shear, to account for anchor groups loaded eccentrically, as defined in Section A.9.9.6.6.2.5;

y_6 = modification factor, for strength in shear, to account for edge distances smaller than $1.5c_1$, as defined in Section A.9.9.6.6.2.6; and

y_7 = modification factor, for strength in shear, to account for cracking, as defined in Section A.9.9.6.6.2.7.

A.9.9.6.2.2 Definitions

Anchor: A metallic element used to transmit applied loads including headed bolts, headed studs, and hooked bolts (J- or L-bolt).

Anchor Group: A number of anchors of approximately equal effective embedment depth with each anchor spaced at less than three times its embedment depth from one or more adjacent anchors.

Anchor Pullout Strength: The strength corresponding to the fastening device sliding out from the concrete without breaking out a substantial portion of the surrounding concrete.

Attachment: The element or assembly, external to the surface of the concrete, that transmits loads to or from the anchor.

Brittle Steel Element: An element with a tensile test elongation of less than 14% over a 2 in. gage length, reduction in area of less than 40%, or both.

Concrete Breakout Strength: The strength corresponding to a volume of concrete surrounding the anchor or group of anchors separating from the member.

Concrete Pryout Strength: The strength corresponding to formation of a concrete spall behind a short, stiff anchor with an embedded base that is displaced in the direction opposite to the applied shear force.

Ductile Steel Element: An element with a tensile test elongation of at least 14% over a 2 in. gage length and reduction in area of at least 40%.

Edge Distance: The distance from the edge of the concrete surface to the center of the nearest anchor.

Effective Embedment Depth: The overall depth through which the anchor transfers force to the surrounding concrete. The effective embedment depth will normally be the depth of the failure surface in tension applications. For headed anchor bolts and headed studs, the effective embedment depth is measured from the bearing contact surface of the head.

5% Fractile: A statistical term meaning 90% confidence that 95% of the actual strengths will exceed the nominal strength. Determination shall include the number of tests when evaluating data.

Hooked Bolt: An anchor anchored mainly by mechanical interlock from the 90-degree bend (L-bolt) or 180-degree bend (J-bolt) at its embedded end.

Projected Area: The area on the free surface of the concrete member that is used to represent the larger base of the assumed rectilinear failure surface.

Side-Face Blowout Strength: The strength of anchors with deeper embedment but thinner side cover corresponding to concrete spalling on the side face around the embedded head while no major breakout occurs at the top concrete surface.

A.9.9.6.3 General Requirements

A.9.9.6.3.1

Anchors and anchor groups shall be designed for critical effects of factored loads as determined by elastic analysis. Plastic analysis approaches are permitted where nominal strength is controlled by ductile steel elements, provided that deformational compatibility is taken into account.

A.9.9.6.3.2

When anchor design includes seismic loads the following additional requirements shall apply.

A.9.9.6.3.2.1: In structures assigned to Seismic Design Category C, D, E or F, as determined in Section 9.4.2, the design strength of anchors shall be taken as $0.75\phi N_n$ and $0.75\phi V_n$, where ϕ is given in Section A.9.9.6.4.4 or A.9.9.6.4.5 and N_n and V_n are determined in accordance with Section A.9.9.6.4.1.

A.9.9.6.3.2.2: In structures assigned to Seismic Design Category C, D, E or F, as determined in Section 9.4.2, anchors shall be designed to be governed by tensile or shear strength of a ductile steel element, unless Section A.9.9.6.3.3.3 is satisfied.

A.9.9.6.3.2.3: In lieu of Section A.9.9.6.3.3.2, the attachment shall be designed so that it will undergo ductile yielding at a load level no greater than 75% of the minimum anchor design strength.

A.9.9.6.3.3

All provisions for anchor axial tension and shear strength apply to normal weight-concrete. When lightweight aggregate concrete is used, provisions for N_n and V_n shall be modified by multiplying all values of $\sqrt{f'_c}$ affecting N_n and V_n by 0.75 for "all-lightweight" concrete and 0.85 for "sand-lightweight" concrete. Linear interpolation shall be permitted when partial sand replacement is used.

A.9.9.6.3.4

The value of f'_c used for calculations in this section shall not exceed 10,000 psi.

A.9.9.6.4 General Requirements for Strength of Anchors

A.9.9.6.4.1

Strength design of anchors shall be based on the computation or test that take all of the following into consideration:

a) steel strength of anchor in tension (Section A.9.9.6.5.1)

b) steel strength of anchor in shear (Section A.9.9.6.6.1)

c) concrete breakout strength of anchor in tension (Sections A.9.9.6.4.2 and A.9.9.6.5.2)

d) concrete breakout strength of anchor in shear (Sections A.9.9.6.4.2 and A.9.9.6.6.2)

e) pullout strength of anchor in tension (Sections A.9.9.6.4.2 and A.9.9.6.5.3)

f) concrete side-face blowout strength of anchor in tension (Sections A.9.9.6.4.2 and A.9.9.6.5.4)

g) concrete pryout strength of anchor in shear (Sections A.9.9.6.4.2 and A.9.9.6.6.3)

h) required edge distances, spacings and thicknesses to preclude splitting failure (Sections A.9.9.6.4.2 and A.9.9.6.8).

A.9.9.6.4.1.1: The design strengths of anchors in tension or shear, except as required in Section A.9.9.6.3.3, shall satisfy:

$$\phi N_n \geq N_u \qquad \text{(Eq. A.9.9.6-1)}$$

$$\phi V_n \geq V_u \qquad \text{(Eq. A.9.9.6-2)}$$

A.9.9.6.4.1.2: When both N_u and V_u are present, interaction effects shall be considered in accordance with Section A.9.9.6.4.3.

A.9.9.6.4.1.3: In Eqs. (A.9.9.6-1) and (A.9.9.6-2), ϕN_n and ϕV_n are the lowest design strengths determined from all appropriate failure modes. ϕN_n is the lowest design strength in tension of an anchor or group of anchors as determined from consideration of ϕN_s, ϕN_{pn}, either ϕN_{sb} or ϕN_{sbg}, and either ϕN_{cb} or ϕN_{cbg}. ϕV_n is the lowest design strength in shear of an anchor or a group of anchors as determined from consideration of ϕV_s, either ϕV_{cb} or ϕV_{cbg}, and ϕV_{cp}.

A.9.9.6.4.2

The nominal strength for any anchor or group of anchors shall be based on design models that result in predictions of strength in substantial agreement with results of comprehensive tests and that account for size effects. The materials used in the tests shall be compatible with the materials to be used in the structure. The nominal strength shall be based on the 5% fractile of the basic individual anchor strength, with modifications made for the number or anchors, the effects of close spacing of anchors, proximity to edges, depth of the concrete member, eccentric load-

ings of anchor groups, and presence or absence of cracking. Limits on edge distances and anchor spacing in the design models shall be consistent with the tests that verified the model.

A.9.9.6.4.2.1: The effect of supplementary reinforcement provided to confine or restrain the concrete breakout, or both, shall be permitted to be included in the design models of Section A.9.9.6.4.2.

A.9.9.6.4.2.2: For anchors with diameters not exceeding 2 in., and tensile embedments not exceeding 25 in. in depth, the concrete breakout strength requirements of Section A.9.9.6.4.2 shall be considered satisfied by the design procedure of Sections A.9.9.6.5.2 and A.9.9.6.6.2.

A.9.9.6.4.3

Resistance to combined tensile and shear loads shall be considered in design using an interaction expression that results in computation of strength in substantial agreement with results of comprehensive tests. This requirement shall be considered satisfied by Section A.9.9.6.7.

A.9.9.6.4.4

Strength reduction factor ϕ for fastening to concrete shall be as follows when the load combinations of Section 9.2 of Ref. 9.9-1 are used:

a) Anchor governed by tensile or shear strength of a ductile steel element 0.90

b) Anchor governed by tensile or shear strength of a brittle steel element 0.75

c) Anchor governed by concrete breakout, blowout, pullout or pryout strength

	Condition A	Condition B
i) Shear Loads.......	0.85	0.75
ii) Tension Loads....	0.85	0.75

Condition A applies where the potential concrete failure surfaces are crossed by supplementary reinforcement proportioned to tie the potential concrete failure prism into the structural member. Condition B applies where such supplementary reinforcement is not provided or where pullout or pryout strength governs.

A.9.9.6.4.5

Strength reduction factor ϕ for fastening to concrete shall be as follows when using the load combinations of Section 2.3 of this Standard:

a) Anchor governed by tensile or shear
   strength of a ductile steel element 0.80
b) Anchor governed by tensile or shear
   strength of a brittle steel element 0.70
c) Anchor governed by concrete breakout,
   blowout, pullout or pryout strength

	Condition A	Condition B
i) Shear Loads.......	0.75	0.70
ii) Tension Loads....	0.75	0.70

Condition A applies where the potential concrete failure surfaces are crossed by supplementary reinforcement proportioned to tie the potential concrete failure prism into the structural member. Condition B applies where such supplementary reinforcement is not provided or where pullout or pryout strength governs.

A.9.9.6.5 Design Requirements for Tensile Loading

A.9.9.6.5.1 Steel strength of anchor in tension

A.9.9.6.5.1.1: The nominal strength of an anchor in tension as governed by the steel, N_s, shall be determined by calculations based on the properties of the anchor material and the physical dimensions of the anchor. Alternatively, it shall be permitted to use values based on the 5% fractile of test results to establish values of N_s.

A.9.9.6.5.1.2: Unless determined by the 5% fractile of test results, nominal strength of an anchor or group of anchors in tension shall not exceed:

For anchor material with a well defined yield point:

$$N_s = nA_{se}f_y \qquad \text{(Eq. A.9.9.6-3)}$$

for anchor material without a well defined yield point where f_{ut} shall not be taken greater than 125,000 psi:

$$N_s = nA_{se}(0.8f_{ut}) \qquad \text{(Eq. A.9.9.6-4)}$$

A.9.9.6.5.2 Concrete breakout strength of anchor in tension

A.9.9.6.5.2.1: Unless determined in accordance with Section A.9.9.6.4.2, nominal concrete breakout strength of an anchor or group of anchors in tension shall not exceed:

(a) for an anchor:

$$N_{cb} = \frac{A_n}{A_{No}}\Psi_2\Psi_3 N_b \qquad \text{(Eq. A.9.9.6-5a)}$$

(b) for a group of anchors:

$$N_{cbg} = \frac{A_N}{A_{No}}\Psi_1\Psi_2\Psi_3 N_b \qquad \text{(Eq. A.9.9.6-5b)}$$

N_b is the basic concrete breakout strength value for a single anchor in tension in cracked concrete. A_N is the projected area of the failure surface for the anchor or group of anchors that shall be approximated as the base of the rectilinear geometrical figure that results from projecting the failure surface outward $1.5h_{ef}$ from the centerlines of the anchor, or in the case of a group of anchors, from a line through a row of adjacent anchors. A_N shall not exceed nA_{No}, where n is the number of tensioned anchors in the group. A_{No} is the projected area of the failure surface of a single anchor remote from edges:

$$A_{No} = 9h_{ef}^2 \qquad \text{(Eq. A.9.9.6-6)}$$

A.9.9.6.5.2.2: Unless determined in accordance with Section A.9.9.6.4.2, the basic concrete breakout strength of a single anchor in tension in cracked concrete shall not exceed:

$$N_b = k\lambda\sqrt{f'_c}h_{ef}^{1.5} \qquad \text{(Eq. A.9.9.6-7a)}$$

where $k = 24$. Alternatively, for headed studs and headed bolts with 11 in. $< h_{ef} < 25$ in., the basic concrete breakout strength of a single anchor in tension in cracked concrete shall not exceed:

$$N_b = k\sqrt{f'_c}h_{ef}^{5/3} \qquad \text{(Eq. A.9.9.6-7b)}$$

where $k = 16$.

A.9.9.6.5.2.3: For the special case of anchors in an application with three or four edges and the largest edge distance $c_{max} < 1.5h_{ef}$, the embedment depth h_{ef} used in Eqs. (A.9.9.6-6), (A.9.9.6-7), (A.9.9.6-8) and (A.9.9.6-9) shall be limited to $c_{max}/1.5$.

A.9.9.6.5.2.4: The modification factor for eccentrically loaded anchor groups where $e'_N < s/2$ is:

$$\Psi_1 = \frac{1}{\left(1 + \dfrac{2e_{N'}}{3h_{ef}}\right)} \leq 1 \qquad \text{(Eq. A.9.9.6-8)}$$

If the loading on an anchor group is such that only some anchors are in tension, only those anchors that are in tension shall be considered when determining the eccentricity, e_N', for use in Eq. (A.9.9.6-8).

In the case where eccentric loading exists about two axes, the modification factor, Ψ_1, shall be computed for each axis individually and the product of these factors used as Ψ_1 in Eq. (A.9.9.6-5).

A.9.9.6.5.2.5: The modification factor for edge effects is:

$$\Psi_2 = 1 \quad \text{if} \quad c_1 \geq 1.5h_{ef} \qquad \text{(Eq. A.9.9.6-9a)}$$

$$\Psi_2 = 0.7 + 0.3 \frac{c_{min}}{1.5h_{ef}} \quad \text{if} \quad c_{min} < 1.5h_{ef}$$
$$\text{(Eq. A.9.9.6-9b)}$$

A.9.9.6.5.2.6: When an anchor is located in a region of a concrete member where analysis indicates no cracking ($f_t < f_r$) at service load levels, the following modification factor shall be permitted:

$$\Psi_3 = 1.25$$

A.9.9.6.5.2.7: When analysis indicates cracking at service load levels, Ψ_3 shall be taken as 1.0. The cracking in the concrete shall be controlled by flexural reinforcement distributed in accordance with Section 10.6.4 of Ref. 9.9-1, or equivalent crack control shall be provided by confining reinforcement.

A.9.9.6.5.2.8: When an additional plate or washer is added under the head of the anchor, it shall be permitted to calculate the projected area of the failure surface by projecting the failure surface outward $1.5h_{ef}$ from the effective perimeter of the plate or washer. The effective perimeter shall not exceed the value at a section projected outward more than t from the outer edge of the head of anchor, where t is the thickness of the washer or plate.

A.9.9.6.5.3 Pullout strength of anchor in tension

A.9.9.6.5.3.1: Unless determined in accordance with Section A.9.9.6.4.2, the nominal pullout strength of an anchor in tension shall not exceed:

$$N_{pn} = \Psi_4 N_p \qquad \text{(Eq. A.9.9.6-10)}$$

A.9.9.6.5.3.2: For single headed studs and headed bolts, it shall be permitted to calculate the pullout strength in tension using Section A.9.9.6.5.3.3. For single J-bolts or L-bolts, it shall be permitted to calculate the pullout strength in tension using Section A.9.9.6.5.3.4. Alternatively, it shall be permitted to use values of N_p based on the 5% fractile of tensile tests performed in such a manner as to exclude the benefit of friction.

A.9.9.6.5.3.3: Unless determined in accordance with Section A.9.9.6.4.2, the pullout strength in tension of a single headed stud or headed bolt, N_p for use in Eq. (A.9.9.6-10), shall not exceed

$$N_p = A_b 8 f_c' \qquad \text{(Eq. A.9.9.6-11a)}$$

A.9.9.6.5.3.4: Unless determined in accordance with Section A.9.9.6.4.2, the pullout strength in tension of a single J-bolt or L-bolt, N_p for use in Eq. (A.9.9.6-10) shall not exceed:

$$N_p = 0.9 f_c' e_h d_o \qquad \text{(Eq. A.9.9.6-11b)}$$

where $3d_o < e_h < 4.5d_o$.

A.9.9.6.5.3.5: For an anchor located in a region of a concrete member where analysis indicates no cracking ($f_t < f_r$) at service load levels, the following modification factor shall be permitted:

$$\Psi_4 = 1.4$$

Otherwise, Ψ_4 shall be taken as 1.0.

A.9.9.6.5.4 Concrete side-face blowout strength of a headed anchor in tension

A.9.9.6.5.4.1: For a single headed anchor with deep embedment close to an edge ($c < 0.4h_{ef}$), unless determined in accordance with Section A.9.9.6.4.2, the nominal side-face blowout strength N_{sb} shall not exceed:

$$N_{sb} = 160c\sqrt{A_b}\sqrt{f_c'} \qquad \text{(Eq. A.9.9.6-12)}$$

If the single headed anchor is located at a perpendicular distance c_2 less than $3c$ from an edge, the value of N_{sb} shall be modified by multiplying it by the factor $(1 + c_2/c)/4$ where $1 < c_2/c < 3$.

A.9.9.6.5.4.2: For multiple headed anchors with deep embedment close to an edge ($c < 0.4h_{ef}$) and spacing between anchors less than $6c$, unless determined in accordance with Section A.9.9.6.4.2, the nominal strength of the group of anchors for a side-face blow-out failure N_{sbg} shall not exceed:

$$N_{sbg} = \left(1 + \frac{s_o}{6c}\right) N_{sb} \quad \text{(Eq. A.9.9.6-13)}$$

where s_o = spacing of the outer anchors along the edge in the group and N_{sb} is obtained from Eq. (A.9.9.6-12) without modification for a perpendicular edge distance.

A.9.9.6.6 Design Requirements for Shear Loading

A.9.9.6.6.1 Steel strength of anchor in shear

A.9.9.6.6.1.1: The nominal strength of an anchor in shear as governed by steel, V_s, shall be determined by calculations based on the properties of the anchor material and the physical dimensions of the anchor. Alternatively, it shall be permitted to use values based on the 5% fractile of test results to establish values of V_s.

A.9.9.6.6.1.2: Unless determined by the 5% fractile of test results, nominal strength of an anchor or group of anchors in shear shall not exceed:

(a) for anchors with a well defined yield point:

$$V_s = nA_{se}f_y \quad \text{(Eq. A.9.9.6-14)}$$

(b) for anchors without a well defined yield point:

$$V_s = n0.6A_{se}f_{ut} \quad \text{(Eq. A.9.9.6-15)}$$

where f_{ut} shall not be taken greater than 125,000 psi

A.9.9.6.1.3: Where anchors are used with built-up grout pads, the nominal strengths of Section A.9.9.6.6.1.2 shall be reduced by 20%.

A.9.9.6.6.2 Concrete breakout strength of anchor in shear

A.9.9.6.6.2.1: Unless determined in accordance with Section A.9.9.6.4.2, nominal concrete breakout strength in shear of an anchor or group of anchors shall not exceed:

(a) for shear force perpendicular to the edge on a single anchor:

$$V_{cb} = \frac{A_V}{A_{Vo}} \Psi_6 \Psi_7 V_b \quad \text{(Eq. A.9.9.6-16a)}$$

(b) for shear force perpendicular to the edge on a group of anchors:

$$V_{cbg} = \frac{A_V}{A_{Vo}} \Psi_5 \Psi_6 \Psi_7 V_b \quad \text{(Eq. A.9.9.6-16b)}$$

(d) for shear force parallel to an edge, V_{cb} or V_{cbg} shall be permitted to be twice the value for shear force determined from Eq. (A.9.9.6-16a or b) respectively with Ψ_6 taken equal to 1.

(e) For anchors located at a corner, the limiting nominal concrete breakout strength shall be determined for each edge and the minimum value shall be used.

V_b is the basic concrete breakout strength value for a single anchor. A_v is the projected area of the failure surface on the side of the concrete member at its edge for a single anchor or a group of anchors. It shall be permitted to evaluate this area as the base of a truncated half pyramid projected on the side face of the member where the top of the half pyramid is given by the axis of the anchor row selected as critical. The value of c_1 shall be taken as the distance from the edge to this axis. A_v shall not exceed nA_{vo}, where n is the number of anchors in the group.

A_{vo} is the projected area for a single anchor in a deep member and remote from edges in the direction perpendicular to the shear force. It shall be permitted to evaluate this area as the base of a half pyramid with a side length parallel to the edge of $3c_1$ and a depth of $1.5c_1$:

$$A_{vo} = 4.5c_1^2 \quad \text{(Eq. A.9.9.6-17)}$$

Where anchors are located at varying distances from the edge and the anchors are welded to the attachment so as to distribute the force to all anchors, it shall be permitted to evaluate the strength based on the distance to the farthest row of anchors from the edge. In this case, it shall be permitted to base the value of c_1 on the distance from the edge to the axis of the farthest anchor row which is selected as critical, and all of the shear shall be assumed to be carried by this critical anchor row alone.

A.9.9.6.6.2.2: Unless determined in accordance with Section A.9.9.6.4.2, the basic concrete breakout strength in shear of a single anchor in cracked concrete shall not exceed:

$$V_b = 7 \left(\frac{\ell}{d_o} \right)^{0.2} \sqrt{d_o} \sqrt{f'_c}\, c_1^{1.5} \quad \text{(Eq. A.9.9.6-18a)}$$

A.9.9.6.6.2.3: For anchors that are rigidly welded to steel attachments having a minimum thickness equal to the greater of 3/8 in. or half of the anchor diameter, unless determined in accordance with Section A.9.9.6.4.2, the basic concrete breakout strength in shear of a single anchor in cracked concrete shall not exceed:

$$V_b = 7 \left(\frac{\ell}{d_o} \right)^{0.2} \sqrt{d_o}\,\lambda\sqrt{f'_c}\, c_1^{1.5} \quad \text{(Eq. A.9.9.6-18b)}$$

provided that:

a) for groups of anchors, the strength is determined based on the strength of the row of anchors farthest from the edge;
b) the center-to-center spacing of the anchors is not less than 2.5 in.;
c) supplementary reinforcement is provided at the corners if $c_2 \leq 1.5h_{ef}$.

A.9.9.6.6.2.4: For the special case of anchors in a thin member influenced by three or more edges, the edge distance c_1 used in Eqs. (A.9.9.6-17), (A.9.9.6-18), (A.9.9.6-19) and (A.9.9.6-20) shall be limited to $h/1.5$.

A.9.9.6.6.2.5: The modification factor for eccentrically loaded anchor groups where $e'_v \leq s/2$ is:

$$\Psi_4 = \frac{1}{\left(1 + \dfrac{2e_v'}{3c_1} \right)} \leq 1 \quad \text{(Eq. A.9.9.6-19)}$$

A.9.9.6.6.2.6: The modification factor for edge effects is:

$$\Psi_6 = 1 \quad \text{if} \quad c_2 \geq 1.5c_1 \quad \text{(Eq. A.9.9.6-20a)}$$

$$\Psi_6 = 0.7 + 0.3\, \frac{c_2}{1.5c_1} \quad \text{if} \quad c_2 < 1.5c_1$$

$$\text{(Eq. A.9.9.6-20b)}$$

A.9.9.6.6.2.7: For anchors located in a region of a concrete member where analysis indicates no cracking ($f_t < f_r$) at service loads, the following modification factor shall be permitted:

$$\Psi = 1.4$$

For anchors located in a region of a concrete member where analysis indicates cracking at service load levels, the following modification factors shall be permitted. In order to be considered as edge reinforcement, the reinforcement shall be designed to intersect the concrete breakout:

$$\Psi_7 = 1.0$$

for anchors in cracked concrete with no edge reinforcement or edge reinforcement smaller than a #4 bar.

$$\Psi_7 = 1.2$$

for anchors in cracked concrete with edge reinforcement of a #4 bar or greater between the anchor and the edge.

$$\Psi_7 = 1.4$$

for anchors in cracked concrete with edge reinforcement of a #4 bar or greater between the anchor and the edge and with the edge reinforcement enclosed within stirrups spaced at not more than 4 in.

A.9.9.6.6.3 Concrete pryout strength of anchor in shear

A.9.9.6.6.3.1: Unless determined in accordance with Section A.9.9.6.4.2, the nominal pryout strength, V_{cp}, shall not exceed:

$$V_{cp} = k_{cp}N_{cb} \quad \text{(Eq. A.9.9.6-21)}$$

where

$$K_{cp} = 1.0 \text{ for } h_{ef} < 2.5 \text{ in.}$$

$$k_{cp} = 2.0 \text{ for } h_{ef} > 2.5 \text{ in.}$$

and N_{cb} shall be determined from Eq. (A.9.9.6-5a), lb.

A.9.9.6.7 Interaction of Tensile and Shear Forces

Unless determined in accordance with Section A.9.9.6.4.3, anchors or groups of anchors that are subjected to both shear and tension shall be designed to satisfy the requirements of Sections A.9.9.6.7.1 through A.9.9.6.7.3. The value of ϕN_n shall be the smallest of the steel strength of the anchor in tension, concrete breakout strength of anchor in tension, pull-out strength of anchor in tension, and side-face blow-out strength. The value of ϕV_n shall be the smallest of the steel strength of anchor in shear, the concrete breakout strength of anchor in shear, and the pryout strength.

A.9.9.6.7.1

If $V_u \le 0.2\phi V_n$, then full design strength in tension shall be permitted: $\phi N_n > N_u$.

A.9.9.6.7.2

If $N_u < 0.2\phi N_n$, then full design strength in shear shall be permitted: $\phi V_n > V_u$.

A.9.9.6.7.3

If $V_u > 0.2\phi V_n$ and $N_u > 0.2\phi N_v$, then:

$$\frac{N_u}{\phi N_n} + \frac{V_u}{V_n} \le 1.2 \quad \text{(Eq. A.9.9.6-22)}$$

A.9.9.6.8 Required Edge Distances, and Spacings to Preclude Splitting Failure

Minimum spacings and edge distances for anchors shall conform to Sections A.9.9.6.8.1 and A.9.9.6.8.2, unless reinforcement is provided to control splitting.

A.9.9.6.8.1

Unless determined in accordance with Section A.9.9.6.8.2, minimum edge distances for headed anchors that will not be torqued shall be based on minimum cover requirements for reinforcement in accordance with Section 7.7 of Ref. 9.9-1. For headed anchors that will be torqued, the minimum edge distances shall be $6d_o$.

A.9.9.6.8.2

For anchors that will remain untorqued, if the edge distance or spacing is less than that specified in Section A.9.9.6.8.1, calculations shall be performed using a fictitious value of d_o that meets the require-ments of Section A.9.9.6.8.1. Calculated forces applied to the anchor shall be limited to the values corresponding to an anchor having that fictitious diameter.

A.9.9.6.8.3

Construction documents shall specify use of anchors with a minimum edge distance as assumed in design.

A.9.9.6.9 Installation of Anchors

A.9.9.6.9.1

Anchors shall be installed in accordance with the construction documents.

A.9.11 SUPPLEMENTARY PROVISIONS FOR MASONRY

A.9.11.1

For the purposes of design of masonry structures using the earthquake loads given in ASCE 7, several amendments to the reference standard are necessary.

A.9.11.2

The references to "Seismic Performance Category" in Section 1.13 and elsewhere of the reference standard shall be replaced by "Seismic Design Category," and the rules given for Category E also apply to Category F.

A.9.11.3

The anchorage forces given in Section 1.13.3.2 of the reference standard shall not be interpreted to replace the anchorage forces given in Section 9 of this standard.

A.9.11.4

To qualify for the R factors given in Section 9 of this Standard, the requirements of the reference standard shall be satisfied and amended as follows:

Ordinary and detailed plain masonry shear walls shall be designed according to Sections 2.1 and 2.2 of the reference standard. Detailed plain masonry shear walls shall be reinforced as a minimum as required in Section 1.13.5.3.3 of the reference Standard.

Ordinary, intermediate and special reinforced masonry shear walls shall be designed according to Sections 2.1 and 2.3 of the reference standard. Ordinary and intermediate reinforced masonry shear walls shall be reinforced as a minimum as required in Section 1.13.5.3.3 of the reference standard. In addition, intermediate reinforced masonry shear walls shall have vertical reinforcing bars spaced no further apart than 48 in. Special reinforced masonry shear walls shall be reinforced as a minimum as required in Sections 1.13.6.3 and 1.13.6.4 of the reference standard.

Special reinforced masonry shear walls shall not have a ratio of reinforcement, ρ, that exceeds that given by either Method A or B below:

A.9.11.4.1 Method A

Method A is permitted to be used where the story drift does not exceed $0.010h_{sx}$ as given in Table 9.5.2.8 and if the extreme compressive fiber strains are less than 0.0035 for clay masonry and 0.0025 for concrete masonry.

1. When walls are subjected to in-plane forces, and for columns and beams, the critical strain condition corresponds to a strain in the extreme tension reinforcement equal to 5 times the strain associated with the reinforcement yield stress, f_y.
2. When walls subjected to out-of-plane forces, the critical strain condition corresponds to a strain in the reinforcement equal to 1.3 times the strain associated with reinforcement yield stress, f_y.

The strain at the extreme compression fiber shall be assumed to be 0.0035 for clay masonry and 0.0025 for concrete masonry.

The calculation of the maximum reinforcement ratio shall include factored gravity axial loads. The stress in the tension reinforcement shall be assumed to be $1.25f_y$. Tension in the masonry shall be neglected. The strength of the compressive zone shall be calculated as 80% of the area of the compressive zone. Stress in reinforcement in the compression zones shall be based on a linear strain distribution.

A.9.11.4.2 Method B

Method B is permitted to be used where story drift does not exceed $0.013h_{sx}$ as given in Table 9.5.2.8.

1. Boundary members shall be provided at the boundaries of shear walls when the compressive strains in the wall exceed 0.002. The strain shall be determined using factored forces and R equal to 1.5.
2. The minimum length of the boundary member shall be three times the thickness of the wall, but shall include all areas where the compressive strain per Section A.9.11.4.2 Item 1, is greater than 0.002.
3. Lateral reinforcement shall be provided for the boundary elements. The lateral reinforcement shall be a minimum of #3 closed ties at a maximum spacing of 8 in. (203 mm) on center within the grouted core, or equivalent approved confinement, to develop an ultimate compressive strain of at least 0.006.
4. The maximum longitudinal reinforcement ratio shall not exceed $0.15f'_m/f_y$.

A.9.11.5

Where allowable stress design is used for load combinations including earthquake, load combinations 3 and 5 of Section 2.4.1 of this standard shall replace combinations (c) and (3) of Section 2.1.1.1.1 of the reference standard. If the increase in stress given in Section 2.1.1.1.3 of the reference standard is used, the restriction on load reduction in Section 2.4.3 of this Standard shall be observed.

A.9.12 SUPPLEMENTARY PROVISIONS FOR WOOD

The conventional light-frame construction braced wall requirements and the conventional construction braced wall requirements specified in Ref. 9.12-8 shall be revised as specified in Tables A.9.12.10.1-1a and 1b. In addition, where Ref. 9.12-8 permits the use of nails in fastening wood diaphragms and wood structural panels, the use of staples shall be permitted if data on factored shear resistance of staples in such application is submitted to and approved by the authority having jurisdiction.

TABLE A.9.12.10.1-1a. Conventional Light-Frame Construction Braced Wall Requirements

Seismic Design Category	Maximum Distance Between Braced Walls (ft)	Maximum Number of Stories Permitted[a]	
		Detached One and Two Family Dwellings	All Other Applications
A[b]	35	3	3
B	35	3	3
C	35	3	2
D	25	2	1
E	25	2	1
E (Seismic Use Group II)	Conventional construction not permitted; conformance with Section 9.11 required		
F			

Note: For conversion to metric, 1 ft = 305 mm.
[a]A cripple wall which exceeds 14 in. in height is considered to be a story above grade; maximum bearing wall height shall not exceed 10 ft.
[b]See exceptions to Section 1.2.

TABLE A.9.12.10.1-1b. Conventional Construction Braced Wall Requirements in Minimum Length of Wall Bracing per 25 Linear Feet of Braced Wall Line[a]

Story Location	Sheathing Type[b]	$0.50g \leq S_{DS} < 0.75g$	$0.75g \leq S_{DS} < 1.0g$[c]	$1.0g \leq S_{DS}$
Top or only story above grade	G-P[d]	14 ft 8 in.	18 ft 8 in.[c]	25 ft 0 in.
	S-W	8 ft 0 in.	9 ft 4 in.[c]	12 ft 0 in.
Story below top story above grade	G-P[d]	NP	NP	NP
	S-W	13 ft 4 in.[c]	17 ft 4 in.[c]	21 ft 4 in.
Bottom story of 3 stories above grade	G-P[d]	Conventional construction not permitted; conformance with Section 9.11 required		
	S-W			

Note: For conversion to metric, 1 in. = 25.4 mm and 1 ft = 304.8 mm.
[a]Minimum length of panel bracing of one face of wall for S-W sheathing or both faces of wall for G-P sheathing; h/w ratio shall not exceed 2/1. For S-W panel bracing of the same material on two faces of the wall, the minimum length is one half the tabulated value but the h/w ratio shall not exceed 2/1 and design for uplift is required if the length is less than the tabulated value.
[b]G-P = gypsumboard, fiberboard, particleboard, lath and plaster, or gypsum sheathing boards; S-W = structural-use panels and diagonal wood sheathing. NP = not permitted.
[c]Applies to one- and two-family detached dwellings only.
[d]Nailing shall be as follows:
  For 1/2-in. gypsum board, 5d (0.086 in. diameter) coolers at 7 in. centers;
  For 5/8-in. gypsum board, 6d (0.092 in. diameter) at 7 in. centers;
  For gypsum sheathing board, 1-3/4-in. long by 7/16-in. head, diamond point galvanized at 4 in. centers;
  For gypsum lath, No. 13 gauge (0.092 in.) by 1-1/8-in. long, 19/64 in. head, plasterboard at 5 in. centers;
  For Portland cement plaster, No. 11 gauge (0.120 in.) by 1-1/2-in. long, 7/16 head at 6 in. centers;
  For fiberboard and particleboard, No. 11 gauge by 1-1/2-in. long, 7/16 head, galvanized at 3 in. centers.
  For structural wood sheathing, the minimum nail size and maximum spacing shall be in accordance with the minimum nails size and maximum spacing allowed for each thickness sheathing in Table A.9.12.9.1-1b.
Nailing as specified above shall occur at all panel edges at studs, at top and bottom plates, and, where occurring, at blocking.

APPENDIX B

B.0 SERVICEABILITY CONSIDERATIONS

This Appendix is not a mandatory part of the standard, but provides guidance for design for serviceability in order to maintain the function of a building and the comfort of its occupants during normal usage. Serviceability limits (e.g., maximum static deformations, accelerations, etc.) shall be chosen with due regard to the intended function of the structure.

Serviceability shall be checked using appropriate loads for the limit state being considered.

B.1 DEFLECTION, VIBRATION AND DRIFT

B.1.1 Vertical Deflections
Deformations of floor and roof members and systems due to service loads shall not impair the serviceability of the structure.

B.1.2 Drift of Walls and Frames
Lateral deflection or drift of structures and deformation of horizontal diaphragms and bracing systems due to wind effects shall not impair the serviceability of the structure.

B.1.3 Vibrations
Floor systems supporting large open areas free of partitions or other sources of damping, where vibration due to pedestrian traffic might be objectionable, shall be designed with due regard for such vibration.

Mechanical equipment that can produce objectionable vibrations in any portion of an inhabited structure shall be isolated to minimize the transmission of such vibrations to the structure.

Building structural systems shall be designed so that wind-induced vibrations do not cause occupant discomfort or damage to the building, its appurtenances or contents.

B.2 DESIGN FOR LONG-TERM DEFLECTION

Where required for acceptable building performance, members and systems shall be designed to accommodate long-term irreversible deflections under sustained load.

B.3 CAMBER

Special camber requirements that are necessary to bring a loaded member into proper relations with the work of other trades shall be set forth in the design documents.

Beams detailed without specified camber shall be positioned during erection so that any minor camber is upward. If camber involves the erection of any member under preload, this shall be noted in the design documents.

B.4 EXPANSION AND CONTRACTION

Dimensional changes in a structure and its elements due to variations in temperature, relative humidity, or other effects shall not impair the serviceability of the structure.

Provision shall be made either to control crack widths or to limit cracking by providing relief joints.

B.5 DURABILITY

Buildings and other structures shall be designed to tolerate long-term environmental effects or shall be protected against such effects.

(This Commentary is not a part of American Society of Civil Engineers Standard *Minimum Design Loads for Buildings and Other Structures*. It is included for information purposes.)

Commentary on American Society of Civil Engineers Standard ASCE 7-98

CONTENTS

COMMENTARY

This Commentary is not a part of the ASCE Standard *Minimum Design Loads for Buildings and Other Structures*. It is included for information purposes.

This Commentary consists of explanatory and supplementary material designed to assist local building code committees and regulatory authorities in applying the recommended requirements. In some cases it will be necessary to adjust specific values in the standard to local conditions; in others, a considerable amount of detailed information is needed to put the provisions into effect. This Commentary provides a place for supplying material that can be used in these situations and is intended to create a better understanding of the recommended requirements through brief explanations of the reasoning employed in arriving at them.

The sections of the commentary are numbered to correspond to the sections of the standard to which they refer. Since it is not necessary to have supplementary material for every section in the standard, there are gaps in the numbering in the Commentary.

C1.0 GENERAL

C1.1 SCOPE

The minimum load requirements contained in this standard are derived from research and service performance of buildings and other structures. The user of this standard, however, must exercise judgement when applying the requirements to "other structures." Loads for some structures other than buildings may be found in Sections 3 to 10 of this standard and additional guidance may be found in the commentary.

Both loads and load combinations are set forth in this document with the intent that they be used together. If one were to use loads from some other source with the load combinations set forth herein or vice versa, the reliability of the resulting design may be affected.

Earthquake loads contained herein are developed for structures that possess certain qualities of ductility and postelastic energy dissipation capability. For this reason, provisions for design, detailing, and construction are provided in Appendix A. In some cases,

these provisions modify or add to provisions contained in design specifications.

C1.3 BASIC REQUIREMENTS

C1.3.1 Strength

Buildings and other structures must satisfy strength limit states in which members are proportioned to carry the design loads safely to resist buckling, yielding, fracture, etc. It is expected that other standards produced under consensus procedures and intended for use in connection with building code requirements will contain recommendations for resistance factors for strength design methods or allowable stresses (or safety factors) for allowable stress design methods.

C1.3.2 Serviceability

In addition to strength limit states, buildings and other structures must also satisfy serviceability limit states which define functional performance and behavior under load and include such items as deflection and vibration. In the United States, strength limit states have traditionally been specified in building codes because they control the safety of the structure. Serviceability limit states, on the other hand, are usually non-catastrophic, define a level of quality of the structure or element and are a matter of judgement as to their application. Serviceability limit states involve the perceptions and expectations of the owner or user and are a contractual matter between the owner or user and the designer and builder. It is for these reasons, and because the benefits themselves are often subjective and difficult to define or quantify, that serviceability limit states for the most part are not included within the three model U.S. Building Codes. The fact that serviceability limit states are usually not codified should not diminish their importance. Exceeding a serviceability limit state in a building or other structure usually means that its function is disrupted or impaired because of local minor damage or deterioration or because of occupant discomfort or annoyance.

C1.3.3 Self-Straining Forces

Constrained structures that experience dimensional changes develop self-straining forces. Examples include moments in rigid frames that undergo differential foundation settlements and shears in bearing wall which support concrete slabs that shrink. Unless provisions are made for self-straining forces, stresses in structural elements, either alone or in combination with stresses from external loads, can be high enough to cause structural distress.

In many cases, the magnitude of self-straining forces can be anticipated by analyses of expected shrinkage, temperature fluctuations, foundation movement, etc. However, it is not always practical to calculate the magnitude of self-straining forces. Designers often provide for self-straining forces by specifying relief joints, suitable framing systems, or other details to minimize the effects of self-straining forces.

This section of the standard is not intended to require the designer to provide for self-straining forces that cannot be anticipated during design. An example is settlement resulting from future adjacent excavation.

C1.4 GENERAL STRUCTURAL INTEGRITY

Through accident or misuse, properly designed structures may be subject to conditions that could lead to either general or local collapse. Except for specially designed protective systems, it is impractical for a structure to be designed to resist general collapse caused by gross misuse of a large part of the system or severe abnormal loads acting directly on a large portion of it. However, precautions can be taken in the design of structures to limit the effects of local collapse, that is, to prevent progressive collapse, which is the spread of an initial local failure from element to element resulting, eventually, in the collapse of an entire structure or a disproportionately large part of it.

Since accidents and misuse are normally unforeseeable events, they cannot be defined precisely. Likewise, general structural integrity is a quality that cannot be stated in simple terms. It is the purpose of Section 1.4 and this Commentary to direct attention to the problem of local collapse, present guidelines for handling it that will aid the design engineer, and promote consistency of treatment in all types of structures and in all construction materials.

Accidents, Misuse, and Their Consequences. In addition to unintentional or willful misuse, some of the

incidents that may cause local collapse are [C1-1][1]: explosions due to ignition of gas or industrial liquids; boiler failures; vehicle impact; impact of falling objects; effects of adjacent excavations; gross construction errors; and very high winds such as tornadoes. Generally, such abnormal events would not be ordinary design considerations.

The distinction between general collapse and limited local collapse can best be made by example.

The immediate demolition of an entire structure by a high-energy bomb is an obvious instance of general collapse. Also, the failure of one column in a one-, two-, three-, or possibly even four-column structure could precipitate general collapse, because the local failed column is a significant part of the total structure at that level. Similarly, the failure of a major bearing element in the bottom story of a two- or three-story structure might cause general collapse of the whole structure. Such collapses are beyond the scope of the provisions discussed herein. There have been numerous instances of general collapse that have occurred as the result of such abnormal events as wartime bombing, landslides, and floods.

An example of limited local collapse would be the containment of damage to adjacent bays and stories following the destruction of one or two neighboring columns in a multibay structure. The restriction of damage to portions of two or three stories of a higher structure following the failure of a section of bearing wall in one story is another example. A prominent case of local collapse that progressed to a disproportionate part of the whole building (and is thus an example of the type of failure of concern here) was the Ronan Point disaster. Ronan Point was a 22-story apartment building of large, precast-concrete, loadbearing panels in Canning Town, England. In March 1968, a gas explosion in an 18th-story apartment blew out a living room wall. The loss of the wall led to the collapse of the whole corner of the building. The apartments above the 18th story, suddenly losing support from below and being insufficiently tied and reinforced, collapsed one after the other. The falling debris ruptured successive floors and walls below the 18th story, and the failure progressed to the ground. Another example is the failure of a one-story parking garage reported in [C1-2]. Collapse of one transverse frame under a concentration of snow led to the later progressive

---

[1]Numbers in brackets refer to references listed at the ends of the major sections in which they appear (that is, Section 1, Section 2, and so forth).

collapse of the whole roof, which was supported by 20 transverse frames of the same type. Similar progressive collapses are mentioned in [C1-3].

There are a number of factors that contribute to the risk of damage propagation in modern structures [C1-4]. Among them are:

1. There can be a lack of awareness that structural integrity against collapse is important enough to be regularly considered in design.
2. In order to have more flexibility in floor plans and to keep costs down, interior walls and partitions are often non-loadbearing and hence may be unable to assist in containing damage.
3. In attempting to achieve economy in structures through greater speed of erection and less site labor, systems may be built with minimum continuity, ties between elements, and joint rigidity.
4. Unreinforced or lightly reinforced load-bearing walls in multistory structures may also have inadequate continuity, ties, and joint rigidity.
5. In roof trusses and arches there may not be sufficient strength to carry the extra loads or sufficient diaphragm action to maintain lateral stability of the adjacent members if one collapses.
6. In eliminating excessively large safety factors, code changes over the past several decades have reduced the large margin of safety inherent in many older structures. The use of higher-strength materials permitting more slender sections compounds the problem in that modern structures may be more flexible and sensitive to load variations and, in addition, may be more sensitive to construction errors.

Experience has demonstrated that the principle of taking precautions in design to limit the effects of local collapse is realistic and can be satisfied economically. From a public-safety viewpoint it is reasonable to expect all multistory structures to possess general structural integrity comparable to that of properly designed, conventional framed structures [C1-4, C1-5].

Design Alternatives. There are a number of ways to obtain resistance to progressive collapse. In [C1-6], a distinction is made between direct and indirect design, and the following approaches are defined:

Direct design: explicit consideration of resistance to progressive collapse during the design process through either:

alternate path method: a method that allows local failure to occur but seeks to provide alternate load paths so that the damage is absorbed and major collapse is averted, or

specific local resistance method: a method that seeks to provide sufficient strength to resist failure from accidents or misuse.

Indirect design: implicit consideration of resistance to progressive collapse during the design process through the provision of minimum levels of strength, continuity, and ductility.

The general structural integrity of a structure may be tested by analysis to ascertain whether alternate paths around hypothetically collapsed regions exist. Alternatively, alternate path studies may be used as guides for developing rules for the minimum levels of continuity and ductility needed in applying the indirect design approach to ensuring general structural integrity. Specific local resistance may be provided in regions of high risk, since it may be necessary for some elements to have sufficient strength to resist abnormal loads in order for the structure as a whole to develop alternate paths. Specific suggestions for the implementation of each of the defined methods are contained in [C1-6].

Guidelines for the Provision of General Structural Integrity. Generally, connections between structural components should be ductile and have a capacity for relatively large deformations and energy absorption under the effect of abnormal conditions. This criterion is met in many different ways, depending on the structural system used. Details that are appropriate for resistance to moderate wind loads and seismic loads often provide sufficient ductility.

Work with large precast panel structures [C1-7, C1-8, C1-9] provides an example of how to cope with the problem of general structural integrity in a building system that is inherently discontinuous. The provision of ties combined with careful detailing of connections can overcome difficulties associated with such a system. The same kind of methodology and design philosophy can be applied to other systems [C1-10]. The ACI Building Code Requirements for Reinforced Concrete [C1-11] includes such requirements in Section 7.13.

There are a number of ways of designing for the required integrity to carry loads around severely damaged walls, trusses, beams, columns, and floors. A few examples of design concepts and details relating particularly, but not solely, to precast and bearing-wall structures are:

1. Good Plan Layout. An important factor in achieving integrity is the proper plan layout of walls (and columns). In bearing-wall structures there should be an arrangement of interior longitudinal walls to support and reduce the span of long sections of crosswall, thus enhancing the stability of individual walls and of the structures as a whole. In the case of local failure this will also decrease the length of wall likely to be affected.

2. Returns on Walls. Returns on interior and exterior walls will make them more stable.

3. Changing Directions of Span of Floor Slab. Where a floor slab is reinforced in order that it can, with a low safety factor, span in another direction if a load-bearing wall is removed, the collapse of the slab will be prevented and the debris loading of other parts of the structure will be minimized. Often, shrinkage and temperature steel will be enough to enable the slab to span in a new direction.

4. Load-Bearing Interior Partitions. The interior walls must be capable of carrying enough load to achieve the change of span direction in the floor slabs.

5. Catenary Action of Floor Slab. Where the slab cannot change span direction, the span will increase if an intermediate supporting wall is removed. In this case, if there is enough reinforcement throughout the slab and enough continuity and restraint, the slab may be capable of carrying the loads by catenary action, though very large deflections will result.

6. Beam Action of Walls. Walls may be assumed to be capable of spanning an opening if sufficient tying steel at the top and bottom of the walls allows them to act as the web of a beam with the slabs above and below acting as flanges (see [C1-7]).

C1.5 CLASSIFICATION OF BUILDINGS AND OTHER STRUCTURES

The categories in Table 1-1 are used to relate the criteria for maximum environmental loads or distortions specified in this standard to the consequence of the loads being exceeded for the structure and its occupants. The category numbering is unchanged from that in the previous edition of the standard (ASCE 7-95). Classification continues to reflect a progression of the anticipated seriousness of the consequence of failure from lowest hazard to human life (Category I) to highest (Category IV).

In Sections 6 and 7, importance factors are presented for the four categories identified. The specific importance factors differ according to the statistical characteristics of the environmental loads and the manner in which the structure responds to the loads. The principle of requiring more stringent loading criteria for situations in which the consequence of failure may be severe has been recognized in previous versions of this standard by the specification of mean recurrence interval maps for wind speed and ground snow load.

This section now recognizes that there may be situations when it is acceptable to assign multiple categories to a structure based on the expected occupancy of the structure during extreme loading events. For instance, a structure that is to be used as a hurricane shelter and has no expected use as a shelter during other extreme events, such as an earthquake, need not be designed for loads consistent with an earthquake shelter. In this case, the structure would be classified in Category IV for wind design and in Category II for seismic design.

Category I contains buildings and other structures that represent a low hazard to human life in the event of failure either because they have a small number of occupants or have a limited period of exposure to extreme environmental loadings. Category II contains all occupancies other than those in Categories I, III and IV and are sometimes referred to as "ordinary" for the purpose of risk exposure. Category III contains those buildings and other structures that have large numbers of occupants, are designated for public assembly or for which the occupants are restrained or otherwise restricted from movement or evacuation. Buildings and other structures in Category III therefore represent a substantial hazard to human life in the event of failure.

Buildings and other structures normally falling into Category III could be classified into Category II if means (secondary containment) are provided to contain the toxic or explosive substances in the case of a spill or to contain a blast. To qualify, secondary containment systems should be designed, installed, and operated to prevent migration of harmful quantities of toxic or explosive substances out of the system to the air, soil, ground water, or surface water at any time during the use of the structure. This requirement is not to be construed as requiring a secondary containment system to prevent a release of any toxic or explosive substance into the air. By recognizing that secondary containment shall not allow releases of "harmful" quantities of contaminants, this standard acknowledges that there are substances that might contaminate ground water but do not produce a sufficient concentration of toxic or explosive material during a vapor release to constitute a health or safety risk to the public.

If the beneficial effect of secondary containment can be negated by external forces, such as the overtopping of dike walls by flood waters, then the buildings or other structures in question should not be classified into Category II. If the secondary containment is to contain a flammable substance, then implementation of a program of emergency response and preparedness combined with an appropriate fire suppression system would be a prudent action associated with a Category II classification. In many jurisdictions, such actions are required by local fire codes.

Buildings and other structures containing toxic or explosive substances also could be classified as Category II for hurricane wind loads when mandatory procedures are used to reduce the risk of release of hazardous substances during and immediately after these predictable extreme loadings. Examples of such procedures include draining hazardous fluids from a tank when a hurricane is predicted or, conversely, filling a tank with fluid to increase its buckling and overturning resistance. As appropriate to minimize the risk of damage to structures containing toxic or explosive substances, mandatory procedures necessary for the Category II classification should include preventative measures such as the removal of objects that might become air-borne missiles in the vicinity of the structure.

Category IV contains buildings and other structures that are designated as essential facilities and are intended to remain operational in the event of extreme environmental loadings. Such occupancies include, but are not limited to hospitals, fire, rescue and other emergency response facilities. Ancillary structures required for the operation of Category IV facilities during an emergency also are included in this category.

C1.7 LOAD TESTS

No specific method of test for completed construction has been given in this standard, since it may be found advisable to vary the procedure according to conditions. Some codes require the construction to sustain a superimposed load equal to a stated multiple of the design load without evidence of serious damage. Others specify that the superimposed load shall be equal to a stated multiple of the live load plus a portion of the dead load. Limits are set on maximum deflection under load and after removal of the load. Recovery of at least three-quarters of the maximum deflection, within 24 hours after the load is removed, is a common requirement [C1-11].

REFERENCES

[C1-1] Leyendecker, E.V., Breen, J.E., Somes, N.F., and Swatta, M. Abnormal loading on buildings and progressive collapse—An annotated bibliography. Washington, D.C.: U.S. Dept. of Commerce, National Bureau of Standards. NBS BSS 67, Jan., 1976.

[C1-2] Granstrom, S., and Carlsson, M. Byggfurskningen T3: Byggnaders beteende vid overpaverkningar [The behavior of buildings at excessive loadings]. Stockholm, Sweden: Swedish Institute of Building Research, 1974.

[C1-3] Seltz-Petrash, A. "Winter roof collapses: Bad luck, bad construction, or bad design." *Civil Engineering*, 42–45, Dec., 1979.

[C1-4] Breen, J.E., Ed. Progressive collapse of building structures [summary report of a workshop held at the Univ. of Texas at Austin, Oct. 1975]. Washington, D.C.: U.S. Dept. of Housing and Urban Development. Rep. PDR-182, Sept., 1976.

[C1-5] Burnett, E.F.P. The avoidance of progressive collapse: Regulatory approaches to the problem. Washington, D.C.: U.S. Dept. of Commerce, National Bureau of Standards. NBS GCR 75-48, Oct., 1975. [Available from: National Technical Information Service, Springfield, Va.]

[C1-6] Leyendecker, E.V., and Ellingwood, B.R. Design methods for reducing the risk of progressive collapse in buildings. Washington, D.C.: U.S. Dept. of Commerce, National Bureau of Standards. NBS BSS 98, 1977.

[C1-7] Schultz, D.M., Burnett, E.F.P., and Fintel, M. A design approach to general structural integrity, design and construction of large-panel concrete structures. Washington, D.C.: U.S. Dept. of Housing and Urban Development, 1977.

[C1-8] PCI Committee on Precast Bearing Walls. "Considerations for the design of precast bearing-wall buildings to withstand abnormal loads." *J. Prestressed Concrete Institute*, 21(2), 46–69, March/April, 1976.

[C1-9] Fintel, M., and Schultz, D.M. "Structural integrity of large-panel buildings." *J. Am. Concrete Inst.*, 76(5), 583–622, May, 1979.

[C1-10] Fintel, M., and Annamalai, G. "Philosophy of structural integrity of multistory load-bearing concrete masonry structures." *Concrete Inst.*, 1(5), 27–35, May, 1979.

[C1-11] Building Code Requirements for Reinforced Concrete, ACI Standard 318–89, American Concrete Institute, Detroit, Mich., 1989.

[C1-12] NEHRP Recommended Provisions for the Development of Seismic Regulations for New Buildings, 1991 Edition, Part 1 Provisions, Building Seismic Safety Council, Washington, D.C., 1991.

C2.0 COMBINATIONS OF LOADS

Loads in this standard are intended for use with design specifications for conventional structural materials, including steel, concrete, masonry, and timber. Some of these specifications are based on allowable stress design, while others employ strength design. In the case of allowable stress design, design specifications define allowable stresses that may not be exceeded by load effects due to unfactored loads, that is, allowable stresses contain a factor of safety. In strength design, design specifications provide load factors and, in some instances, resistance factors. Structural design specifications based on limit states design have been adopted by a number of specification-writing groups. Therefore, it is desirable to include herein common load factors that are applicable to these new specifications. It is intended that these load factors be used by all material-based design specifications that adopt a strength design philosophy in conjunction with nominal resistances and resistance factors developed by individual material-specification-writing groups. Load factors given herein were developed using a first-order probabilistic analysis and a broad survey of the reliabilities inherent in contemporary design practice. References [C2-9, C2-19, C2-10] also provide guidelines for materials-specification-writing groups to aid them in developing resistance factors that are compatible, in terms of inherent reliability, with load factors and statistical information specific to each structural material.

C2.2 SYMBOLS AND NOTATION

Self-straining forces can be caused by differential settlement of foundations, ground movements from expensive soils, creep in concrete members, shrinkage in members after placement, expansion of shrinkage-compensating concrete, and changes in temperature of members during the service life of the structure. In some cases, these forces may be a significant design consideration. In concrete or masonry structures, the reduction in stiffness that occurs upon cracking may relieve these self-straining forces, and the assessment of loads should consider this reduced stiffness.

Some permanent loads, such as landscaping loads on plaza areas, may be more appropriately considered as live loads for purposes of design.

C2.3 COMBINING LOADS USING STRENGTH DESIGN

C2.3.1 Applicability

Load factors and load combinations given in this section apply to limit states or strength design criteria (referred to as "Load and Resistance Factor Design" by the steel and wood industries) and they should not be used with allowable stress design specifications.

C2.3.2 Basic Combinations

Unfactored loads to be used with these load factors are the nominal loads of Sections 3 through 9 of this standard. Load factors are from NBS SP 577 with the exception of the factor 1.0 for E, which is based on the more recent NEHRP research on seismic-resistant design [C2-21]. The basic idea of the load combination scheme is that in addition to dead load, which is considered to be permanent, one of the variable loads takes on its maximum lifetime value while the other variable loads assume "arbitrary point-in-time" values, the latter being loads that would be measured at any instant of time [C2-23]. This is consistent with the manner in which loads actually combine in situations in which strength limit states may be approached. However, nominal loads in Sections 3 through 9 are substantially in excess of the arbitrary point-in-time values. To avoid having to specify both a maximum and an arbitrary point-in-time value for each load type, some of the specified load factors are less than unity in combinations (2) through (6).

Load factors in Section 2.3.2 are based on a survey of reliabilities inherent in existing design practice. The load factor on wind load in combinations (4) and (6) has been increased to 1.6 in the current standard from the value of 1.3 appearing in ASCE 7-95. The reasons for this increase are two fold.

First, the previous wind load factor, 1.3, incorporated a factor of 0.85 to account for wind directionality, that is, the reduced likelihood that the maximum windspeed occurs in a direction that is most unfavorable for building response [C2-12]. This directionality effect was not taken into account in ASD. Recent wind engineering research has made it possible to identify wind directionality factors explicity for a number of common structures. Accordingly, new wind directionality factors, K_d, are presented in Table 6-6 of this standard; these factors now are reflected in the nominal wind forces, W, used in both strength design and allowable stress design. This change alone mandates an increase in the wind load factor to approximately 1.53.

Second, the previous value, 1.3, was based on a statistical analysis of wind forces on buildings at sites not exposed to hurricane winds [C2-12]. Studies have shown that, owing to differences between statistical characteristics of wind forces in hurricane-prone coastal areas of the United States [C2-16, C2-17, C2-15] the probability of exceeding the factored (or design-basis) wind force 1.3W is higher in hurricane-prone coastal areas than in the interior regions. Two recent studies [C2-13, C2-14] have shown that the wind load factor in hurricane-prone areas should be increased to approximately 1.5 to 1.8 (depending on site) to maintain comparable reliability.

To move toward uniform risk in coastal and interior areas across the country, two steps were taken. First, the windspeed contours in hurricane-prone areas were adjusted to take the differences in extreme hurricane wind speed probability distributions into account (as explained in C6.5.4); these differences previously were accounted for in ASCE 7-95 by the "importance factor." Second, the wind load factor was increased from 1.3 to 1.6. This approach (a) reflects the removal of the directionality factor, and (b) avoids having to specify separate load criteria for coastal and interior areas.

Load combination (6) and (7) apply specifically to the case in which the structural actions due to lateral forces and gravity loads counteract one another.

Load factors given herein relate only to strength limit states. Serviceability limit states and associated load factors are covered in Appendix B of this standard.

This standard historically has provided specific procedures for determining magnitudes of dead, occupancy live, wind, snow, and earthquake loads. Other loads not traditionally considered by this standard may also require consideration in design. Some of these loads may be important in certain material specifications and are included in the load criteria to enable uniformity to be achieved in the load criteria for different materials. However, statistical data on these loads are limited or nonexistent, and the same procedures used to obtain load factors and load combinations in Section 2.3.2 cannot be applied at the present time. Accordingly, load factors for fluid load (F), lateral pressure due to soil and water in soil (H), and self-straining forces and effects (T) have been chosen to yield designs that would be similar to those obtained with existing specifications, if appro-

priate adjustments consistent with the load combinations in Section 2.3.2 were made to the resistance factors. Further research is needed to develop more accurate load factors because the load factors selected for H and F are probably conservative.

Fluid load, F, defines structural actions in structural supports, framework, or foundations of a storage tank, vessel or similar container due to stored liquid products. The product in a storage tank shares characteristics of both dead and live load. It is similar to a dead load in that its weight has a maximum calculated value, and the magnitude of the actual load may have a relatively small dispersion. However, it is not permanent; emptying and filling causes fluctuating forces in the structure, the maximum load may be exceeded by overfilling; and densities of stored products in a specific tank may vary. Adding F to combination 1 provides additional conservatism for situations in which F is the dominant load.

It should be emphasized that uncertainties in lateral forces from bulk materials, included in H, are higher than those in fluids, particularly when dynamic effects are introduced as the bulk material is set in motion by filling or emptying operations. Accordingly, the load factor for such loads is set equal to 1.6.

C2.3.3 Load Combinations Including Flood Load

The nominal flood load, F_a, is based on the 100-year flood (Section 5.3.3.1). The recommended flood load factor of 2.0 in V Zones and coastal A Zones is based on a statistical analysis of flood loads associated with hydrostatic pressures, pressures due to steady overland flow, and hydrodynamic pressures due to waves, as specified in Section 5.3.3.

The flood load criteria were derived from an analysis of hurricane-generated storm tides produced along the U.S. East and Gulf coasts [C2-14], where storm tide is defined as the water level above mean sea level resulting from wind-generated storm surge added to randomly phased astronomical tides. Hurricane wind speeds and storm tides were simulated at 11 coastal sites based on historical storm climatology and on accepted wind speed and storm surge models. The resulting wind speed and storm tide data were then used to define probability distributions of wind loads and flood loads using wind and flood load equations specified in Sections 6.5, 5.3.3 and in other publications (U.S. Army Corps of Engineers). Load factors for these loads were then obtained using established reliability methods [C2-12] and achieve approximately the same level of reliability as do combinations involving wind loads acting without floods. The relatively high flood load factor stems from the high variability in floods relative to other environmental loads. The presence of $2.0 F_a$ in combinations (4) and (6) in V Zones and coastal A Zones is the result of high stochastic dependence between extreme wind and flood in hurricane-prone coastal zones. The $2.0 F_a$ also applies in coastal areas subject to northeasters, extra tropical storms or coastal storms other than hurricanes, where a high correlation exists between extreme wind and flood.

Flood loads are unique in that they are initiated only after the water level exceeds the local ground elevation. As a result, the statistical characteristics of flood loads vary with ground elevation. The load factor 2.0 is based on calculations (including hydrostatic, steady flow, and wave forces) with still-water flood depths ranging from approximately 4–9 ft (average still-water flood depth of approximately 6 ft), and applies to a wide variety of flood conditions. For lesser flood depths, load factors exceed 2.0 because of the wide dispersion in flood loads relative to the nominal flood load. As an example, load factors appropriate to water depths slightly less than 4 ft equal 2.8 [C2-14]. However, in such circumstances, the flood load itself generally is small. Thus, the load factor 2.0 is based on the recognition that flood loads of most importance to structural design occur in situations where the depth of flooding is greatest.

C2.4 COMBINING LOADS USING ALLOWABLE STRESS DESIGN

C2.4.1 Basic Combinations

The load combinations listed cover those loads for which specific values are given in other parts of this standard. However, these combinations are not all-inclusive, and designers will need to exercise judgement in some situations. Design should be based on the load combination causing the most unfavorable effect. In some cases this may occur when one or more loads are not acting. No safety factors have been applied to these loads, since such factors depend on the design philosophy adopted by the particular material specification.

Wind and earthquake loads need not be assumed to act simultaneously. However, the most unfavorable effects of each should be considered separately in design, where appropriate. In some instances, forces due to wind might exceed those due to earthquake, while ductility requirements might be determined by earthquake loads.

Load combinations (4) and (5) are new in ASCE 7-98. They address the situation in which the effects of lateral or uplift forces counteract the effect of gravity loads. This eliminates an inconsistency in the treatment of counteracting loads in allowable stress design and strength design, and emphasizes the importance of checking stability. The earthquake load effect is multiplied by 0.7 to align allowable stress design for earthquake effects with the definition of E in Section 9.2.2.6 which is based on strength principles.

C2.4.2 Load Combinations Including Flood Load

The basis for the load combinations involving flood load is presented in detail in Section C2.3.3 on strength design. Consistent with the treatment of flood loads for strength design, F_a has been added to load combinations (3) and (4); the multiplier on F_a aligns allowable stress design for flood load with strength design.

C2.4.3 Load Reduction

Most loads, other than dead loads, vary significantly with time. When these variable loads are combined with dead loads, their combined effect should be sufficient to reduce the risk of unsatisfactory performance to an acceptably low level. However, when more than one variable load is considered, it is extremely unlikely that they will all attain their maximum value at the same time. Accordingly, some reduction in the total of the combined load effects is appropriate. This reduction is accomplished through the 0.75 load combination factor. The 0.75 factor applies only to the sum of the transient loads, not to the dead load.

Some material design standards that permit a one-third increase in allowable stress for certain load combinations have justified that increase by this same concept. Where that is the case, simultaneous use of both the one-third increase in allowable stress and the 25% reduction in combined loads is unsafe and is not permitted. In contrast, allowable stress increases that are based upon duration of load or loading rate effects, which are independent concepts may be combined with the reduction factor for combining multiple transient loads. Effects apply to the total stress; that is, the stress resulting from the combination of all loads. Load combination reduction factors for combined transient loads are different in that they apply only to the transient loads, and they do not affect the permanent loads nor the stresses caused by permanent loads. This explains why the 0.75 factor applied to the sum of all loads, to this edition, in which the 0.75 factor applies only to the sum of the transient loads, not the dead load.

Certain material design standards permit a one-third increase in allowable stress for load combinations with one transient load where that transient is earthquake load. This standard handles allowable stress design for earthquake loads in a fashion to give results comparable to the strength design basis for earthquake loads as explained in the C9 Commentary section titled "Use of Allowable Stress Design Standards."

C2.5 LOAD COMBINATIONS FOR EXTRAORDINARY EVENTS

ASCE Standard 7 Commentary C1.4 recommends approaches to providing general structural integrity in building design and construction. Commentary C2.5 explains the basis for the load combinations that the designer should use if the Direct Design alternative in Commentary C1.4 is selected. If the authority having jurisdiction requires the Indirect Design alternative, that authority may use these load requirements as one basis for determining minimum required levels of strength, continuity and ductility. Generally, extraordinary events with a probability of occurrence in the range 10^{-6}–10^{-4}/year or greater should be identified, and measures should be taken to ensure that the performance of key loading-bearing structural systems and components is sufficient to withstand such events.

Extraordinary events arise from extraordinary service or environmental conditions that traditionally are not considered explicitly in design of ordinary buildings and structures. Such events are characterized by a low probability of occurrence and usually a short duration. Few buildings are ever exposed to such events and statistical data to describe their magnitude and structural effects are rarely available. Included in the category of extraordinary events would be fire, explosions of volatile liquids or natural gas in building service systems, sabotage, vehicular impact, misuse by building occupants, subsidence (not settlement) of subsoil, and tornadoes. The occurrence of any of these events is likely to lead to structural damage or failure. If the structure is not properly designed and detailed, this local failure may initiate a chain reaction of failures that propagates throughout a major portion of the structure and leads to a potentially catastrophic collapse. Approximately 15–20% of building collapses occur in this way [C2-1]. Although all buildings are susceptible to progressive failures in varying degrees, types of construction that lack inherent continuity and ductility are particularly vulnerable [C2-2, C2-22].

Good design practice requires that structures be robust and that their safety and performance not be sensitive to uncertainties in loads, environmental influences and other situations not explicitly considered in design. The structural system should be designed in such a way that if an extraordinary event occurs, the probability of damage disproportionate to the original event is sufficiently small [C2-7]. The philosophy of designing to limit the spread of damage rather than to prevent damage entirely is different from the traditional approach to designing to withstand dead, live, snow and wind loads, but is similar to the philosophy adopted in modern earthquake-resistant design [C2-21].

In general, structural systems should be designed with sufficient continuity and ductility that alternate load paths can develop following individual member failure so that failure of the structure as a whole does not ensue. At a simple level, continuity can be achieved by requiring development of a minimum tie force—say 20 kN/m—between structural elements [C2-3]. Member failures may be controlled by protective measures that ensure that no essential load-bearing member is made ineffective as a result of an accident, although this approach may be more difficult to implement. Where member failure would inevitably result in a disproportionate collapse, the member should be designed for a higher degree of reliability [C2-20]. In either approach, an enhanced quality assurance and maintenance program may be required.

Design limit states include loss of equilibrium as a rigid body, large deformations leading to significant second-order effects, yielding or rupture of members of connections, formation of a mechanism, instability of members or the structure as a whole. These limit states are the same as those considered for other load events, but the load-resisting mechanisms in a damaged structure may be different and sources of load-carrying capacity that normally would not be considered in ordinary ultimate limit states design, such as a membrane or catenary action, may be included. The use of elastic analysis vastly underestimates the load-carrying capacity of the structure. Materially or geometrically nonlinear or plastic analyses may be used, depending on the response of the structure to the actions.

Specific design provisions to control the effect of extraordinary loads and risk of progressive failure can be developed with a probabilistic basis [C2-8, C2-11]. One can either attempt to reduce the likelihood of the extraordinary event or design the structure to withstand or absorb damage from the event if it occurs. Let F be the event of failure and A be the event that a structurally damaging event occurs. The probability of failure due to event A is,

$$P_f = P(F|A)P(A) \qquad \text{(Eq. C2-1)}$$

in which $P(F|A)$ is the conditional probability of failure of a damaged structure and $P(A)$ is the probability of occurrence of event A. The separation of $P(F|A)$ and $P(A)$ allows one to focus on strategies for reducing risk. $P(A)$ depends on siting, controlling the use of hazardous substances, limiting access, and other actions that are essentially independent of structural design. In contrast, $P(F|A)$ depends on structural design measures ranging from minimum provisions for continuity to a complete post-damage structural evaluation.

The probability, $P(A)$, depends on the specific hazard. Limited data for severe fires, gas explosions, bomb explosions and vehicular collisions indicate that the event probability depends on building size, measured in dwelling units or square footage, and ranges from about 0.23×10^{-6}/dwelling unit/year to about 7.8×10^{-6}/dwelling unit/year [C2-6, C2-8]. Thus, the probability that a building structure is affected depends on the number of dwelling units (or square footage) in the building. If one were to set the conditional limit state probability, $P(F|A) = 0.1-0.2$/ yr, however, the annual probability of structural failure from Eq. C2-1 would be on the order of $10^{-7}-10^{-6}$, placing the risk in the low-magnitude background along with risks from rare accidents [C2-24].

Design requirements corresponding to this desired $P(F|A) = 0.1-0.2$ can be developed using first-order reliability analysis if the limit state function describing structural behavior is available [C2-19, C2-10]. As an alternative, one can leave material and structural behavior considerations to the responsible material specifications and consider only the load combination aspect of the safety check, which is more straightforward.

For checking a structure to determine its residual load-carrying capacity following occurrence of a damaging extraordinary event, selected load-bearing elements should be notionally removed and the capacity of the remaining structure evaluated using the following load combination:

$$(0.9 \text{ or } 1.2)D + (0.5L \text{ or } 0.2S) + 0.2W$$
$$\text{(Eq. C2-2)}$$

For checking the capacity of a structure or structural element to withstand the effect of an extraordinary event, the following load combinations should be used:

$$1.2D + A_k + (0.5L \text{ or } 0.2S) \qquad \text{(Eq. C2-3)}$$

$$(0.9 \text{ or } 1.2)D + A_k + 0.2W \qquad \text{(Eq. C2-4)}$$

The value of the load or load effect resulting from extraordinary event A used in design is denoted as A_k. Only limited data are available to define the frequency distribution of the load, and A_k must be specified by the authority having jurisdiction [C2-4]. The uncertainty in the load due to the extraordinary event is encompassed in the selection of a conservative A_k and thus the load factor on A_k is set equal to 1.0, as is done in the earthquake load combinations in Section 2.3. Load factors less than 1.0 on the companion actions reflect the small probability of a joint occurrence of the extraordinary load and the design live, snow or wind load. The companion action $0.5L$ corresponds, approximately, to the mean of the yearly maximum load [C2-5]. Companion actions $0.2S$ and $0.2W$ are interpreted similarly. A similar set of load combinations for extraordinary events appears in Eurocode 1 [C2-18].

REFERENCES

[C2-1] Allen, D.E., and Schriever, W. "Progressive collapse, abnormal loads and building codes." *Structural Failures: Modes, Causes, Responsibilities*, ASCE, New York, N.Y., 21–48, 1973.

[C2-2] Breen, J.E., and Siess, C.P. "Progressive collapse—symposium summary." *ACI Journal*, 76(9), 997–1004, 1979.

[C2-3] British standard structural use of steelwork in building (BS 5950: Part 8). British Standards Institution, London, U.K., 1990.

[C2-4] Burnett, E.F.P. "The avoidance of progressive collapse: regulatory approaches to the problem." National Bureau of Standards Report GCR 75-48, Washington, D.C. (Available from NTIS, Springfield, Va.), 1975.

[C2-5] Chalk, P.L., and Corotis, R.B. "Probability models for design live loads." *J. Struct. Div.*, ASCE, 106(10), 2017–2033, 1980.

[C2-6] CIB W14. "A conceptual approach towards a probability based design guide on structural fire safety." *Fire Safety Journal*, 6(1), 1–79, 1983.

[C2-7] "Commentary C Structural Integrity." National Building Code of Canada, National Research Council of Canada, Ottawa, Ontario, 1990.

[C2-8] Ellingwood, B., and Leyendecker, E.V. "Approaches for design against progressive collapse." *J. Struct. Div.*, ASCE, 104(3), 413–423, 1978.

[C2-9] Ellingwood, B., Galambos, T.V., MacGregor, J.G., and Cornell, C.A. "Development of a probability-based load criterion for American National Standard A58." Washington, D.C.: U.S. Dept. of Commerce, National Bureau of Standards. NBS SP 577, 1980.

[C2-10] Ellingwood, B., Galambos, T.V., MacGregor, J.G., and Cornell, C.A. "Probability-based load criteria: load factors and load combinations." *J. Struct. Div.*, ASCE, 108(5), 978–997, 1982.

[C2-11] Ellingwood, B., and Corotis, R.B. "Load combinations for buildings exposed to fires." *Engrg. J.*, ASIC, 28(1), 37–44, 1991.

[C2-12] Ellingwood, B. "Wind and snow load statistics for probabilistic design." *J. Struct. Engrg.*, ASCE, 107(7), 1345–1349, 1981.

[C2-13] Ellingwood, B., and Tekie, P.B. "Wind load statistics for probability-based structural design." Report prepared for NAHB Research Center, Upper Marlboro, Md. (Available through NTIS; also submitted to *Journal of Structural Engineering*, ASCE), 1997.

[C2-14] Mehta, K.C., et al. "An investigation of load factors for flood and combined wind and flood." Report prepared for Federal Emergency Management Agency, Washington, D.C., 1998.

[C2-15] Peterka, J.A., and Shahid, S. "Design gust wind speeds in the United States." *J. Struct. Engrg.*, ASCE, 124(2), 207–214, 1998.

[C2-16] Vickery, P.J., and Twisdale, L.A. "Prediction of hurricane windspeeds in the United States." *J. Struct. Engrg.*, ASCE, 121(11), 1691–1699, 1995.

[C2-17] Whalen, T. "Probabilistic estimates of design load factors for wind-sensitive structures using the 'peaks over threshold' approach." National Institute of Standards and Technology, NIST TN 1418, Gaithersburg, Md, 1966.

[C2-18] Eurocode 1. "Common unified rules for different types of construction and material." Commission of the European Communities, Brussels, 1990.

[C2-19] Galambos, T.V., Ellingwood, B., MacGregor, J.G., and Cornell, C.A. "Probability-based load criteria: assessment of current design practice." *J. Struct. Div.*, ASCE, 108(5), 959–977, 1982.

[C2-20] NKB (1987). "Guidelines for loading and safety regulations for structural design." NKB (Nordic Committee on Building Regulations) Report No. 55E, Copenhagen, Denmark.

[C2-21] "NEHRP recommended provisions for the development of seismic regulations for buildings." FEMA Report 222, Federal Emergency Management Agency, Washington, D.C., 1992.

[C2-22] Taylor, D.A. "Progressive collapse." *Canadian J. Civ. Engrg.*, 2(4), 517–529, 1975.

[C2-23] Turkstra, C.J., and Madsen, H. "Load combinations for codified structural design." *J. Struct. Div.*, ASCE, 106(12), 2527–2543, 1980.

[C2-24] Wilson, R., and Crouch, E.A. "Risk assessment and comparisons: An introduction." *Science*, 236, 267–270.1, 1987.

C3.0 DEAD LOADS

C3.2 WEIGHTS OF MATERIALS AND CONSTRUCTIONS

To establish uniform practice among designers, it is desirable to present a list of materials generally used in building construction, together with their proper weights. Many building codes prescribe the minimum weights for only a few building materials, and in other instances no guide whatsoever is furnished on this subject. In some cases the codes are so drawn up as to leave the question of what weights to use to the discretion of the building official, without providing any authoritative guide. This practice, as well as the use of incomplete lists, has been subjected to much criticism. The solution chosen has been to present, in this Commentary, an extended list that will be useful to designer and official alike. However, special cases will unavoidably arise, and authority is therefore granted in the standard for the building official to deal with them.

For ease of computation, most values are given in terms of pounds per square foot (lb/ft^2) (kN/m^2) of given thickness [see Table C3-1 (given in English and Standard International Units)]. Pounds-per-cubic-foot (lb/ft^3) (kN/m^3) values, consistent with the pounds-per-square foot (kilonewtons per square meter) values, are also presented in some cases [see Table C3-2 (given in English and Standard International Units)]. Some constructions for which a single figure is given actually have a considerable range in weight. The average figure given is suitable for general use, but when there is reason to suspect a considerable deviation from this, the actual weight should be determined.

Engineers, architects, and building owners are advised to consider factors that result in differences between actual and calculated loads.

Engineers and architects cannot be responsible for circumstances beyond their control. Experience has shown, however, that conditions are encountered which, if not considered in design, may reduce the future utility of a building or reduce its margin of safety. Among them are:

1. Dead Loads. There have been numerous instances in which the actual weights of members and construction materials have exceeded the values used in design. Care is advised in the use of tabular values. Also, allowances should be made for such factors as the influence of formwork and support deflections on the actual thickness of a concrete slab of prescribed nominal thickness.
2. Future Installations. Allowance should be made for the weight of future wearing or protective surfaces where there is a good possibility that such may be applied. Special consideration should be given to the likely types and position of partitions, as insufficient provision for partitioning may reduce the future utility of the building.

Attention is directed also to the possibility of temporary changes in the use of a building, as in the case of clearing a dormitory for a dance or other recreational purpose.

TABLE C3-1. Minimum Design Dead Loads* (English Units)

Component	Load (psf)	Component	Load (psf)	Component	Load (psf)
CEILINGS		FLOOR FILL		Clay brick wythes:	
Acoustical Fiber Board	1	Cinder concrete, per inch	9	4 in.	39
Gypsum board (per mm thickness)	0.55	Lightweight concrete, per inch	8	8 in.	79
Mechanical duct allowance	4	Sand, per inch	8	12 in.	115
Plaster on tile or concrete	5	Stone concrete, per inch	12	16 in.	155
Plaster on wood lath	8	FLOORS AND FLOOR FINISHES			
Suspended steel channel system	2	Asphalt block (2-in.), 1/2-in. mortar	30	Hollow concrete masonry unit wythes:	
Suspended metal lath and cement plaster	15	Cement finish (1-in.) on stone-concrete fill	32	Wythe thickness (in inches)	4	6	8	10
Suspended metal lath and gypsum plaster	10	Ceramic or quarry tile (3/4-in.) on 1/2-in. mortar bed	16	Density of unit (16.49 kN/m³)	12
Wood furring suspension system	2.5	Ceramic or quarry tile (3/4-in.) on 1-in. mortar bed	23	No grout	22	24	31	37	43
COVERINGS, ROOF, AND WALL		Concrete fill finish (per inch thickness)	12	48" o.c.	29	38	47	55
Asbestos-cement shingles	4	Hardwood flooring, 7/7-in.	4	40" o.c. grout	30	40	49	57
Asphalt shingles	2	Linoleum or asphalt tile, 1/4-in.	1	32" o.c. spacing	32	42	52	61
Cement tile	16	Marble and mortar on stone-concrete fill	33	24" o.c.	34	46	57	67
Clay tile (for mortar add 10 psf)		Slate (per mm thickness)	15	16" o.c.	40	53	66	79
Book tile, 2-in.	12	Solid flat tile on 1-in. mortar base	23	Full grout	55	75	95	115
Book tile, 3-in.	20	Subflooring, 3/4-in.	3		
Ludowici	10	Terrazzo (1-1/2-in.) directly on slab	19	Density of unit (125 pcf):	
Roman	12	Terrazzo (1-in.) on stone-concrete fill	32	No grout	26	28	36	44	50
Spanish	19	Terrazzo (1-in.), 2-in. stone concrete	32	48" o.c.	33	44	54	62
Composition:		Wood block (3-in.) on mastic, no fill	10	40" o.c. grout	34	45	56	65
Three-ply ready roofing	1	Wood block (3-in.) on 1/2-in. mortar base	16	32" o.c. spacing	36	47	58	68
Four-ply felt and gravel	5.5	FLOORS, WOOD-JOIST (NO PLASTER)		24" o.c.	39	51	63	75
Five-ply felt and gravel	6	DOUBLE WOOD FLOOR		16" o.c.	44	59	73	87

		12-in. spacing (lb/ft²)	16-in. spacing (lb/ft²)	24-in. spacing (lb/ft²)
Copper or tin	1			
Corrugated asbestos-cement roofing	4			
Deck, metal, 20 gage	2.5			
Deck, metal, 18 gage	3			
Decking, 2-in. wood (Douglas fir)	5			
Decking, 3-in. (Douglas fir)	8			
Fiberboard, 1/2-in.	0.75			
Gypsum sheathing, 1/2-in.	2			
Insulation, roof boards (per in. thickness)				
Cellular glass	0.7			
Fibrous glass	1.1			
Fiberboard	1.5			
Perlite	0.8			
Polystyrene foam	0.2			
Urethane foam with skin	0.5			
Plywood (per 1/8-in. thickness)	0.4			
Rigid insulation, 1/2-in.	0.75			
Skylight, metal frame, 3/8-in. wire glass	8			
Slate, 3/16-in.	7			
Slate, 1/4-in.	10			
Waterproofing membranes:				
Bituminous, gravel-covered	5.5			
Bituminous, smooth surface	1.5			
Liquid applied	1			
Single-ply, sheet	0.7			
Wood sheathing (per in. thickness)	3			
Wood shingles	3			

Joist sizes (in.):

	12-in. spacing (lb/ft²)	16-in. spacing (lb/ft²)	24-in. spacing (lb/ft²)
2 × 6	6	5	5
2 × 8	6	6	5
2 × 10	7	6	6
2 × 12	8	7	6

FRAME PARTITIONS

Movable steel partitions	4
Wood or steel studs, 1/2-in. gypsum board each side	8

FRAME WALLS

Wood studs, 2 × 4, unplastered	4
Wood studs, 2 × 4, plastered one side	12
Wood studs, 2 × 4, plastered two sides	20
Exterior stud walls:	
2 × 4 @ 16-in., 5/8-in. gypsum, insulated, 3/8-in. siding	11
2 × 6 @ 16-in., 5/8-in. gypsum, insulated, 3/8-in. siding	12
Exterior stud walls with brick veneer	48
Windows, glass, frame and sash	8

CONCRETE MASONRY UNIT WYTHES

Density of Unit (21.21 kN/m³)

Density of Unit (Full Grout)	29	59	81	102	123
No grout		30	39	47	54
48" o.c.		36	47	57	66
40" o.c. grout		37	48	59	69
32" o.c. spacing		38	50	62	72
24" o.c.		41	54	67	78
16" o.c.		46	61	76	90
Full grout		62	83	105	127

Solid concrete masonry unit wythes (incl. concrete brick)

Wythe thickness (in mm)	4	6	8	10	12
Density of unit (105 pcg):	32	51	69	87	105
Density of unit (125 pcf):	38	60	81	102	124
Density of unit (135 pcf.):	41	64	87	110	133

*Weights of masonry include mortar but not plaster. For plaster, add 5 lb/ft² for each face plastered. Values given represent averages. In some cases there is a considerable range of weight for the same construction.

TABLE C3-1. Minimum Design Dead Loads* (Standard International Units)

Panel 1

Component	Load (kN/m²)
CEILINGS	
Acoustical Fiber Board	0.05
Gypsum board (per mm thickness)	0.008
Mechanical duct allowance	0.19
Plaster on tile or concrete	0.24
Plaster on wood lath	0.38
Suspended steel channel system	0.10
Suspended metal lath and cement plaster	0.72
Suspended metal lath and gypsum plaster	0.48
Wood furring suspension system	0.12
COVERINGS, ROOF, AND WALL	
Asbestos-cement shingles	0.19
Asphalt shingles	0.10
Cement tile	0.77
Clay tile (for mortar add 0.48kN/m²)	
Book tile, 51 mm	0.57
Book tile, 76 mm	0.96
Ludowici	0.48
Roman	0.57
Spanish	0.91
Composition:	
Three-ply ready roofing	0.05
Four-ply felt and gravel	0.26
Five-ply felt and gravel	0.29
Copper or tin	0.05
Corrugated asbestos-cement roofing	0.19
Deck, metal, 20 gage	0.12
Deck, metal, 18 gage	0.14
Decking, 51 mm wood (Douglas fir)	0.24
Decking, 76 mm wood (Douglas fir)	0.38
Fiberboard, 13 mm	0.04

Panel 2

Component	Load (kN/m²)
FLOOR FILL	
Cinder concrete, per mm	0.017
Lightweight concrete, per mm	0.015
Sand, per mm	0.015
Stone concrete, per mm	0.023
FLOORS AND FLOOR FINISHES	
Asphalt block (51 mm), 13 mm mortar	1.44
Cement finish (25 mm) on stone-concrete fill	1.53
Ceramic or quarry tile (19 mm) on 13 mm mortar bed	0.77
Ceramic or quarry tile (19 mm) on 25 mm mortar bed	1.10
Concrete fill finish (per mm thickness)	0.023
Hardwood flooring, 22 mm	0.19
Linoleum or asphalt tile, 6 mm	0.05
Marble and mortar on stone-concrete fill	1.58
Slate (per mm thickness)	0.028
Solid flat tile on 25 mm mortar base	1.10
Subflooring, 19 mm	0.14
Terrazzo (38 mm) directly on slab	0.91
Terrazzo (25 mm) on stone-concrete fill	1.53
Terrazzo (25 mm), 51 mm stone concrete	1.53
Wood block (76 mm) on mastic, no fill	0.48
Wood block (76 mm) on 13 mm mortar base	0.77

FLOORS, WOOD-JOIST (NO PLASTER)
DOUBLE WOOD FLOOR

Joist sizes (mm)	305 mm spacing (kN/m²)	406 mm spacing (kN/m²)	610 mm spacing (kN/m²)
51 × 152	0.29	0.24	0.24
51 × 203	0.29	0.29	0.24
51 × 254	0.34	0.29	0.29
51 × 305	0.38	0.34	0.29

Panel 3

Component	Load (kN/m²)
Clay brick wythes:	
102 mm	1.87
203 mm	3.78
305 mm	5.51
406 mm	7.42

Hollow concrete masonry unit wythes:

	Wythe thickness (in mm)				
	102	152	203	254	305
Density of unit (16.49 kN/m³)					
No grout	1.05	1.29	1.68	2.01	2.35
1,219 mm		1.48	1.92	2.35	2.78
1,016 mm grout		1.58	2.06	2.54	3.02
813 mm spacing		1.63	2.15	2.68	3.16
610 mm		1.77	2.35	2.92	3.45
406 mm		2.01	2.68	3.35	4.02
Full grout		2.73	3.69	4.69	5.70
Density of unit (125 pcf):					
No grout	1.25	1.34	1.72	2.11	2.39
1,219 mm		1.58	2.11	2.59	2.97
1,016 mm grout		1.63	2.15	2.68	3.11
813 mm spacing		1.72	2.25	2.78	3.26
610 mm		1.87	2.44	3.02	3.59
406 mm		2.11	2.78	3.50	4.17
Full Grout		2.82	3.88	4.88	5.89
Density of unity (21.21 kN/m³)					
No grout	1.39	1.68	2.15	2.59	3.02
1,219 mm		1.58	2.39	2.92	3.45
1,016 mm grout		1.72	2.54	3.11	3.69
813 mm spacing		1.82	2.63	3.26	3.83

610 mm	1.96	2.82	3.50	4.12
406 mm	2.25	3.16	3.93	4.69
Full Grout	3.06	4.17	5.27	6.37

Solid concrete masonry unit wythes (incl. concrete brick):

Wythe thickness (in mm)	102	152	203	254	305
Density of unit (16.49 kN/m³):	1.53	2.35	3.21	4.02	4.88
Density of unit (19.64 kN/m³):	1.82	2.82	3.78	4.79	5.79
Density of unit (21.21 kN/m³):	1.96	3.02	4.12	5.17	6.27

Gypsum sheathing, 13 mm	0.10
Insulation, roof boards (per mm thickness)	
Cellular glass	0.0013
Fibrous glass	0.0021
Fiberboard	0.0028
Perlite	0.0015
Polystyrene foam	0.0004
Urethane foam with skin	0.0009
Plywood (per mm thickness)	0.006
Rigid insulation, 13 mm	0.04
Skylight, metal frame, 10 mm wire glass	0.38
Slate, 5 mm	0.34
Slate, 6 mm	0.48
Waterproofing membranes:	
Bituminous, gravel-covered	0.26
Bituminous, smooth surface	0.07
Liquid applied	0.05
Single-ply, sheet	0.03
Wood sheathing (per mm thickness)	0.0057
Wood shingles	0.14

FRAME PARTITIONS

Movable steel partitions	0.19
Wood or steel studs, 13 mm gypsum board each side	0.38
Wood studs, 51 × 102, unplastered	0.19
Wood studs, 51 × 102, plastered one side	0.57
Wood studs, 51 × 102, plastered two sides	0.96

FRAME WALLS

Exterior stud walls:	
51 mm × 102 mm @ 406 mm, 16 mm gypsum, insulated, 10 mm siding	0.53
51 mm × 152 mm @ 406 mm, 16 mm gypsum, insulated, 10 mm siding	0.57
Exterior stud walls with brick veneer	2.30
Windows, glass, frame and sash	0.38

*Weights of masonry include mortar but not plaster. For plaster, add 0.24 kN/m² for each face plastered. Values given represent averages. In some cases there is a considerable range of weight for the same construction.

229

TABLE C3-2. Minimum Densities for Design Loads from Materials (English Units)

Material	Load (lb/ft³)	Material	Load (lb/ft³)
Aluminum	170	Lead	710
Bituminous products		Lime	
Asphaltum	81	Hydrated, loose	32
Graphite	135	Hydrated, compacted	45
Parafin	56	Masonry, Ashlar Stone	
Petroleum, crude	55	Granite	165
Petroleum, refined	50	Limestone, crystalline	165
Petroleum, benzine	46	Limestone, oolitic	135
Petroleum, gasoline	42	Marble	173
Pitch	69	Sandstone	144
Tar	75	Masonry, Brick	
Brass	526	Hard (low absorption)	130
Bronze	552	Medium (medium absorption)	115
Cast-stone masonry (cement, stone, sand)	144	Soft (high absorption)	100
Cement, portland, loose	90	Masonry, Concrete[1]	
Ceramic tile	150	Lightweight units	105
Charcoal	12	Medium weight units	125
Cinder fill	57	Normal weight units	135
Cinders, dry, in bulk	45	Masonry Grout	140
Coal		Masonry, Rubble Stone	
Anthracite, piled	52	Granite	153
Bituminous, piled	47	Limestone, crystalline	147
Lignite, piled	47	Limestone, oolitic	138
Peat, dry, piled	23	Marble	156
Concrete, plain		Sandstone	137
Cinder	108	Mortar, cement or lime	130
Expanded-slag aggregate	100	Particleboard	45
Haydite (burned-clay aggregate)	90	Plywood	36
Slag	132	Riprap (Not submerged)	
Stone (including gravel)	144	Limestone	83
Vermiculite and perlite aggregate, nonload-bearing	25–50	Sandstone	90
Other light aggregate, load-bearing	70–105	Sand	
Concrete, Reinforced		Clean and dry	90
Cinder	111	River, dry	106
Slag	138	Slag	
Stone (including gravel)	150	Bank	70
Copper	556	Bank screenings	108
Cork, compressed	14	Machine	96
Earth (not submerged)		Sand	52
Clay, dry	63	Slate	172
Clay, damp	110	Steel, cold-drawn	492
Clay and gravel, dry	100	Stone, Quarried, Piled	
Silt, moist, loose	78	Basalt, granite, gneiss	96
Silt, moist, packed	96	Limestone, marble, quartz	95
Silt, flowing	108	Sandstone	82
Sand and gravel, dry, loose	100	Shale	92
Sand and gravel, dry, packed	110	Greenstone, hornblende	107
Sand and gravel, wet	120	Terra Cotta, Architectural	
Earth (submerged)		Voids filled	120
Clay	80	Voids unfilled	72
Soil	70	Tin	459
River mud	90	Water	

TABLE C3-2. Minimum Densities for Design Loads from Materials (English Units) (*Continued*)

Material	Load (lb/ft³)	Material	Load (lb/ft³)
Sand or gravel	60	Fresh	62
Sand or gravel and clay	65	Sea	64
Glass	160	Wood, Seasoned	
Gravel, dry	104	Ash, commercial white	41
Gypsum, loose	70	Cypress, southern	34
Gypsum, wallboard	50	Fir, Douglas, coast region	34
Ice	57	Hem fir	28
Iron		Oak, commercial reds and whites	47
Cast	450	Pine, southern yellow	37
Wrought	480	Redwood	28
		Spruce, red, white, and Stika	29
		Western hemlock	32
		Zinc, rolled sheet	449

¹Tabulated values apply to solid masonry and to the solid portion of hollow masonry.

TABLE C3-2. Minimum Densities for Design Loads from Materials (Standard International Units)

Material	Load (kN/m³)	Material	Load (kN/m³)
Aluminum	26.7	Lead	111.5
Bituminous products		Lime	
Asphaltum	12.7	Hydrated, loose	5.0
Graphite	21.2	Hydrated, compacted	7.1
Parafin	8.8	Masonry, Ashlar Stone	
Petroleum, crude	8.6	Granite	25.9
Petroleum, refined	7.9	Limestone, crystalline	25.9
Petroleum, benzine	7.2	Limestone, oolitic	21.2
Petroleum, gasoline	6.6	Marble	27.2
Pitch	10.8	Sandstone	22.6
Tar	11.8	Masonry, Brick	
Brass	82.6	Hard (low absorption)	20.4
Bronze	86.7	Medium (medium absorption)	18.1
Cast-stone masonry (cement, stone, sand)	22.6	Soft (high absorption)	15.7
Cement, portland, loose	14.1	Masonry, Concrete¹	
Ceramic tile	23.6	Lightweight units	16.5
Charcoal	1.9	Medium weight units	19.6
Cinder fill	9.0	Normal weight units	21.2
Cinders, dry, in bulk	7.1	Masonry Grout	22.0
Coal		Masonry, Rubble Stone	
Anthracite, piled	8.2	Granite	24.0
Bituminous, piled	7.4	Limestone, crystalline	23.1
Lignite, piled	7.4	Limestone, oolitic	21.7
Peat, dry, piled	3.6	Marble	24.5
Concrete, plain		Sandstone	21.5
Cinder	17.0	Mortar, cement or lime	20.4
Expanded-slag aggregate	15.7	Particleboard	7.1
Haydite (burned-clay aggregate)	14.1	Plywood	5.7
Slag	20.7	Riprap (Not submerged)	

231

TABLE C3-2. Minimum Densities for Design Loads from Materials (Standard International Units)
(Continued)

Material	Load (kN/m³)	Material	Load (kN/m³)
Stone (including gravel)	22.6	Limestone	13.0
Vermiculite and perlite aggregate, nonload-bearing	3.9–7.9	Sandstone	14.1
Other light aggregate, load-bearing	11.0–16.5	Sand	
Concrete, Reinforced		Clean and dry	14.1
Cinder	17.4	River, dry	16.7
Slag	21.7	Slag	
Stone (including gravel)	23.6	Bank	11.0
Copper	87.3	Bank screenings	17.0
Cork, compressed	2.2	Machine	15.1
Earth (not submerged)		Sand	8.2
Clay, dry	9.9	Slate	27.0
Clay, damp	17.3	Steel, cold-drawn	77.3
Clay and gravel, dry	15.7	Stone, Quarried, Piled	
Silt, moist, loose	12.3	Basalt, granite, gneiss	15.1
Silt, moist, packed	15.1	Limestone, marble, quartz	14.9
Silt, flowing	17.0	Sandstone	12.9
Sand and gravel, dry, loose	15.7	Shale	14.5
Sand and gravel, dry, packed	17.3	Greenstone, hornblende	16.8
Sand and gravel, wet	18.9	Terra Cotta, Architectural	
Earth (submerged)		Voids filled	18.9
Clay	12.6	Voids unfilled	11.3
Soil	11.0	Tin	72.1
River mud	14.1	Water	
Sand or gravel	9.4	Fresh	9.7
Sand or gravel and clay	10.2	Sea	10.1
Glass	25.1	Wood, Seasoned	
Gravel, dry	16.3	Ash, commercial white	6.4
Gypsum, loose	11.0	Cypress, southern	5.3
Gypsum, wallboard	7.9	Fir, Douglas, coast region	5.3
Ice	9.0	Hem fir	4.4
Iron		Oak, commercial reds and whites	7.4
Cast	70.7	Pine, southern yellow	5.8
Wrought	75.4	Redwood	4.4
		Spruce, red, white, and Stika	4.5
		Western hemlock	5.0
		Zinc, rolled sheet	70.5

[1]Tabulated values apply to solid masonry and to the solid portion of hollow masonry.

C4.0 LIVE LOADS

C4.2 UNIFORMLY DISTRIBUTED LOADS

C4.2.1 Required Live Loads

A selected list of loads for occupancies and uses more commonly encountered is given in Section 4.2.1, and the authority having jurisdiction should approve on occupancies not mentioned. Tables C4-1 and C4-2 are offered as a guide in the exercise of such authority.

In selecting the occupancy and use for the design of a building or a structure, the building owner should consider the possibility of later changes of occupancy involving loads heavier than originally contemplated. The lighter loading appropriate to the first occupancy should not necessarily be selected. The building owner should ensure that a live load greater than that for which a floor or roof is approved by the authority having jurisdiction is not placed, or caused or permitted to be placed, on any floor or roof of a building or other structure.

In order to solicit specific informed opinion regarding the design loads in Table 4-1, a panel of 25 distinguished structural engineers was selected. A Delphi [C4-1] was conducted with this panel in which design values and supporting reasons were requested for each occupancy type. The information was summarized and recirculated back to the panel members for a second round of responses; those occupancies for which previous design loads were reaffirmed, as well as those for which there was consensus for change, were included.

It is well known that the floor loads measured in a live-load survey usually are well below present design values [C4-2–C4-5]. However, buildings must be designed to resist the maximum loads they are likely to be subjected to during some reference period T, frequently taken at 50 years. Table C4-2 briefly summarizes how load survey data are combined with a theoretical analysis of the load process for some common occupancy types and illustrates how a design load might be selected for an occupancy not specified in Table 4-1 [C4-6]. The floor load normally present for the intended functions of a given occupancy is referred to as the sustained load. This load is modeled as constant until a change in tenant or occupancy type occurs. A live-load survey provides the statistics of the sustained load. Table C4-2 gives the mean, m_s, and standard deviation, σ_x, for particular reference areas. In addition to the sus-

tained load, a building is likely to be subjected to a number of relatively short-duration, high-intensity, extraordinary or transient loading events (due to crowding in special or emergency circumstances, concentrations during remodeling, and the like). Limited survey information and theoretical considerations lead to the means, m_r, and standard deviations, σ_r, of single transient loads shown in Table C4-2.

Combinations of the sustained load and transient load processes, with due regard for the probabilities of occurrence, leads to statistics of the maximum total load during a specified reference period T. The statistics of the maximum total load depend on the average duration of an individual tenancy, τ, the mean rate of occurrence of the transient load, v_e, and the reference period, T. Mean values are given in Table C4-2. The mean of the maximum load is similar, in most cases, to the Table 4-1 values of minimum uniformly distributed live loads and, in general, is a suitable design value.

C4.3 CONCENTRATED LOADS

C4.3.1 Accessible Roof-Supporting Members

The provision regarding concentrated loads supported by roof trusses or other primary roof members is intended to provide for a common situation for which specific requirements are generally lacking.

C4.4 LOADS ON HANDRAILS, GUARDRAIL SYSTEMS, GRAB BAR SYSTEMS AND VEHICLE BARRIER SYSTEMS

C4.4.2 Loads

(a) Loads that can be expected to occur on handrail and guardrail systems are highly dependent on the use and occupancy of the protected area. For cases in which extreme loads can be anticipated, such as long straight runs of guardrail systems against which crowds can surge, appropriate increases in loading shall be considered.

(b) When grab bars are provided for use by persons with physical disabilities the design is governed by CABO A117 Accessible and Usable Buildings and Facilities.

(c) Vehicle barrier systems may be subjected to horizontal loads from moving vehicles. These horizontal loads may be applied normal to the plane of the barrier system, parallel to the plane of the barrier system, or at any intermediate angle.

Loads in garages accommodating trucks and buses may be obtained from the provisions contained in Standard Specifications for Highway Bridges, 1989, The American Association of State Highway and Transportation Officials.

(d) This provision is introduced to the standard in 1998 and is consistent with the provisions for stairs.

(e) Side rail extensions of fixed ladders are often flexible and weak in the lateral direction. OSHA (CFR 1910) requires side rail extensions, with specific geometric requirements only. The load provided is introduced to the standard in 1998, and has been determined on the basis of a 250 lb person standing on a rung of the ladder, and accounting for reasonable angles of pull on the rail extension.

C4.6 PARTIAL LOADING

It is intended that the full intensity of the appropriately reduced live load over portions of the structure or member be considered, as well as a live load of the same intensity over the full length of the structure or member.

Partial-length loads on a simple beam or truss will produce higher shear on a portion of the span than a full-length load. "Checkerboard" loadings on multistoried, multipanel bents will produce higher positive moments than full loads, while loads on either side of a support will produce greater negative moments. Loads on the half span of arches and domes or on the two central quarters can be critical. For roofs, all probable load patterns should be considered. Cantilevers cannot rely on a possible live load on the anchor span for equilibrium.

C4.7 IMPACT LOADS

Grandstands, stadiums, and similar assembly structures may be subjected to loads caused by crowds swaying in unison, jumping to its feet, or stomping. Designers are cautioned that the possibility of such loads should be considered.

C4.8 REDUCTION IN LIVE LOADS

C4.8.1 General

The concept of, and methods for, determining member live load reductions as a function of a

loaded member's influence area, A_I, was first introduced into this standard in 1982 and was the first such change since the concept of live load reduction was introduced over 40 years ago. The revised formula is a result of more extensive survey data and theoretical analysis [C4-7]. The change in format to a reduction multiplier results in a formula that is simple and more convenient to use. The use of influence area, now defined as a function of the tributary area, A_T, in a single equation has been shown to give more consistent reliability for the various structural effects. The influence area is defined as that floor area over which the influence surface for structural effects is significantly different from zero.

The factor K_{LL} is the ratio of the influence area (A_I) of a member to its tributary area (A_T), i.e., $K_{LL} = A_I/A_T$, and is used to better define the influence area of a member as a function of its tributary area. Fig. C4-1 illustrates typical influence areas and tributary areas for a structure with regular bay spacings. Table 4-2 has established K_{LL} values (derived from calculated K_{LL} values) to be used in Eq. 4-1 for a variety of structural members and configurations. Calculated K_{LL} values vary for column and beam members having adjacent cantilever construction, as is shown in Fig. C4-1, and the Table 4-2 values have been set for these cases to result in live load reductions which are slightly conservative. For unusual shapes, the concept of significant influence effect should be applied.

An example of a member without provisions for continuous shear transfer normal to its span would be a precast T-beam or double-T beam which may have an expansion joint along one or both flanges, or which may have only intermittent weld tabs along the edges of the flanges. Such members do not have the ability to share loads located within their tributary areas with adjacent members, thus resulting in $K_{LL} = 1$ for these types of members.

Reductions are permissible for two-way slabs and for beams, but care should be taken in defining the appropriate influence area. For multiple floors, areas for members supporting more than one floor are summed.

The formula provides a continuous transition from unreduced to reduced loads. The smallest allowed value of the reduction multiplier is 0.4 (providing a maximum 60% reduction), but there is a minimum of 0.5 (providing a 50% reduction) for members with a contributory load from just one floor.

C4.8.2 Heavy Live Loads

In the case of occupancies involving relatively heavy basic live loads, such as storage buildings, several adjacent floor panels may be fully loaded. However, data obtained in actual buildings indicate that rarely is any story loaded with an average actual live load of more than 80% of the average rated live load. It appears that the basic live load should not be reduced for the floor-and-beam design, but that it could be reduced a flat 20% for the design of members supporting more than one floor. Accordingly, this principle has been incorporated in the recommended requirement.

C4.9 MINIMUM ROOF LIVE LOADS

C4.9.1 Flat, Pitched, and Curved Roofs

The values specified in Eq. 4-1 that act vertically upon the projected area have been selected as minimum roof live loads, even in localities where little or no snowfall occurs. This is because it is considered necessary to provide for occasional loading due to the presence of workers and materials during repair operations.

C4.9.2 Special Purpose Roofs

Designers should consider any additional dead loads that may be imposed by saturated landscaping materials. Special purpose or occupancy roof live loads may be reduced in accordance with the requirements of Section 4.8.

C4.10 CRANE LOADS

All support components of moving bridge cranes and monorail cranes, including runway beams, brackets, bracing, and connections, shall be designed to support the maximum wheel load of the crane and the vertical impact, lateral, and longitudinal forces induced by the moving crane. Also, the runway beams shall be designed for crane stop forces. The methods for determining these loads vary depending on the type of crane system and support. References [C4-8–C4-11] describe types of bridge cranes and monorail cranes. Cranes described in these references include top running bridge cranes with top running trolley, underhung bridge cranes, and underhung monorail cranes. Reference [C4-12] gives more stringent re-

quirements for crane runway design that are more appropriate for higher capacity or higher speed crane systems.

REFERENCES

[C4-1] Corotis, R.B., Fox, R.R., and Harris, J.C. Delphi methods: Theory and design load application. *J. Struct. Div.*, ASCE, 107(6), 1095–1105, June, 1981.

[C4-2] Peir, J.C., and Cornell, C.A. Spatial and temporal variability of live loads. *J. Struct. Div.*, ASCE, 99(5), 903–922, May, 1973.

[C4-3] McGuire, R.K., and Cornell, C.A. Live load effects in office buildings. *J. Struct. Div.*, ASCE, 100(7), 1351–1366, July, 1974.

[C4-4] Ellingwood, B.R., and Culver, C.G. Analysis of live loads in office buildings. *J. Struct. Div.*, ASCE, 103(8), 1551–1560, Aug., 1977.

[C4-5] Sentler, L. A stochastic model for live loads on floors in buildings. Lund Institute of Technology, Division of Building Technology, Report 60, Lund, Sweden, 1975.

[C4-6] Chalk, P.L., and Corotis, R.B. A probability model for design live loads. *J. Struct. Div.*, ASCE, 106(10), 2017–2030, Oct., 1980.

[C4-7] Harris, M.E., Corotis, R.B., and Bova, C.J. Area-dependent processes for structural live loads. *J. Struct. Div.*, ASCE, 107(5), 857–872, May, 1981.

[C4-8] Specifications for Underhung Cranes and Monorail Systems, ANSI MH 27.1, Material Handling Industry, Charlotte, N.C., 1981.

[C4-9] Specifications for Electric Overhead Traveling Cranes, No. 70, Material Handling Industry, Charlotte, N.C., 1994.

[C4-10] Specifications for Top Running and Under Running Single Girder Electric Overhead Traveling Cranes, No. 74, Material Handling Industry, Charlotte, N.C., 1994.

[C4-11] Metal Building Manufacturers Association. Low Rise Building Systems Manual, MBMA, Inc., Cleveland, Oh., 1986.

[C4-12] Association of Iron and Steel Engineers, Technical Report No. 13, Pittsburgh, Pa., 1979.

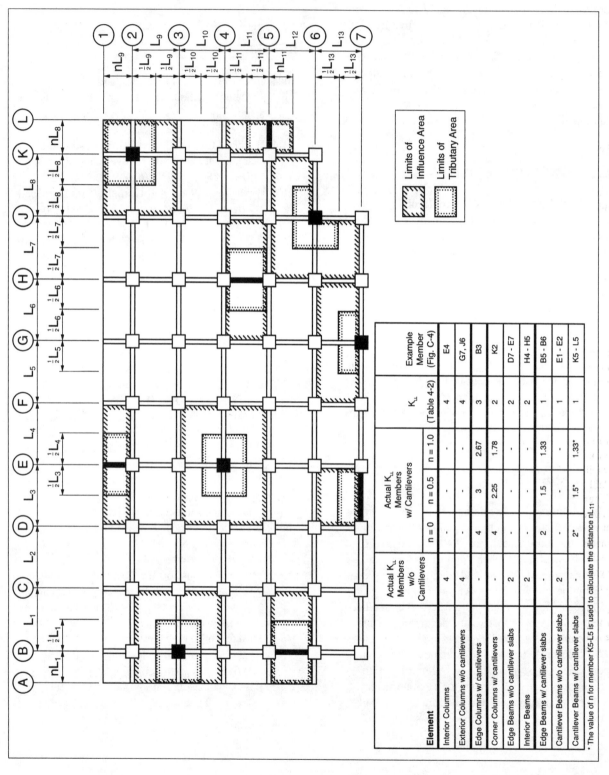

FIGURE 4-1. Typical Tributary and Influence Areas

Element	Actual K_{LL} Members w/o Cantilevers	Actual K_{LL} Members w/ Cantilevers			K_{LL} (Table 4-2)	Example Member (Fig. C-4)
		n = 0	n = 0.5	n = 1.0		
Interior Columns	4	-	-	-	4	E4
Exterior Columns w/o cantilevers	4	-	-	-	4	G7, J6
Edge Columns w/ cantilevers	-	4	3	2.67	3	B3
Corner Columns w/ cantilevers	-	4	2.25	1.78	2	K2
Edge Beams w/o cantilever slabs	2	-	-	-	2	D7 - E7
Interior Beams	2	-	-	-	2	H4 - H5
Edge Beams w/ cantilever slabs	-	2	1.5	1.33	1	B5 - B6
Cantilever Beams w/o cantilever slabs	2	-	-	-	1	E1 - E2
Cantilever Beams w/ cantilever slabs	-	2*	1.5*	1.33*	1	K5 - L5

* The value of n for member K5-L5 is used to calculate the distance nL_{11}

TABLE C4-1. Minimum Uniformly Distributed Live Loads

Occupancy or Use	Live Load lb/ft² (kN/m²)	Occupancy or Use	Live Load lb/ft² (kN/m²)
Air-conditioning	200[1] (9.58)	Kitchens, other than domestic	150[2] (7.18)
(machine space)		Laboratories, scientific	100[1] (4.79)
Amusement park	100[1] (4.79)	Laundries	150[1] (7.18)
structure		Libraries, corridors	80[1] (3.83)
Attic, nonresidental		Manufacturing, ice	300 (14.36)
Nonstorage	25 (1.20)	Morgue	125 (6.00)
Storage	80[1] (3.83)	Office Buildings	
Bakery	150 (7.18)	Business machine equipment	100[1] (4.79)
Exterior	100 (4.79)	Files (see file room)	
Interior (fixed seats)	60 (2.87)	Printing Plants	
Interior (movable seats)	100 (4.79)	Composing rooms	100 (4.79)
Boathouse, floors	100[1] (4.79)	Linotype rooms	100 (4.79)
Boiler room, framed	300[1] (14.36)	Paper storage	—[4]
Broadcasting studio	100 (4.79)	Press rooms	150[1] (7.18)
Catwalks	25 (1.20)	Public rooms	100 (4.79)
Ceiling, accessible furred	10[6] (0.48)	Railroad tracks	—[5]
Cold storage		Ramps	
No overhead system	250[2] (11.97)	Driveway (see garages)	
Overhead system		Pedestrian (see sidewalks and	
Floor	150 (7.18)	corridors in Table 4–1)	
Roof	250 (11.97)	Seaplane (see hangars)	
Computer equipment	150[1] (7.18)	Rest rooms	60 (2.87)
Courtrooms	50–100 (2.40–4.79)	Rinks	
Dormitories		Ice skating	250 (11.97)
Nonpartitioned	80 (3.83)	Roller skating	100 (4.79)
Partitioned	40 (1.92)	Storage, hay or grain	300[1] (14.36)
Elevator machine room	150[1] (7.18)	Telephone exchange	150[1] (7.18)
Fan room	150[1] (7.18)	Theaters:	
File room		Dressing rooms	40 (1.92)
Duplicating equipment	150[1] (7.18)	Grid-iron floor or fly gallery:	
Card	125[1] (6.00)	Grating	60 (2.87)
Letter	80[1] (3.83)	Well beams, 250 lb/ft per pair	
Foundries	600[1] (28.73)	Header beams, 1,000 lb/ft	
Fuel rooms, framed	400 (19.15)	Pin rail, 250 lb/ft	
Garages—trucks	—[3]	Projection room	100 (4.79)
Greenhouses	150 (7.18)	Toilet rooms	60 (2.87)
Hangars	150[3] (7.18)	Transformer rooms	200[1] (9.58)
Incinerator charging floor	100 (4.79)	Vaults, in offices	250[1] (11.97)

[1]Use weight of actual equipment or stored material when greater.

[2]Plus 150 lb/ft² (7.18 kN/m²) for trucks.

[3]Use American Association of State Highway and Transportation Officials lane loads. Also subject to not less than 100% maximum axle load.

[4]Paper storage 50 lb/ft (2.40 kN/m²) of clear story height.

[5]As required by railroad company.

[6]Accessible ceilings normally are not designed to support persons. The value in this table is intended to account for occasional light storage or suspension of items. If it may be necessary to support the weight of maintenance personnel, this shall be provided for.

TABLE C4-2. Typical Live Load Statistics

| Occupancy or Use | Survey Load | | Transient Load | | Temporal Constants | | | Mean Maximum Load[1] |
	m_s lb/ft^2 (kN/m^2)	σ_s[1] lb/ft^2 (kN/m^2)	m_t[1] lb/ft^2 (kN/m^2)	σ_t[1] lb/ft^2 (kN/m^2)	τ_s[2] (years)	ν_e[3] (per year)	T[4] (years)	lb/ft^2 (kN/m^2)
Office buildings								
offices	10.9 (0.52)	5.9 (0.28)	8.0 (0.38)	8.2 (0.39)	8	1	50	55 (2.63)
Residential								
renter occupied	6.0 (0.29)	2.6 (0.12)	6.0 (0.29)	6.6 (0.32)	2	1	50	36 (1.72)
owner occupied	6.0 (0.29)	2.6 (0.12)	6.0 (0.29)	6.6 (0.32)	10	1	50	38 (1.82)
Hotels								
guest rooms	4.5 (0.22)	1.2 (0.06)	6.0 (0.29)	5.8 (0.28)	5	20	50	46 (2.2)
Schools								
classrooms	12.0 (0.57)	2.7 (0.13)	6.9 (0.33)	3.4 (0.16)	1	1	100	34 (1.63)

[1]For 200-ft^2 (18.58 m^2) area, except 1,000 ft^2 (92.9 m^2) for schools.
[2]Duration of average sustained load occupancy.
[3]Mean rate of occurrence of transient load.
[4]Reference period.

C5.0 SOIL AND HYDROSTATIC PRESSURE AND FLOOD LOADS

C5.1 PRESSURE ON BASEMENT WALLS

Table 5-1 includes high earth pressures, 85 pcf (13.36 kN/m^2) or more, to show that certain soils are poor backfill material. In addition, when walls are unyielding the earth pressure is increased from active pressure toward earth pressure at rest, resulting in 60 pcf (9.43 kN/m^2) for granular soils and 100 pcf (15.71 kN/m^2) for silt and clay type soils (see [C5-4]). Examples of light floor systems supported on shallow basement walls mentioned in Table 5-1 are floor systems with wood joists and flooring, and cold formed steel joists without cast in place concrete floor attached.

Expansive soils exist in many regions of the United States and may cause serious damage to basement walls unless special design considerations are provided. Expansive soils should not be used as backfill because they can exert very high pressures against walls. Special soil testing is required to determine the magnitude of these pressures. It is preferable to excavate expansive soil and backfill with non-expansive freely-draining sands or gravels. The excavated backslope adjacent to the wall should be no steeper than 45 degrees from the horizontal in order to minimize the transmission of swelling pressure from the expansive soil through the new backfill. Other special details are recommended, such as a cap of non-pervious soil on top of the backfill and provision of foundation drains. Refer to current reference books on geotechnical engineering for guidance.

C5.2 UPLIFT ON FLOORS AND FOUNDATIONS

If expansive soils are present under floors or footings large pressures can be exerted and must be resisted by special design. Alternatively, the expansive soil can be excavated to a depth of at least 2 ft (0.60 m) and backfilled with non-expansive freely-draining sands or gravel. A geotechnical engineer should make recommendations in these situations.

C5.3 FLOOD LOADS

This section presents information for the design of buildings and other structures in areas prone to flooding. Much of the impetus for flood-resistant design has come about from the federal government sponsored initiatives of flood insurance and flood-damage mitigation.

The National Flood Insurance Program (NFIP) is based on an agreement between the federal government and participating communities that have been identified as being floodprone. The Federal Emergency Management Agency (FEMA) through the Federal Insurance Administration (FIA), makes flood insurance available to the residents of communities provided that the community adopts and enforces adequate floodplain management regulations that meet the minimum requirements. Included in the NFIP requirements, found under Title 44 of the U.S. Code of Federal Regulations [C5-3], are minimum building design and construction standards for buildings and other structures located in Special Flood Hazard Areas (SFHA).

Special Flood Hazards Areas (SFHA) are those identified by FEMA's Mitigation Directorate as being subject to inundation during the 100-year flood. SFHA are shown on Flood Insurance Rate Maps (FIRM), which are produced for floodprone communities. SFHA are identified on FIRM as Zones A, A1-30, AE, AR, AO, AH and coastal high hazard areas as V1-30, V and VE. The SFHA is the area in which communities must enforce NFIP-complaint, flood damage resistant design and construction practices.

Prior to designing a structure in a floodprone area, design professionals should contact the local building official to determine if the site in question is located in a SFHA or other floodprone area that is regulated under the community's floodplain management regulations. If the proposed structure is located within the regulatory floodplain, local building officials can explain the regulatory requirements.

Answers to specific questions on flood-resistant design and construction practices may be directed to the Mitigation Division of each of FEMA's regional offices. FEMA has regional offices that are available to assist design professionals.

C5.3.1 Definitions

Three new concepts are added with ASCE 7-98. First, the concept of the Design Flood is introduced. The Design Flood will, at a minimum, be equivalent to the flood having a 1% chance of being equaled or exceeded in any given year (i.e., the Base Flood or 100-year flood, which served as the load basis in ASCE 7-95). In some instances, the Design Flood may exceed the Base Flood in elevation or spatial extent—this will occur where a community has designated a greater flood (lower frequency, higher return period) as the flood to which the community will regulate new construction.

Many communities have elected to regulate to a flood standard higher than the minimum requirements of the National Flood Insurance Program (NFIP). Those communities may do so in a number of ways. For example: a community may require new construction to be elevated a specific vertical distance above the Base Flood Elevation (this is referred to as freeboard); a community may select a lower frequency flood as its regulatory flood; a community may conduct hydrologic and hydraulic studies, upon which Flood Hazard Maps are based, in a manner different than the Flood Insurance Study prepared by the NFIP (the community may complete flood hazard studies based upon development conditions at build-out, rather than following the NFIP procedure which uses conditions in existence at the time the studies are completed; the community may include watersheds smaller than 1 mi^2 in size in its analysis, rather than following the NFIP procedure which neglects watersheds smaller than 1 mi^2).

Use of the Design Flood concept in ASCE 7-98 will ensure that the requirements of this standard are not less restrictive than a community's requirements where that community has elected to exceed minimum NFIP requirements. In instances where a community has adopted the NFIP minimum requirements, the Design Flood described in this standard will default to the Base Flood.

Second, this standard also introduces the terms, Flood Hazard Area and Flood Hazard Map, to correspond to and show the areas affected by the Design Flood. Again, in instances where a community has adopted the minimum requirements of the NFIP, the Flood Hazard Area defaults to the NFIP's Special Flood Hazard Area and the Flood Hazard Map defaults to the Flood Insurance Rate Map.

Third, the concept of a coastal A Zone is introduced, to facilitate application of revised load combinations contained in Section 2. Coastal A Zones lie landward of V Zones, or landward of an open coast shoreline where V Zones have not been mapped (e.g., the shorelines of the Great Lakes), coastal A Zones are subject to the effects of waves, high-velocity flows and erosion, although not to the extent that V Zones are. Like V Zones, flood forces in Coastal A Zones will be highly correlated with coastal winds or coastal seismic activity.

C5.3.2.1 Design Loads

Wind loads and flood loads may act simultaneously at coastlines, particularly during hurricanes and coastal storms. This may also be true during severe storms at the shorelines of large lakes, and during riverine flooding of long duration.

C5.3.3.1 Load Basis

Water loads are the loads or pressures on surfaces of buildings and structures caused and induced by the presence of floodwaters. These loads are of two basic types: hydrostatic and hydrodynamic. Impact loads result from objects transported by floodwaters striking against buildings and structures or part thereof. Wave loads can be considered a special type of hydrodynamic load.

C5.3.3.2 Hydrostatic Loads

Hydrostatic loads are those caused by water either above or below the ground surface, free or confined, which is either stagnant or moves at velocities less than 5 ft/sec (1.52 m/s). These loads are equal to the product of the water pressure multiplied by the surface area on which the pressure acts.

Hydrostatic pressure at any point is equal in all directions and always acts perpendicular to the surface on which it is applied. Hydrostatic loads can be subdivided into vertical downward loads, lateral loads and vertical upward loads (uplift or buoyancy). Hydrostatic loads acting on inclined, rounded or irregular surfaces may be resolved into vertical downward or upward loads and lateral loads based on the geometry of the surfaces and the distribution of hydrostatic pressure.

C5.3.3.3 Hydrodynamic Loads

Hydrodynamic loads are those loads induced by the flow of water moving at moderate to high velocity above the ground level. They are usually lateral loads caused by the impact of the moving mass of water and the drag forces as the water flows around the obstruction. Hydrodynamic loads are computed by recognized engineering methods. In the coastal high hazard area the loads from high velocity currents due to storm surge and overtopping are of particular importance. Reference [C5-1] is one source of design information regarding hydrodynamic loadings.

Potential sources of information regarding velocities of floodwaters include local, state and federal government agencies and consulting engineers specializing in coastal engineering, stream hydrology or hydraulics.

C5.3.3.4 Wave Loads

The magnitude of wave forces (lb/ft^2) (kN/m^2) acting against buildings or other structures can be 10 or more times higher than wind forces and other forces under design conditions. Thus, it should be readily apparent that elevating above the wave crest elevation is crucial to the survival of buildings and other structures. Even elevated structures, however, must be designed for large wave forces that can act over a relatively small surface area of the foundation and supporting structure.

Wave load calculation procedures in Section 5.3.3.4 are taken from [C5-1] and [C5-5]. The analytical procedures described by Eqs. 5-2–5-10 should be used to calculate wave heights and wave loads unless more advanced numerical or laboratory procedures permitted by this Standard are used.

Wave load calculations using the analytical procedures described in this Standard all depend upon the initial computation of the wave height, which is determined using Eqs. 5-2 and 5-3. These equations result from the assumptions that waves propagating into shallow water break when the wave height equals 78% of the local stillwater depth and that 70% of the wave height lies above the local stillwater level. These assumptions are identical to those used by FEMA in its mapping of coastal flood hazard areas on FIRMs.

It should be pointed out that present NFIP mapping procedures distinguish between A Zones and V Zones by the wave heights expected in each zone. Generally speaking, A Zones are designated where wave heights less than 3 ft (0.91 m) are expected; V Zones are designated where wave heights equal to or greater than 3 ft (0.91 m) are expected. Designers should proceed cautiously, however. Large wave forces can be generated in some A Zones, and wave force calculations should not be restricted to V Zones. Present NFIP mapping procedures do not designate V Zones in all areas where wave heights greater than 3 ft (0.91 m) can occur during base flood conditions. Rather than rely exclusively on flood hazard maps, designers should investigate historical flood damages near a site to determine whether or not wave forces can be significant.

C5.3.3.4.2 Breaking wave loads on vertical walls

Equations used to calculate breaking wave loads on vertical walls contain a coefficient, C_p. Reference [C5-5] provides recommended values of the coefficient as a function of probability of exceedance. The probabilities given by [C5-5] are not annual probabilities of exceedance, but probabilities associated with a distribution of breaking wave pressures measured during laboratory wave tank tests. Note that the distribution is independent of water depth. Thus, for any water depth, breaking wave pressures can be expected to follow the distribution described by the probabilities of exceedance in Table 5-2.

This Standard assigns values for C_p according to Building Category, with the most important buildings having the largest values of C_p. Category II buildings are assigned a value of C_p corresponding to a 1% probability of exceedance, which is consistent with wave analysis procedures used by FEMA in mapping coastal flood hazard areas and in establishing minimum floor elevations. Category I buildings are assigned a value of C_p corresponding to a 50% probability of exceedance, but designers may wish to choose a higher value of C_p. Category III buildings are assigned a value of C_p corresponding to a 0.2% probability of exceedance, while Category IV buildings are assigned a value of C_p corresponding to a 0.1% probability of exceedance.

Breaking wave loads on vertical walls reach a maximum when the waves are normally incident (direction of wave approach perpendicular to the face of the wall; wave crests are parallel to the face of the wall). As guidance for designers of coastal buildings or other structures on normally dry land (i.e., flooded only during coastal storm or flood events), it can be assumed that the direction of wave approach will be approximately perpendicular to the shoreline. Therefore, the direction of wave approach relative to a vertical wall will depend upon the orientation of the wall relative to the shoreline. Section 5.3.3.4.4 provides a method for reducing breaking wave loads on vertical walls for waves not normally incident.

C5.3.3.5 Impact Loads

Normal impact loads are those which result from isolated occurrences of logs, ice floes and other objects normally encountered striking buildings, structures or parts thereof. Special impact loads are those which result from large objects, such as broken up ice floats and accumulations of debris, either striking or resting against a building, structures or parts thereof. Extreme impact loads are those which result from very large objects such as boats, barges, or collapsed buildings striking the building, structure or component under consideration [C5-2].

Impact load is calculated as follows:

$$F_i = ma \qquad \text{(Eq. C5-1)}$$

where

F_i = impact load in pounds (kN);

$$m = w/g = \text{mass in slugs (kg)} \qquad \text{(Eq. C5-2)}$$

w = weight of object in pounds (kN);
g = acceleration due to gravity 32.2 ft/s^2 (9.81 m/s^2);

$$a = \Delta V \Delta / t \qquad \text{(Eq. C5-3)}$$

ΔV = change from V_b to zero velocity;
V_b = velocity of object in feet per second (m/s); and
Δt = time to decelerate object in seconds.

Assume that the velocity of the object is reduced to 0 in 1 s resulting in a minimum impact load:

$F_i = 31 \times V_b$ in pounds

$$[\text{ in SI: } F_i = 0.453 \times V_b \text{ in kN }] \qquad \text{(Eq. C5-4)}$$

Where larger than normal impact loads are likely to occur, the minimum impact load presented in this section is inadequate and special consideration in the design of the building or other structure is required. For example, rigid structures of concrete or steel may reduce the velocity of the object to zero within a time interval of 0.1–0.5 s.

C5.3.4 Special Flood Hazard Areas—A Zone

Note that the elevation requirements specified in Table 5-3 should apply to A Zones identified on FIRMs, and to other community-identified flood hazard areas with a 1% or greater chance of flooding in any given year.

Exception: Community-identified V Zones not identified as such on FIRMs.

C5.3.5 Coastal High Hazards Areas—V Zones

In coastal flood hazard areas the velocity of flow of the water erodes soil supporting the foundation elements, such as piles, piers, and columns. Adequate embedment must be provided to resist the effects of the design flood as well as the accumulated scour from a number of smaller flood events. Similar design requirements should be met in other flood hazard areas, including Coastal A Zones, and along river areas and lakefronts where water depths, wind speeds and fetch lengths are sufficient to generate damaging waves and high velocity flows capable of causing erosion and scour.

Note that the elevation requirements specified in Table 5-4 should apply to V Zones identified on FIRMs, and to other community-identified flood hazard areas subject to high velocity wave action, and with a 1% or greater chance of flooding in any given year.

C5.3.5.1 Elevation

Shear walls have not been included as a means of elevating buildings or other structures in coastal high hazard areas. They can be very efficient in resisting loads in the plane of the shear wall from wind, etc. There is concern about serious damage or collapse from wave forces normal to the plane of the shear wall. However, it is possible to design shear walls to resist such wave forces and special impact forces. When shear walls are used, they should be oriented to minimize the total surface area exposed to potential hydrodynamic and impact loads.

C5.3.5.3 Erosion and Scour

Scour is an important consideration in the design of foundations in coastal high hazard areas or V Zone. In coastal areas, scour can be significant due to area erosion resulting from the effects of storm surge and wave action. Local scour around foundation elements such as a pile or a column, mat foundation or grade beam must also be considered in determining the required depth of embedment or anchorage. Along the coastline the erosion from many small to large storms must be considered through evaluation of historical records. The required embedment of piles and foundations must include consideration of both areal erosion and local scour at the foundation.

REFERENCES

[C5-1] U.S. Army Corps of Engineers, Coastal Engineering Research Center, Waterways Experiment Station, Shore Protection Manual, 2 Vols., 4th Ed., 1984.

[C5-2] U.S. Army Corps of Engineers, Office of the Chief of Engineers, Flood Proofing Regulations, EP 1165-2-314, March, 1992.

[C5-3] Federal Emergency Management Agency, National Flood Insurance Program, 44 CFR Ch. 1, Parts 59 and 60, (10-1-90 Edition), 1990.

[C5-4] Terzaghi, K., and Peck, R.B., *Soil mechanics in engineering practice*, Wiley, 2nd Ed., 1967.

[C5-5] Walton, T.L., Jr., J.P. Ahrens, C.L. Truitt, and R.G. Dean, *Criteria for evaluating coastal flood protection structures*, Technical Report CERC 89-15, U.S. Army Corps of Engineers, Waterways Experiment Station, 1989.

C6.0 WIND LOADS

C6.1 GENERAL

The ASCE 7-98 version of the wind load standard provides three methods from which the designer can choose. A new "simplified method" (Method 1) for which the designer can select wind pressures directly without any calculation when the building meets all the requirements for application of the procedure; and two other methods (Analytical Method and Wind Tunnel Procedure) which are essentially the same methods as previously given in the standard except for some changes that are noted.

The ASCE 7-98 version of the standard has been entirely reformatted into a more "user friendly" format. Specific step-by-step design procedures are listed for application of Methods 1 and 2 to aid the user in applying the standard.

Temporary bracing should be provided to resist wind loading on structural components and structural assemblages during erection and construction phases.

C6.2 DEFINITIONS

New definitions have been added for "approved," "building envelope," "regular shaped building," "rigid building and other structure," "simple diaphragm building," "escarpment," "glazing," "impact resistant glazing," "impact resistant covering," "hill," "hurricane prone regions," "mean roof height," "openings," "ridge," and "wind borne debris region." These terms are used throughout the standard and are provided to clarify application of the standard provisions.

Main wind-force resisting system can consist of a structural frame or an assemblage of structural elements that work together to transfer wind loads acting on the entire structure to the ground. Structural elements such as cross-bracing, shear walls, roof trusses and roof diaphragms are part of the main wind-force resisting system when they assist in transferring overall loads [C6-87].

Building, enclosed, open, partially enclosed: These definitions relate to the proper selection of internal pressure coefficients, GC_{pi}. Building, open and Building, partially enclosed are specifically defined. All other buildings are considered to be enclosed by definition, although there may be large openings in two or more walls. An example of this is a parking garage through which the wind can pass. The internal pressure coefficient for such a building would be ± 0.18, and the internal pressures would act on the solid areas of the walls and roof.

Components and cladding: Components receive wind loads directly or from cladding and transfer the load to the main wind-force resisting system. Cladding receives wind loads directly. Examples of components include fasteners, purlins, girts, studs, roof decking and roof trusses. Components can be part of the main wind-force resisting system when they act as shear walls or roof diaphragms, but they may also be loaded as individual components. The engineer needs to use appropriate loadings for design of components, which may require certain components to be designed for more than one type of loading, e.g., long span roof trusses should be designed for loads associated with main wind-force resisting systems, and individual members of trusses should also be designed for component and cladding loads [C6-87]. Examples of cladding include wall coverings, curtain walls, roof coverings, exterior windows (fixed and operable) and doors, and overhead doors.

Effective wind area is the area of the building surface used to determine GC_p. This area does not necessarily correspond to the area of the building surface contributing to the force being considered. Two cases arise. In the usual case the effective wind area does correspond to the area tributary to the force component being considered. For example, for a cladding panel, the effective wind area may be equal to the total area of the panel; for a cladding fastener, the effective wind area is the area of cladding secured by a single fastener. A mullion may receive wind from several cladding panels; in this case, the effective wind area is the area associated with the wind load that is transferred to the mullion.

The second case arises where components such as roofing panels, wall studs or roof trusses are spaced closely together; the area served by the component may become long and narrow. To better approximate the actual load distribution in such cases, the width of the effective wind area used to evaluate GC_p need not be taken as less than one third the length of the area. This increase in effective wind area has the effect of reducing the average wind pressure acting on the component. Note however that this effective wind area should only be used in determining the GC_p in Figs. 6-5 through 6-8. The induced wind load should be applied over the actual area tributary to the component being considered.

For membrane roof systems, the effective wind area is the area of an insulation board (or deck panel if insulation is not used) if the boards are fully adhered (or the membrane is adhered directly to the deck). If the insulation boards or membrane are mechanically attached or partially adhered, the effective wind area is the area of the board or membrane secured by a single fastener or individual spot or row of adhesive.

Flexible buildings and other structures: A building or other structure is considered flexible if it contains a significant dynamic resonant response. Resonant response depends on the gust structure contained in the approaching wind, on wind loading pressures generated by the wind flow about the building, and on the dynamic properties of the building or structure. Gust energy in the wind is smaller at frequencies above about 1 Hz; therefore the resonant response of most buildings and structures with lowest natural frequency above 1 Hz will be sufficiently small that resonant response can often be ignored. When buildings or other structures have a height exceeding four times the least horizontal dimension or when there is reason to believe that the natural frequency is less than 1 Hz (natural period greater than 1 s), natural frequency for it should be investigated. A useful calculation procedure for natural frequency or period for various building types is contained in Section 9.

Regular shaped buildings and other structures: Defining the limits of applicability of the analytical procedures within the standard is a difficult process, requiring a balance between the practical need to use the provisions past the range for which data has been obtained and restricting use of the provisions past the range of realistic application. Wind load provisions are based primarily on wind-tunnel tests on shapes shown in Figs. 6-3 through 6-8. Extensive wind-tunnel tests on actual structures under design show that relatively large changes from these shapes can, in many cases, have minor changes in wind load, while in other cases seemingly small changes can have relatively large effects, particularly on cladding pressures. Wind loads on complicated shapes are frequently smaller than those on the simpler shapes of Figs. 6-3 through 6-8, and so wind loads determined from these provisions reasonably envelope most structure shapes. Buildings which are clearly unusual should use the provisions of Section 6.4 for wind-tunnel tests.

Rigid buildings and other structures: The defining criteria for rigid, in comparison to flexible, is that the natural frequency is greater than or equal to 1 Hz. A general guidance is that most rigid buildings

and structures have height to minimum width less than 4. Where there is concern about whether or not a building or structure meets this requirement, the provisions of Section 9 provide a method for calculating natural frequency (period = 1/natural frequency).

C6.3 SYMBOLS AND NOTATION

The following additional symbols and notation are used herein:

A_{ob} = average area of open ground surrounding each obstruction;

n = reference period, in years;

P_a = annual probability of wind speed exceeding a given magnitude (see Eq. C6-1);

P_n = probability of exceeding design wind speed during n years (see Eq. C6-1);

s_{ob} = average frontal area presented to the wind by each obstruction;

V_t = wind speed averaged over t seconds (see Fig. C6-1), in mi/h (m/s);

V_{3600} = mean wind speed averaged over 1 h (see Fig. C6-1), in mi/h (m/s);

β = structural damping coefficient (percentage of critical damping)

C6.4 METHOD 1—SIMPLIFIED PROCEDURE

Method 1 has been added to the standard for a designer having the relatively common low-rise ($h \leq$ 30 ft) regular shaped, simple diaphragm building case (see new definitions for "simple diaphragm building" and "regular shaped building") where pressures for the roof and walls can be selected directly from a table. Two tables are provided; Table 6-2 for the main wind force resisting system and Table 6-3A and 6-3B for components and cladding. For components and cladding, values are provided for enclosed and partially enclosed buildings. Note that for the main wind force resisting system in a diaphragm building, the internal pressure cancels for loads on the walls, but must be considered for the roof. This is true because when wind forces are transferred by horizontal diaphragms (such as floors and roofs) to the vertical elements of the main wind force resisting system (such as shear walls, X-bracing, or moment frames), the collection of wind forces from windward and leeward sides of the building occurs in the horizontal diaphragms. Once transferred into the horizontal diaphragms by the wall systems, the wind forces become a net horizontal wind force that is delivered to the vertical elements. The equal and opposite internal pressures on the walls cancel in the horizontal diaphragm. Method 1 combines the windward and leeward pressures into a net horizontal wind pressure, with the internal pressures canceled. The user is cautioned to consider the precise application of windward and leeward wall loads to members of the roof diaphragm where openings may exist and where particular members such as drag struts are designed. The design of the roof members of the main wind force resisting system is still influenced by internal pressures, but for the limitations imposed on the simple diaphragm building type, it can be assumed that the maximum uplift, produced by a positive internal pressure, is the controlling load case. For the designer to use Method 1, the building must conform to all seven requirements in Section 6.4.1 otherwise Method 2 or 3 must be used. Values are tabulated for Exposure B; multiplying factors are provided for other common exposures. The use of the Simplified Procedure for low-rise buildings in Exposure A is not recommended because of the greater uncertainty of wind load distribution in such an environment. The following values have been used in preparation of the tables:

$h = 30$ ft, Exposure B $K_z = 0.70$;
$K_d = 0.85$ $G = 0.85$ $K_{zt} = 1.0$ $I = 1.0$;
$GC_{pi} = \pm 0.18$ (enclosed building);
$GC_{pi} = \pm 0.55$ (partially enclosed building);
Pressure coefficients from Figs. 6-3 and 6-5.

C6.5 METHOD 2—ANALYTICAL PROCEDURE

C6.5.1 Scope

The analytical procedure provides wind pressures and forces for the design of main wind-force resisting systems and for the design of components and cladding of buildings and other structures. The procedure involves the determination of wind directionality and a velocity pressure, the selection or determination of an appropriate gust effect factor, and the selection of appropriate pressure or force coefficients. The procedure allows for the level of structural reliability required, the effects of differing wind exposures, the speed-up effects of certain topographic features such as hills and escarpments, and the size and geometry

of the building or other structure under consideration. The procedure differentiates between rigid and flexible buildings and other structures, and the results generally envelope the most critical load conditions for the design of main wind-force resisting systems as well as components and cladding.

C6.5.2 Limitations of Analytical Procedure

The provisions given under Section 6.5.2 apply to the majority of site locations and buildings and structures, but for some locations, these provisions may be inadequate. Examples of site locations and buildings and structures (or portions thereof) that require use of recognized literature for documentation pertaining to wind effects, or the use of the wind tunnel procedure of Section 6.6 include:

1. Site locations which have channeling effects or wakes from upwind obstructions. Channeling effects can be caused by topographic features (e.g., mountain gorge) or buildings (e.g., a cluster of tall buildings). Wakes can be caused by hills or by buildings or other structures.
2. Buildings with unusual or irregular geometric shape, including domes, barrel vaults, and other buildings whose shape (in plan or profile) differs significantly from a uniform or series of superimposed prisms similar to those indicated in Figs. 6-3 through 6-8. Unusual or irregular geometric shapes include buildings with multiple setbacks, curved facades, irregular plan resulting from significant indentations or projections, openings through the building, or multi-tower buildings connected by bridges.
3. Buildings with unusual response characteristics, which result in across-wind and/or dynamic torsional loads, loads caused by vortex shedding, or loads resulting from instabilities such as flutter or galloping. Examples of buildings and structures which may have unusual response characteristics include flexible buildings with natural frequencies normally below 1 Hz, tall slender buildings (building height-to-width ratio exceeds 4), and cylindrical buildings or structures. Note that vortex shedding occurs when wind blows across a slender prismatic or cylindrical body. Vortices are alternately shed from one side of the body and then the other side, which results in a fluctuating force acting at right angles to the wind direction (across-wind) along the length of the body.
4. Bridges, cranes, electrical transmission lines, guyed masts, telecommunication towers and flagpoles.

C6.5.2.1 Shielding

Due to the lack of reliable analytical procedures for predicting the effects of shielding provided by buildings and other structures or by topographic features, reductions in velocity pressure due to shielding are not permitted under the provisions of Section 6.5. However, this does not preclude the determination of shielding effects and the corresponding reductions in velocity pressure by means of the wind tunnel procedure in Section 6.6.

C6.5.2.2 Air-Permeable Cladding

Air-permeable roof or wall claddings allow partial air pressure equalization between their exterior and interior surfaces. Examples include siding, pressure-equalized rain screen walls, shingles, tiles, concrete roof pavers, and aggregate roof surfacing.

The design wind pressures derived from Section 6.5 represent the pressure differential between the exterior and interior surfaces of the exterior envelope (wall or roof system). Because of partial air-pressure equalization provided by air permeable claddings, the pressures derived from Section 6.5 can overestimate the load on air-permeable cladding elements. The designer may elect either to use the loads derived from Section 6.5, or to use loads derived by an approved alternative method. If the designer desires to determine the pressure differential across the air-permeable cladding element, appropriate full-scale pressure measurements should be made on the applicable cladding element, or reference be made to recognized literature [C6-9, C6-16, C6-37, C6-73], for documentation pertaining to wind loads.

C6.5.4 Basic Wind Speed

The ASCE 7 wind map proposed for the 1998 Standard has been updated from the map in ASCE 7-95 based on a new and more complete analysis of hurricane wind speeds [C6-78, C6-79]. This new hurricane analysis yields predictions of 50 and 100 year return period peak gust wind speeds along the coast which are generally similar to those given in [C6-88] and [C6-89]. The decision within the Task Committee on Wind Loads to update the map relied on several factors important to an accurate wind specification:

1. The new hurricane results include many more predictions for sites away from the coast than have been available in the past. It is desirable to include the best available decrease in speeds with inland distance. Significant reductions in wind speeds occur in inland Florida for the new analysis.

2. The distance inland to which hurricanes can influence wind speed increases with return period. It is desirable to include this distance in the map for design of ultimate events (working stress multiplied by an appropriate load factor).

3. A hurricane coast importance factor of 1.05 acting on wind speed was included explicitly in past ASCE 7 Standards (1993 and earlier) to account for the more rapid increase of hurricane speeds with return period in comparison to non-hurricane winds. The hurricane coast importance factor actually varies in magnitude and position along the coast and with distance inland. In order to produce a more uniform risk of failure, it is desirable to include the effect of the importance factor in the map by first mapping an ultimate event and then reducing the event to a design basis.

The Task Committee on Wind Loads chose to use a map which includes the hurricane importance factor in the map contours. The map is specified so that the loads calculated from the Standard, after multiplication by the load factor, represent an ultimate load having approximately the same return period as loads for non-hurricane winds. (An alternative not selected was to use an ultimate wind speed map directly in the Standard, with a load factor of 1.0.)

The approach required selection of an ultimate return period. A return period of about 500 years has been used previously for earthquake loads. This return period can be derived from the non-hurricane speeds in ASCE 7-95. A factor of 0.85 is included in the load factor of ASCE 7-95 to account for wind and pressure coefficient directionality [C6-76]. Removing this from the load factor gives an effective load factor F of $1.30/0.85 = 1.529$ (round to 1.5). Of the uncertainties affecting the wind load factor, the variability in wind speed has the strongest influence [C6-77], such that changes in the coefficient of variation in all other factors by 25% gives less than a 5% change in load factor. The non-hurricane multiplier of 50-year wind speed for various return periods averages $Fc = 0.36 + 0.1 \ln(12T)$, with T in years [C6-74]. Setting $Fc = \sqrt{F} = \sqrt{1.5} = 1.225$ yields $T = 476$ years. On this basis, a 500-year speed can reasonably represent an approximate ultimate limit state event.

A set of design level hurricane speed contours, which include the hurricane importance factor, were obtained by dividing 500-year hurricane wind speed contours by $\sqrt{F} = 1.225$. The implied importance factor ranges from near 1.0 up to about 1.25 (the explicit value in ASCE 7-93 is 1.05).

The design level speed map has several advantages. First, a design using the map results in an ultimate load (loads inducing the design strength after use of the load factor) which has a more uniform risk for buildings than occurred with earlier versions of the map. Second, there is no need for a designer to use and interpolate a hurricane coast importance factor. It is not likely that the 500-year event is the actual speed at which engineered structures are expected to fail, due to resistance factors in materials, due to conservative design procedures which do not always analyze all load capacity, and due to a lack of a precise definition of "failure."

The wind speed map of Fig. 6-1 presents basic wind speeds for the contiguous United States, Alaska and other selected locations. The wind speeds correspond to 3-second gust speeds at 33 ft (10 m) above ground for Exposure Category C. Because the National Weather Service has phased out the measurement of fastest-mile wind speeds, the basic wind speed has been redefined as the peak gust which is recorded and archived for most NWS stations. Given the response characteristics of the instrumentation used, the peak gust is associated with an averaging time of approximately 3 s. Because the wind speeds of Fig. 6-1 reflect conditions at airports and similar open-country exposures, they do not account for the effects of significant topographic features such as those described in Section 6.5.7. Note that the wind speeds shown in Fig. 6-1 are not representative of speeds at which ultimate limit states are expected to occur. Allowable stresses or load factors used in the design equation(s) lead to structural resistances and corresponding wind loads and speeds that are substantially higher than the speeds shown in Fig. 6-1.

The hurricane wind speeds given in Fig. 6-1 replace those given in ASCE-7-95 which were based on a combination of the data given in [C6-5, C6-15, C6-20, C6-54, and C6-88], supplemented with some judgment. The non-hurricane wind speeds of Fig. 6-1 were prepared from peak gust data collected at 485 weather stations where at least 5 years of data were available [C6-29, C6-30, C6-74]. For non-hurricane regions, measured gust data were assembled from a number of stations in state-sized areas to decrease sampling error, and the assembled data were fit using a Fisher-Tippett Type I extreme value distribution. This procedure gives the same speed as does area-averaging the 50-year speeds from the set of stations. There was insufficient variation in 50-year speeds over the eastern 3/4 of the lower 48 states to justify contours. The division between the 90 and 85 mph

(40.2 and 38.0 m/s) regions, which follows state lines, was sufficiently close to the 85 mph (38.0 m/s) contour that there was no statistical basis for placing the division off political boundaries. This data is expected to follow the gust factor curve of Fig. C6-1 [C6-13].

Limited data were available on the Washington and Oregon coast; in this region, existing fastest-mile wind speed data were converted to peak gust speeds using open-country gust factors [C6-13]. This limited data indicates that a speed of 100 mph is appropriate in some portions of the special coastal region in Washington and 90 mph in the special coastal region in Oregon; these speeds do not include that portion of the special wind region in the Columbia River Gorge where higher speeds may be justified. Speeds in the Aleutian Islands and in the interior of Alaska were established from gust data. Contours in Alaska are modified slightly from ASCE 7-88 based on measured data, but insufficient data were available for a detailed coverage of the mountainous regions.

C6.5.4.1 Special Wind Regions

Although the wind-speed map of Fig. 6-1 is valid for most regions of the country, there are special regions in which wind-speed anomalies are known to exist. Some of these special regions are noted in Fig. 6-1. Winds blowing over mountain ranges or through gorges or river valleys in these special regions can develop speeds that are substantially higher than the values indicated on the map. When selecting basic wind speeds in these special regions, use of regional climatic data and consultation with a wind engineer or meteorologist is advised.

It is also possible that anomalies in wind speeds exist on a micrometeorological scale. For example, wind speed-up over hills and escarpments is addressed in Section 6.5.7. Wind speeds over complex terrain may be better determined by wind-tunnel studies as described in Section 6.6. Adjustments of wind speeds should be made at the micrometeorological scale on the basis of wind engineering or meteorological advice and used in accordance with the provisions of Section 6.5.4.2 when such adjustments are warranted.

C6.5.4.2 Estimation of Basic Wind Speeds from Regional Climatic Data

When using regional climatic data in accordance with the provisions of Section 6.5.4.2 and in lieu of the basic wind speeds given in Fig. 6-1, the user is cautioned that the gust factors, velocity pressure exposure coefficients, gust effect factors, pressure coefficients, and force coefficients of this Standard are intended for use with the 3-second gust speed at 33 ft (10 m) above ground in open country. It is necessary, therefore, that regional climatic data based on a different averaging time, for example hourly mean or fastest mile, be adjusted to reflect peak gust speeds at 33 ft (10 m) above ground in open country. The results of statistical studies of wind-speed records, reported by [C6-13] for extratropical winds and for hurricanes [C6-78], are given in Fig. C6-1 which defines the relation between wind speed averaged over t seconds, V_t, and over 1 h, V_{3600}. New research cited in [C6-78] indicates that the old Krayer Marshall curve [C6-20] does not apply in hurricanes. Therefore it has been removed in Fig. C6-1 in ASCE 7-98. The gust factor adjustment to reflect peak gust speeds is not always straightforward and advice from a wind engineer or meteorologist may be needed.

In using local data, it should be emphasized that sampling errors can lead to large uncertainties in specification of the 50-year wind speed. Sampling errors are the errors associated with the limited size of the climatological data samples (years of record of annual extremes). It is possible to have a 20 mph (8.9 m/s) error in wind speed at an individual station with a record length of 30 years. It was this type of error that led to the large variations in speed in the non-hurricane areas of the ASCE 7-88 wind map. While local records of limited extent often must be used to define wind speeds in special wind areas, care and conservatism should be exercised in their use.

If meteorological data are used to justify a wind speed lower than 85-mph 50-year peak gust at 10 m, an analysis of sampling error is required to demonstrate that the wind record could not occur by chance. This can be accomplished by showing that the difference between predicted speed and 85 mph contains 2 to 3 standard deviations of sampling error [C6-67]. Other equivalent methods may be used.

C6.5.4.3 Limitation

In recent years, advances have been made in understanding the effects of tornadoes on buildings. This understanding has been gained through extensive documentation of building damage caused by tornadic storms and through analysis of collected data. It is recognized that tornadic wind speeds have a significantly lower probability of occurrence at a point than the probability for basic wind speeds. In addition, it is found that in approximately one-half of

the recorded tornadoes, gust speeds are less than the gust speeds associated with basic wind speeds. In intense tornadoes, gust speeds near the ground are in the range of 150–200 mph (67–89 m/s). Sufficient information is available to implement tornado resistant design for above-ground shelters and for buildings that house essential facilities for post-disaster recovery. This information is in the form of tornado risk probabilities, tornadic wind speeds, and associated forces. Several references provide guidance in developing wind load criteria for tornado-resistant design [C6-1, C6-2, C6-24–C6-28, C6-57].

Tornadic wind speeds, which are gust speeds, associated with an annual probability of occurrence of 1×10^{-5} (100,000 year mean recurrence interval) are shown in Fig. C6-1A. This map was developed by the American Nuclear Society committee ANS 2.3 in the early 1980s. Tornado occurrence data of the last 15 years can provide a more accurate tornado hazard wind speed for a specific site.

C6.5.4.4 Wind Directionality Factor

The existing wind load factor 1.3 in ASCE 7-95 includes a "wind directionality factor" of 0.85 [C6-76, C6-77]. This factor accounts for two effects: (1) the reduced probability of maximum winds coming from any given direction; and (2) the reduced probability of the maximum pressure coefficient occurring for any given wind direction. The wind directionality factor (identified as K_d in the new standard) has been hidden in previous editions of the standard and has generated renewed interest in establishing the design values for wind forces determined by using the standard. Accordingly, the Task Committee on Wind Loads, working with the Task Committee on Load Combinations, has decided to separate the wind directionality factor from the load factor and include its effect in the equation for velocity pressure. This has been done by developing a new factor, K_d, that is tabulated in the new Table 6-6 for different structure types. As new research becomes available, this factor can be directly modified without changing the wind load factor. Values for the factor were established from references in the literature and collective committee judgment. It is noted that the k_d value for round chimneys, tanks and similar structures is given as 0.95 in recognition of the fact that the wind load resistance may not be exactly the same in all directions as implied by a value of 1.0. A value of 0.85 might be more appropriate if a triangular trussed frame is shrouded in a round cover. 1.0 might be more appropriate for a round chimney having a lateral load resistance equal in all directions. The designer is cautioned by the footnote to Table 6-6 and the statement in Section 6.5.4.4 where reference is made to the fact that this factor is only to be used in conjunction with the load combination factors specified in Sections 2.3 and 2.4.

C6.5.5 Importance Factor

The importance factor is used to adjust the level of structural reliability of a building or other structure to be consistent with the building classifications indicated in Table 1-1. The importance factors given in Table 6-1 adjust the velocity pressure to different annual probabilities of being exceeded. Importance-factor values of 0.87 and 1.15 are, for the non-hurricane winds, associated, respectively, with annual probabilities of being exceeded of 0.04 and 0.01 (mean recurrence intervals of 25 and 100 years). In the case of hurricane winds, the annual exceedance probabilities implied by the use of the importance factors of 0.77 and 1.15 will vary along the coast; however, the resulting risk levels associated with the use of these importance factors when applied to hurricane winds will be approximately consistent with those applied to the non-hurricane winds.

The probability P_n that the wind speed associated with a certain annual probability P_a will be equaled or exceeded at least once during an exposure period of n years is given by

$$P_n = 1 - (1 - P_a)^n \qquad \text{(Eq. C6-1)}$$

and values of P_n for various values of P_a and n are listed in Table C6-2. As an example, if a design wind speed is based upon $P_a = 0.02$ (50 year mean recurrence interval), there exists a probability of 0.40 that this speed will be equaled or exceeded during a 25-year period, and a 0.64 probability of being equaled or exceeded in a 50-year period.

For applications of serviceability, design using maximum likely events, or other applications, it may be desired to use wind speeds associated with mean recurrence intervals other than 50 years. To accomplish this, the 50-year speeds of Fig. 6-1 are multiplied by the factors listed in Table C6-3. Table C6-3 is strictly valid for the non-hurricane winds only ($V < 100$ mph for continental U.S. and all speeds in Alaska), where the design wind speeds have a nominal annual exceedance probability of 0.02. Using the factors given in Table C6-3 to adjust the hurricane wind speeds will yield wind speeds and resulting wind loads that are approximately risk consistent

with those derived for the non-hurricane prone regions. The true return periods associated with the hurricane wind speeds cannot be determined using the information given in this standard.

The difference in wind speed ratios between continental U.S. ($V < 100$ mph) and Alaska were determined by data analysis and probably represent a difference in climatology at different latitudes.

C6.5.6 Exposure Categories

Revisions have been made to definitions for Exposure C and Exposure D in recognition of new research [C6-86]. The definitions of Exposures A and B remain unchanged. Proper application of exposure requires the designer to consider the following:

1. Terrain roughness for the particular surface area surrounding the site including height and density of topographic features and other structures.
2. For each wind direction assumed, the vertical frontal area of each obstruction to the wind for any selected upwind fetch surface area.

The Task Committee on Wind Loads has made a judgment that 1,500 ft or 10 times the structure height, whichever is greater, is the appropriate fetch distance to consider for Exposure B. One half mile or 10 times the structure height, whichever is greater, remains the fetch distance to consider for Exposure A.

In an effort to assist the designer in the proper selection of the Exposure B Category, the following guidance is provided. Terrain roughness corresponding to Exposure B may be defined by consideration of the frontal area of each obstruction to the wind in the upwind fetch surface area [C6-75]. For Exposure B to apply, on average the vertical frontal area of each obstruction to the wind in the upwind fetch surface area should be at least 5% of the area of open ground surrounding it. Mathematically this may be expressed as follows:

$$\frac{s_{ob}}{A_{ob}} \geq 0.05 \qquad \text{(Eq. C6-2)}$$

where

s_{ob} = average frontal area presented to the wind by each obstruction; and

A_{ob} = average area of open ground surrounding each obstruction.

Vertical frontal area is defined as the area of the

EXPOSURE A: Large city center with at least 50% of the buildings having a height in excess of 70 ft. The subject building must have this terrain upwind for at least one-half mile or 10 times the height of the building, whichever is greater

EXPOSURE B: Suburban residential area with mostly single family dwellings. Structures in the center of the photograph have an Exposure B terrain greater than 1,500 ft or ten times the height of the structure, whichever is greater, in any wind direction

EXPOSURE B: Urban area with numerous closely spaced obstructions having the size of single family dwellings or larger. For all structures shown, terrain representative of Exposure B extends more than 10 times the height of the structure or 1,500 ft, whichever is greater, in the upwind direction

EXPOSURE B: Structures in the foreground are subjected to an Exposure B terrain. Structures in the center top of the photograph adjacent to the clearing to the left, which is greater than 600 ft in length, are subjected to Exposure C when wind comes from the left over the clearing

EXPOSURE C: Flat open grassland with scattered obstructions having heights generally less than 30 ft

EXPOSURE C: Open terrain with scattered obstructions having heights generally less than 30 ft. For most wind directions, all structures in the photograph are less than 1,500 ft or 10 times the height of the structure, whichever is greater, from an open field which prevents the use of Exposure B

EXPOSURE D: A building at the shoreline (excluding shorelines in hurricane prone regions) with wind flowing over open water for a distance of at least one mile. Shorelines in Exposure D include inland waterways, the Great Lakes, and coastal areas of California, Oregon, Washington, and Alaska

projection of the obstruction onto a vertical plane normal to the wind direction. The area s_{ob} may be estimated by summing the vertical frontal areas of all obstructions within a selected area of upwind fetch and dividing the sum by the number of obstructions in the area. Likewise A_{ob} may be estimated by dividing the size of the selected area of upwind fetch by the number of obstructions in it.

Trees and bushes are porous and are deformed by strong winds, which reduces their effective frontal area. For conifers and other evergreens no more than 50% of their gross frontal area can be taken to be effective in obstructing the wind. For deciduous trees and bushes, no more than 15% of their gross frontal area can be taken to be effective in obstructing the wind. Gross frontal area is defined in this context as the projection onto a vertical plane (normal to the wind) of the area enclosed by the perimeter envelope of the tree or bush.

The "upwind fetch surface area" for evaluation is currently left to engineering judgment. For compo-

nent and cladding design, where a single exposure is to be selected representing the most severe condition (highest wind loads), it is suggested that 30° sectors be considered in assigning the most severe terrain exposure for the site.

A recent study [C6-66] has estimated that the majority of buildings (perhaps as much as 60–80%) have a terrain exposure corresponding to Exposure B.

A new requirement in ASCE 7-98 states that if a site is located in a transition zone between categories, the category resulting in the largest wind forces shall apply.

Exposure C has been expanded to include the shoreline in hurricane prone regions based on a recent study of the subject [C6-86]. It is recommended that Exposure C be used for the shoreline in non-hurricane prone regions where open water extends at least 600 ft but less than one mile upwind from the building or other structure. If less than 600 ft of open water extends upwind then the terrain further upwind becomes a factor in exposure selection.

Aerial photographs, representative of each exposure type, have been included in the commentary to aid the user in establishing the proper exposure for a given site (see Exposures A–D). Obviously, the proper assessment of exposure is a matter of good engineering judgment.

A significant improvement in ASCE 7-98 is that wind forces for both the main wind force resisting system and components and cladding are now based on the actual exposure that the structure is judged to be in rather than the concept of using only Exposure C or using a modifying exposure factor for Exposure B. (See Commentary discussion under Section C6.5.11.) In the case of component and cladding elements and for low-rise buildings designed using Fig. 6-4, wind profiles for Exposures A and B have been truncated in the bottom 100 ft and 30 ft, respectively. The truncation accounts for increased wind loading coefficients caused by local turbulence and from increased wind speeds near the surface associated with openings in the surface roughness such as parking lots, wide roads, road intersections, underdeveloped lots and tree clearings. Where such clearings adjacent to the building exceed 600 ft (183 m) the use of Exposure C is recommended. This effect has been accounted for by adjusting the velocity pressure exposure coefficient values in Table 6-5. The use of a truncated profile eliminates the underestimation in the wind loads which would occur in the absence of the truncation now that the actual exposures are used in the design.

In the case of component and cladding loads on high rise buildings, the negative pressures will not be affected by the truncation since these loads are referenced to the dynamic pressure at roof height. In the case of the positive pressures near the base of a high rise building ($h > 60$ ft) in an Exposure B environment, the component and cladding loads will increase in the lowest 30 ft. This increase is observed in wind tunnel tests, and the estimate of the positive cladding pressures near the base of taller buildings is, and has in the past, been underestimated in most cases using a non-truncated profile.

In the case of overall loads on buildings and structures, the net effect of the changes in the characteristics of the turbulence is less important as the loads are integrated over the entire structure. These overall loads are reasonably well estimated when normalized by the peak gust wind speed obtained using the non-truncated profile.

In general, when using Fig. 6-3, it is reasonable to base main wind force resisting system loads on specific wind directions since the frame loads typically can be correlated with a particular wind direction and act in a direction very close to the wind direction. However, it should also be recognized that the pressure coefficients in Fig. 6-3 are based on two perpendicular wind directions along the building axis. Pressures on components and cladding required for design have, from experience on many wind tunnel tests, not been consistently identified with a particular wind direction. In defining the exposure category for components and cladding in Sections 6.5.6.3.1 and 6.5.6.3.2, the phrase ". . . based on the exposure resulting in the highest wind loads for any wind direction at the site'' has been added to alert the user to the fact that there is no way to know, except for wind tunnel testing, which pressure zones are associated with which wind directions. Wind tunnel tests have shown high pressure zones to occur at unexpected wind directions. Therefore, a single exposure representing the most severe condition (highest wind loads) from any wind direction is required for use in design of component and cladding elements. However, in the design of the main wind force resisting system using Fig. 6-3, the designer may consider one or more wind directions and assign a terrain exposure to each direction for use in calculating wind loads for that direction. For example the designer may select a terrain exposure for each 90° quadrant and calculate wind loads for each quadrant. It is common in wind tunnel studies to consider 36 ten-degree sectors.

C6.5.6.4 Velocity Pressure Exposure Coefficient

The velocity pressure exposure coefficient K_z can be obtained using the equation:

$$K_z = \begin{cases} 2.01 \left(\dfrac{z}{z_g}\right)^{2/\alpha} & \text{for } 15 \text{ ft} \leq z \leq z_g \quad \text{(Eq. C6-3a)} \\ 2.01 \left(\dfrac{15}{z_g}\right)^{2/\alpha} & \text{for } z < 15 \text{ ft} \quad \text{(Eq. C6-3b)} \end{cases}$$

in which values of α and z_g are given in Table 6-4. These equations are now given in Table 6-5 to aid the user.

The values of the gradient height, z_g, listed in Table 6-4 are consistent with those values used in ASCE 7-93. However, because the shape of the wind speed profile changes when the reference wind speed is changed from fastest mile to the 3-second gust speed, it was necessary to modify the values of α in the power-law representation of the wind speed profile. The new values of α in Table 6-4 are based on a comprehensive review of existing data and representations of gust speed profiles. Correspondingly, the multiplier in Eq. C6-3 for the exposure coefficient, K_z, changes from 2.58 to 2.01. Values of K_z are assumed to be constant for heights less than 15, and for heights greater than the gradient height. For Case 1 in Exposures A and B, the value of K_z has been truncated at 100 ft and 30 ft, respectively (refer to Table 6-5).

The new values of α (Table 6-4) make the wind speed profiles flatter than they are in ASCE 7-93; this is consistent with theory as wind speed averaging time is changed from fastest-mile to 3-second gust speed [C6-67, C6-68]. Field data collected recently in flat terrain in 1992 on a 160 ft high tower with anemometers at several levels also indicated that α value changes as averaging time of wind speed is varied. In order to determine α values related to 3-second gust speed for Exposure Categories A, B, C and D, a published atmospheric boundary layer turbulence model [C6-67] was used. Mean velocity profiles, defined using power-law exponent from ASCE 7-93, were combined with the turbulence profiles to produce profiles of peak gust speed. Mean velocity profile power law exponents consistent with the gust profiles are listed in Table 6-4 for reference. In the profile calculations, boundary layer heights are the same ones used for ASCE 7-93 as indicated in Table C6-2 of that Standard. The gust profile was fitted to determine α values. As an additional check, log-law profiles were fitted with power-law profiles in the bottom 500 ft (152.4 m) to obtain α values. The roughness length z_o selected to represent Exposure Categories A, B, C and D were 0.3, 0.1, 0.01, and 0.003 m consistent with gust speed [C6-68]. The α values are rounded for use in the Standard. The α values of Table 6-4 increase K_z values for Exposure Categories A and B while decreasing the values for Exposure Category D as compared to values K_z in ASCE 7-93. In addition, differences in K_z values between exposure categories are reduced; e.g. for Exposure Category B at 30 ft (9 m) the values of $K_z = 0.7$ in Table 6-5 as compared to the value of $K_z = 0.5$ in ASCE 7-93. The new values of K_z in Table 6-5 are very close to the values specified in the Australian Standard [C6-34], which also uses the 3-second gust speed format; this agreement lends credence to the new values in Table 6-4.

C6.5.7 Wind Speed-Up over Hills and Escarpments

As an aid to the designer, this section has been rewritten to specify when topographic effects need to be applied to a particular structure rather than when they do not as in the previous version. In addition, the upwind distance to consider has been lengthened from 50 to 100 times the height of the topographic feature (100 H) and from one to two miles. In an effort to exclude situations where little or no topographic effect exists, condition (2) has been added to include the fact that the topographic feature should protrude significantly above (by a factor of two or more) upwind terrain features before it becomes a factor. For example, if a significant upwind terrain feature has a height of 35 ft above its base elevation and has a top elevation of 100 ft above mean sea level then the topographic feature (hill, ridge or escarpment) must have at least the H specified and extend to elevation 170 mean sea level (100 ft + 2 × 35 ft) within the two mile radius specified.

A recent wind tunnel study [C6-81] and observation of actual wind damage has shown that the affected height H is less than previously specified. Accordingly, condition (5) has been changed to 15 ft in Exposure C.

Buildings sited on the upper half of an isolated hill or escarpment may experience significantly higher wind speeds than buildings situated on level ground. To account for these higher wind speeds, the velocity pressure exposure coefficients in Table 6-5 are multiplied by a topographic factor, K_{zt}, defined in Eq. 6-13 of Section 6.5.10. The topographic feature (2-D ridge or escarpment, or 3-D axisymmetrical hill) is described by two parameters, H and L_h. H is the height of the hill or difference in elevation between the crest and that of the upwind terrain. L_h is the distance upwind of the crest to where the ground elevation is equal to half the height of the hill. K_{zt} is determined from three multipliers, K_1, K_2 and K_3, which are obtained from Fig. 6-2, respectively. K_1 is related to the shape of the topographic feature and the maximum speed-up near the crest, K_2 accounts for the reduction in speed-up with distance upwind or downwind of the crest, and K_3 accounts for the reduction in speed-up with height above the local ground surface.

The multipliers listed in Fig. 6-2 are based on the assumption that the wind approaches the hill along the direction of maximum slope, causing the greatest speed-up near the crest. The average maximum upwind slope of the hill is approximately $H/2L_h$, and measurements have shown that hills with slopes of less than about 0.10 ($H/L_h < 0.20$) are unlikely to produce significant speed-up of the wind. For values of $H/L_h > 0.5$ the speed-up effect is assumed to be independent of slope. The speed-up principally affects the mean wind speed rather than the amplitude of the turbulent fluctuations and this fact has been accounted for in the values of K_1, K_2 and K_3 given in Fig. 6-2. Therefore, values of K_{zt} obtained from Table 6-2 are intended for use with velocity pressure exposure coefficients, K_h and K_z, which are based on gust speeds.

It is not the intent of Section 6.5.7 to address the general case of wind flow over hilly or complex terrain for which engineering judgment, expert advice, or wind tunnel tests as described in Section 6.6 may be required. Background material on topographic speed-up effects may be found in the literature [C6-18, C6-21, C6-56].

The designer is cautioned that, at present, the standard contains no provision for vertical wind speed-up because of a topographic effect, even though this phenomenon is known to exist and can cause additional uplift on roofs. Additional research is required to quantify this effect before it can be incorporated into the standard.

C6.5.8 Gust Effect Factors

ASCE 7-98 contains a single gust effect factor of 0.85 for rigid buildings. As an option, the designer can incorporate specific features of the wind environment and building size to more accurately calculate a gust effect factor. One such procedure, previously contained in the Commentary, is now located in the body of the Standard [C6-63, C6-64]. A suggested procedure is also included for calculating the gust effect factor for flexible structures. The rigid structure gust factor is typically 0–4% higher than in ASCE 7-95 and is 0–10% lower than the simple, but conservative, value of 0.85 permitted in the Standard without calculation. The procedures for both rigid and flexible structures have been changed from the previous version to: (1) keep the rigid gust factor calculation within a few percent of the previous model; (2) provide a superior model for flexible structures which displays the peak factors g_Q and g_R; and (3) causes the flexible structure value to match the rigid structure as resonance is removed (an advantage not included in the previous version). A designer is free to use any other rational procedure in the approved literature, as stated in Section 6.5.8.3.

The gust effect factor accounts for the loading effects in the along-wind direction due to wind turbulence-structure interaction. It also accounts for along-wind loading effects due to dynamic amplification for flexible buildings and structures. It does not include

allowances for across-wind loading effects, vortex shedding, instability due to galloping or flutter, or dynamic torsional effects. For structures susceptible to loading effects that are not accounted for in the gust effect factor, information should be obtained from recognized literature [C6-60–C6-65] or from wind tunnel tests.

Along-Wind Response. Based on the preceding definition of the gust effect factor, predictions of along-wind response, e.g., maximum displacement, rms and peak acceleration, can be made. These response components are needed for survivability and serviceability limit states. In the following, expressions for evaluating these along-wind response components are given.

Maximum Along-Wind Displacement. The maximum along-wind displacement $X_{max}(z)$ as a function of height above the ground surface is given by

$$X_{max}(z) = \frac{\phi(z)\rho B h C_{fx} \bar{V}_{\bar{z}}^2}{2m_1(2\pi n_1)^2} KG \quad \text{(Eq. C6-4)}$$

where

$\phi(z)$ = the fundamental model shape = $(z/h)^\xi$;
ξ = the mode exponent;
ρ = aair density; and
C_{fx} = mean alongwind force coefficient;

$$m_1 = \text{modal mass} = \int_o^h \mu(z)\phi^2(z)\,dz \quad \text{(Eq. C6-5)}$$

$\mu(z)$ = mass per unit height

$$K = (1.65)^{\hat{\alpha}}/(\hat{\alpha} + \xi + 1) \quad \text{(Eq. C6-6)}$$

$\hat{V}_{\bar{z}}$ is the 3-second gust speed at height $\bar{z}$. This can be evaluated $\bar{V}_{\bar{z}} = \hat{b}(z/33)^{\hat{\alpha}}V$, where V is the 3-second gust speed in Exposure C at the reference height (obtained from Fig. 6-1); $\hat{b}$ and $\hat{\alpha}$ are given in Table 6-4.

RMS Along-Wind Acceleration. The rms along-wind acceleration $\sigma_{\ddot{x}}(z)$ as a function of height above the ground surface is given by

$$\sigma_{\ddot{x}}(z) = \frac{0.85\phi(z)\rho B h C_{fx} \bar{V}_{\bar{z}}^2}{m_1} I_{\bar{z}} KR \quad \text{(Eq. C6-7)}$$

where $\bar{V}_{\bar{z}}$ is the mean hourly wind speed at height $\bar{z}$, ft/s

$$\bar{V}_{\bar{z}} = \hat{b} \left(\frac{\bar{z}}{33} \right)^{\hat{\alpha}} V \qquad \text{(Eq. C6-8)}$$

where $\hat{b}$ and $\hat{\alpha}$ are defined in Table 6-4.

Maximum Along-Wind Acceleration. The maximum along-wind acceleration as a function of height above the ground surface is given by

$$\ddot{X}_{\max}(z) = g_{\ddot{x}}\sigma_{\ddot{x}}(z) \qquad \text{(Eq. C6-9)}$$

$$g_{\ddot{x}} = \sqrt{2 \ln(n_1 T)} + \frac{0.5772}{\sqrt{2 \ln(n_1 T)}} \qquad \text{(Eq. C6-10)}$$

where T = the length of time over which the minimum acceleration is computed, usually taken to be 3,600 s to represent 1 h.

Example:

This example is presented to illustrate the calculation of the gust effect factor. Tables C6-4, C6-5 and C6-6 may be found at the end of Section C6.

Table C6-4 uses the given information to obtain values from Table 6-4. Table C6-5 presents the calculated values. Table C6-6 summarizes the calculated displacements and accelerations as a function of the height, z.

Given Values:

Basic wind speed at reference height in Exposure
  C = 90 mph
Type of exposure = A
Building height h = 600 ft
Building width B = 100 ft
Building depth L = 100 ft
Building natural frequency n_1 = 0.2 Hz
Damping ratio = 0.01
C_{fx} = 1.3
Mode exponent = 1.0
Building density = 12 lb/cu ft = 0.3727 slugs/cu ft
Air density = 0.0024 slugs/cu ft

C6.5.9 Enclosure Classifications

The magnitude and sense of internal pressure is dependent upon the magnitude and location of openings around the building envelope with respect to a given wind direction. Accordingly, the Standard requires that a determination be made of the amount of openings in the envelope in order to assess enclosure classification (enclosed, partially enclosed, or open).

"Openings" are specifically defined in this version of the Standard as "apertures or holes in the building envelope which allow air to flow through the building envelope and which are designed as "open" during design winds." Examples include doors, operable windows, air intake exhausts for air conditioning and/or ventilation systems, gaps around doors, deliberate gaps in cladding and flexible and operable louvers. Once the enclosure classification is known, the designer enters Table 6-7 to select the appropriate internal pressure coefficient.

This version of the Standard creates four new definitions applicable to enclosure: "wind borne debris regions," "glazing," "impact resistant glazing," and "impact resistant covering." "Wind borne debris regions" are defined to alert the designer to areas requiring consideration of missile impact design and potential openings in the building envelope. "Glazing" is defined as "any glass or transparent or translucent plastic sheet used in windows, doors, or skylights." "Impact resistant glazing" is specifically defined as "glazing which has been shown by a test method, acceptable to the authority having jurisdiction, to withstand impact of wind borne missiles likely to be generated in wind borne debris regions during design winds." "Impact resistant coverings" over glazing can be shutters or screens designed for wind borne debris impact. While impact resistance can now be tested using the new ASTM Standard E 1886 [C6-70], no ASTM specification yet exists (one is currently under development) for defining specific missiles and impact speeds that can be referenced in the body of the Standard. The designer can refer to SBCCI SSTD 12 [C6-71] which does define specific missiles and corresponding impact speeds. Origins of missile impact provisions contained in these test standards are summarized in references [C6-72, C6-80].

Attention is made to Section 6.5.9.3 which requires the designer to assume glazing not designed to be impact resistant glazing or protected by an impact resistant covering on surfaces receiving positive pressure in wind borne debris regions to be treated as openings when assessing enclosure classification.

C6.5.10 Velocity Pressure

The basic wind speed is converted to a velocity pressure q_z in pounds per square foot (newtons per square meter) at height z by the use of Eq. 6-13.

The constant 0.00256 (or 0.613 in SI) reflects the mass density of air for the standard atmosphere, i.e., temperature of 59°F (15°C) and sea level pressure of 29.92 in. of mercury (101.325 kPa), and dimensions associated with wind speed in mi/h (m/s). The constant is obtained as follows:

$$\text{constant} = 1/2[(0.0765 \text{ lb/ft}^3)/(32.2 \text{ ft/s}^2)]$$
$$\times [(\text{mi/h})(5,280 \text{ ft/mi})$$
$$\times (1 \text{ h}/3,600 \text{ s})]^2$$
$$= 0.00256$$

$$\text{constant} = 1/2[(1.225 \text{ kg/m}^3)/(9.81 \text{ m/s}^2)]$$
$$\times [(\text{m/s})]^2[9.81 \text{ N/kg}]$$
$$= 0.613 \qquad \text{(Eq. C6-11)}$$

The numerical constant of 0.00256 should be used except where sufficient weather data are available to justify a different value of this constant for a specific design application. The mass density of air will vary as a function of altitude, latitude, temperature, weather, and season. Average and extreme values of air density are given in Table C6-1.

C6.5.11 Pressure and Force Coefficients

The pressure and force coefficients provided in Figs. 6-3 through 6-8 and in Tables 6-4 through 6-10 have been assembled from the latest boundary-layer wind-tunnel and full-scale tests and from previously available literature. Since the boundary-layer wind tunnel results were obtained for specific types of building such as low- or high-rise buildings and buildings having specific types of structural framing systems, the designer is cautioned against indiscriminate interchange of values among the figures and tables.

Loads on Main Wind-Force Resisting Systems. Figs. 6-3 and 6-4. The pressure coefficients for main wind-force resisting systems are separated into two categories:

1. Buildings of all heights (Fig. 6-3); and
2. Low-rise buildings having a height less than or equal to 60 ft (18 m) (Fig. 6-4).

In generating these coefficients, two distinctly different approaches were used. For the pressure coefficients given in Fig. 6-3, the more traditional approach was followed and the pressure coefficients reflect the actual loading on each surface of the building as a function of wind direction; namely, winds perpendicular or parallel to the ridge line.

For low-rise buildings having a height less than or equal to 60 ft (18 m), however, the values of GC_{pf} represent "pseudo" loading conditions (Case A and Case B) which, when applied to the building, envelope the desired structural actions (bending moment, shear, thrust) independent of wind direction and exposure. To capture all appropriate structural actions, the building must be designed for all wind directions by considering in turn each corner of the building as the windward corner shown in the sketches of Fig. 6-4. Note also that for all roof slopes, load Case A and B must be considered individually in order to determine the critical loading for a given structural assemblage or component thereof. These two separate loading conditions are required for each of the windward corners to generate the wind actions, including torsion, to be resisted by the structural systems. Note that the building "end zones" must be aligned in accordance with the assumed windward corner (see Fig. C6-2).

To develop the appropriate "pseudo" values of GC_{pf}, investigators at the University of Western Ontario [C6-11] used an approach which consisted essentially of permitting the building model to rotate in the wind tunnel through a full 360° while simultaneously monitoring the loading conditions on each of the surfaces (see Fig. C6-3). Both Exposures B and C were considered. Using influence coefficients for rigid frames, it was possible to spatially average and time average the surface pressures to ascertain the maximum induced external force components to be resisted. More specifically, the following structural actions were evaluated:

1. total uplift
2. total horizontal shear
3. bending moment at knees (2-hinged frame)
4. bending moment at knees (3-hinged frame), and
5. bending moment at ridge (2-hinged frame)

The next step involved developing sets of "pseudo" pressure coefficients to generate loading conditions which would envelope the maximum induced force components to be resisted for all possible wind directions and exposures. Note, for example, that the wind azimuth producing the maximum bending moment at the knee would not necessarily produce the maximum total uplift. The maximum induced external force components determined for each of the above five categories were used to develop the coefficients. The end result was a set of coefficients which represent fictitious loading conditions, but which conservatively envelope the maximum induced force components (bending moment, shear, and thrust) to be resisted, independent of wind direction.

The original set of coefficients was generated for the framing of conventional pre-engineered buildings, i.e., single story moment-resisting frames in one of the principal directions and bracing in the other principal direction. The approach was later extended to single story moment-resisting frames with interior columns [C6-19].

Subsequent wind tunnel studies [C6-69] have shown that the GC_{pf} values of Fig. 6-4 are also applicable to low-rise buildings with structural systems other than moment-resisting frames. That work examined the instantaneous wind pressures on a low-rise building with a 4:12 pitched gable roof and the resulting wind-induced forces on its main wind-force resisting system. Two different main wind-force resisting systems were evaluated. One consisted of shear walls and roof trusses at different spacings. The other had moment resisting frames in one direction, positioned at the same spacings as the roof trusses, and diagonal wind bracing in the other direction. Wind tunnel tests were conducted for both Exposures B and C. The findings of this study showed that the GC_{pf} values of Fig. 6-4 provided satisfactory estimates of the wind forces for both types of structural systems. This work confirms the validity of Fig. 6-4, which reflects the combined action of wind pressures on different external surfaces of a building and thus take advantage of spatial averaging.

In the original wind tunnel experiments, both B and C exposure terrains were checked. In these early experiments, Exposure B did not include nearby buildings. In general, the force components, bending moments, etc. were found comparable in both exposures, although GC_{pf} values associated with Exposure B terrain would be higher than that for Exposure C terrain because of reduced velocity pressure in Exposure B terrain. The GC_{pf} values given in Fig. 6-4 (also in Figs. 6-5, 6-6 and 6-7) are derived from wind tunnel studies modeled with Exposure C terrain. However, they may also be used in other exposures when the velocity pressure representing the appropriate exposure is used.

In recent comprehensive wind tunnel studies conducted by [C6-66] at the University of Western Ontario, it was determined that when low buildings ($h < 60$ ft) are embedded in suburban terrain (Exposure B which included nearby buildings), the pressures in most cases are lower than those currently used in existing standards and codes, although the values show a very large scatter because of high turbulence and many variables. The results seem to indicate that some reduction in pressures for buildings located in

Exposure B is justified. The Task Committee on Wind Loads believes it is desirable to design buildings for the exposure conditions consistent with the exposure designations defined in the Standard. In the case of low buildings, the effect of the increased intensity of turbulence in rougher terrain (i.e., Exposure A or B vs. C) increases the local pressure coefficients. In ASCE 7-95 this effect was accounted for by allowing the designer of a building situated in Exposure A or B to use the loads calculated as if the building were located in Exposure C, but to reduce the loads by 15%. In ASCE 7-98 the effect of the increased turbulence intensity on the loads is treated with the truncated profile. Using this approach, the actual building exposure is used and the profile truncation corrects for the underestimate in the loads that would be obtained otherwise. The resulting wind loads on components and cladding obtained using this approach are much closer to the true values than those obtained using Exposure C loads combined with a 15% reduction in the resulting pressures.

Fig. 6-4 is most appropriate for low buildings with a width greater than twice their height and a mean roof height that does not exceed 33 ft (10 m). The original data base included low buildings with width no greater than 5 times their eave height, and eave height did not exceed 33 ft (10 m). In the absence of more appropriate data, Fig. 6-4 may also be used for buildings with mean roof height which does not exceed the least horizontal dimension and is less than or equal to 60 ft (18 m). Beyond these extended limits, Fig. 6-3 should be used.

Internal pressure coefficients (GC_{pi}) to be used for loads on main wind-force resisting systems are given in Table 6-4. The internal pressure load can be critical in one-story moment resisting frames and in the top story of a building where the main wind-force resisting system consists of moment resisting frames. Loading cases with positive and negative internal pressures should be considered. The internal pressure load cancels out in the determination of total lateral load and base shear. The designer can use judgment in the use of internal pressure loading for the main wind-force resisting system of high rise buildings.

Loads on Components and Cladding. In developing the set of pressure coefficients applicable for the design of components and cladding as given in Figs. 6-5, 6-6 and 6-7, an envelope approach was followed but using different methods than for the main wind-force resisting systems of Fig. 6-4. Because of the small effective area which may be involved in the design of a particular component (consider, for example, the effective area associated with the design of a fastener), the point-wise pressure fluctuations may be highly correlated over the effective area of interest. Consider the local purlin loads shown in Fig. C6-2. The approach involved spatial averaging and time averaging of the point pressures over the effective area transmitting loads to the purlin while the building model was permitted to rotate in the wind tunnel through 360°. As the induced localized pressures may also vary widely as a function of the specific location on the building, height above ground level, exposure, and more importantly, local geometric discontinuities and location of the element relative to the boundaries in the building surfaces (walls, roof lines), these factors were also enveloped in the wind tunnel tests. Thus, for the pressure coefficients given in Figs. 6-5, 6-6 and 6-7, the directionality of the wind and influence of exposure have been removed and the surfaces of the building "zoned" to reflect an envelope of the peak pressures possible for a given design application.

As indicated in the discussion for Fig. 6-4, the wind tunnel experiments checked both B and C exposure terrains. Basically GC_p values associated with Exposure B terrain would be higher than that for Exposure C terrain because of reduced velocity pressure in Exposure B terrain. The GC_p values given in Figs. 6-5, 6-6 and 6-7 are associated with Exposure C terrain as obtained in the wind tunnel. However, they also may be used for any exposure when the correct velocity pressure representing the appropriate exposure is used (see Commentary discussion in Section C6.5.11 under Loads on Main Wind Force Resisting Systems).

The wind tunnel studies conducted by [C6-66] determined that when low buildings ($h < 60$ ft) are embedded in suburban terrain (Exposure B), the pressures on components and cladding in most cases are lower than those currently used in the standards and codes, although the values show a very large scatter because of high turbulence and many variables. The results seem to indicate that some reduction in pressures for components and cladding of buildings located in Exposure B is justified.

The pressure coefficients given in Fig. 6-8 for buildings with mean height greater than 60 feet were developed following a similar approach, but the influence of exposure was not enveloped [C6-42]. Therefore, Exposure Categories A, B, C or D may be used with the values of GC_p in Fig. 6-8 as appropriate.

Fig. 6-5. The pressure coefficient values provided in this figure are to be used for buildings with a mean roof height of 60 ft (18 m) or less. The values were obtained from wind-tunnel tests conducted at the University of Western Ontario [C6-10, C6-11], at the James Cook University of North Queensland [C6-6], and at Concordia University [C6-39, C6-40, C6-44, C6-45, C6-47]. These coefficients have been refined to reflect results of full-scale tests conducted by the National Bureau of Standards [C6-22] and the Building Research Station, England [C6-14]. Pressure coefficients for hemispherical domes on ground or on cylindrical structures have been reported [C6-52]. Some of the characteristics of the values in the figure are as follows:

1. The values are combined values of GC_p; the gust effect factors from these values should not be separated.
2. The velocity pressure q_h evaluated at mean roof height should be used with all values of GC_p.
3. The values provided in the figure represent the upper bounds of the most severe values for any wind direction. The reduced probability that the design wind speed may not occur in the particular direction for which the worst pressure coefficient is recorded has not been included in the values shown in the figure.
4. The wind-tunnel values, as measured, were based on the mean hourly wind speed. The values provided in the figures are the measured values divided by $(1.53)^2$ (see Fig. C6-1) to reflect the reduced pressure coefficient values associated with a 3-second gust speed.

Each component and cladding element should be designed for the maximum positive and negative pressures (including applicable internal pressures) acting on it. The pressure coefficient values should be determined for each component and cladding element on the basis of its location on the building and the effective area for the element. As recent research has shown [C6-41, C6-43], the pressure coefficients provided generally apply to facades with architectural features such as balconies, ribs and various facade textures.

Figs. 6-6 and 6-7A. These figures present values of GC_p for the design of roof components and cladding for buildings with multispan gable roofs and buildings with monoslope roofs. The coefficients are based on wind tunnel studies reported by [C6-46, C6-47, C6-51].

ASCE WIND FORCES ON STRUCTURES

Fig. 6-7B. The values of GC_p in this figure are for the design of roof components and cladding for buildings with sawtooth roofs and mean roof height, h, less than or equal to 60 ft (18 m). Note that the coefficients for corner zones on Segment A differ from those coefficients for corner zones on the segments designated as B, C, and D. Also, when the roof angle is less than or equal to 10°, values of GC_p for regular gables roofs (Fig. 6-5B) are to be used. The coefficients included in Fig. 6-7B are based on wind tunnel studies reported by [C6-35].

Fig. 6-8. The pressure coefficients shown in this figure have been revised to reflect the results obtained from comprehensive wind tunnel studies carried out by [C6-42]. In general, the loads resulting from these coefficients are lower than those required by ASCE 7-93. However, the area averaging effect for roofs is less pronounced when compared with the requirements of ASCE 7-93. The availability of more comprehensive wind tunnel data has also allowed a simplification of the zoning for pressure coefficients; flat roofs are now divided into three zones, and walls are represented by two zones.

The external pressure coefficients and zones given in Fig. 6-8 were established by wind tunnel tests on isolated "box-like" buildings [C6-2, C6-31]. Boundary-layer wind tunnel tests on high-rise buildings (mostly in downtown city centers) show that variations in pressure coefficients and the distribution of pressure on the different building facades are obtained [C6-53]. These variations are due to building geometry, low attached buildings, non-rectangular cross sections, setbacks, and sloping surfaces. In addition, surrounding buildings contribute to the variations in pressure. Wind-tunnel tests indicate that pressure coefficients are not distributed symmetrically and can give rise to torsional wind loading on the building.

Boundary-layer wind tunnel tests that include modeling of surrounding buildings permit the establishment of more exact magnitudes and distributions of GC_p for buildings that are not isolated or "box-like" in shape.

Tables 6-8 to 6-13. With the exception of Table 6-13, the pressure and force coefficient values in these tables are unchanged from ANSI A58.1-1972 and 1982, and ASCE 7-88 and 7-93. The coefficients specified in these tables are based on wind-tunnel tests conducted under conditions of uniform flow and low turbulence, and their validity in turbulent boundary layer flows has yet to be completely established. Additional pressure coefficients for conditions not

specified herein may be found in two references [C6-3, C6-36]. With regard to Table 6-10, local maximum and minimum peak pressure coefficients for cylindrical structures with $h/D < 2$ are $GC_p = 1.1$ and $GC_p = -1.1$, respectively, for Reynolds numbers ranging from 1.1×10^5 to 3.1×10^5 [C6-23]. The latter results have been obtained under correctly simulated boundary layer flow conditions.

With regard to Table 6-13, the force coefficients are a refinement of the coefficients specified in ANSI A58.1-1982 and in ASCE 7-93. The force coefficients specified are offered as a simplified procedure that may be used for trussed towers and are consistent with force coefficients given in ANSI/EIA/TIA-222-E-1991, Structural Standards for Steel Antenna Towers and Antenna Supporting Structures, and force coefficients recommended by Working Group No. 4 (Recommendations for Guyed Masts), International Association for Shell and Spatial Structures (1981).

It is not the intent of the Standard to exclude the use of other recognized literature for the design of special structures such as transmission and telecommunications towers. Recommendations for wind loads on tower guys are not provided as in previous editions of the Standard. Recognized literature should be referenced for the design of these special structures as is noted in Section C6.4.2.1. For the design of flagpoles, see ANSI/NAAMM FP1001-97, 4th Ed., Guide Specifications for Design of Metal Flagpoles.

C6.5.11.1 Internal Pressure Coefficients

The internal pressure coefficient values in Table 6-7 were obtained from wind tunnel tests [C6-38] and full scale data [C6-59]. Even though the wind-tunnel tests were conducted primarily for low-rise buildings, the internal pressure coefficient values are assumed to be valid for buildings of any height. The values $GC_{pi} = +0.18$ and -0.18 are for enclosed buildings. It is assumed that the building has no dominant opening or openings and that the small leakage paths that do exist are essentially uniformly distributed over the building's envelope. The internal pressure coefficient values for partially enclosed buildings assume that the building has a dominant opening or openings. For such a building, the internal pressure is dictated by the exterior pressure at the opening and is typically increased substantially as a result. Net loads, i.e., the combination of the internal and exterior pressures, are therefore also significantly increased on the building surfaces that do not contain the opening. Therefore, higher GC_{pi} values of +0.55 and −0.55 are applicable to this case. These values include a reduction factor to account for the lack of perfect correlation between the internal pressure and the external pressures on the building surfaces not containing the opening [C6-82, C6-83]. Taken in isolation, the internal pressure coefficients can reach values of ±0.8, (or possibly even higher on the negative side).

For partially enclosed buildings containing a large unpartitioned space the response time of the internal pressure is increased and this reduces the ability of the internal pressure to respond to rapid changes in pressure at an opening. The gust factor applicable to the internal pressure is therefore reduced. Eq. 6-14 which is based on references [C6-84, C6-85] is provided as a means of adjusting the gust factor for this effect on structures with large internal spaces such as stadiums and arenas.

Glazing in the bottom 60 ft of buildings that are sited in hurricane-prone regions that is not impact resistant glazing or is not protected by impact resistant coverings should be treated as openings. Because of the nature of hurricane winds [C6-27], glazing in buildings sited in hurricane areas is very vulnerable to breakage from missiles, unless the glazing can withstand reasonable missile loads and subsequent wind loading, or the glazing is protected by suitable shutters. Glazing above 60 ft (18 m) is also somewhat vulnerable to missile damage, but because of the greater height, this glazing is typically significantly less vulnerable to damage than glazing at lower levels. When glazing is breached by missiles, development of high internal pressure results, which can overload the cladding or structure if the higher pressure was not accounted for in the design. Breaching of glazing can also result in a significant amount of water infiltration, which typically results in considerable damage to the building and its contents [C6-33, C6-49, C6-50].

If the option of designing for higher internal pressure (versus designing glazing protection) is selected, it should be realized that if glazing is breached, significant damage from overpressurization to interior partitions and ceilings is likely. The influence of compartmentation on the distribution of increased internal pressure has not been researched. If the space behind breached glazing is separated from the remainder of the building by a sufficiently strong and reasonably air-tight compartment, the increased internal pressure would likely be confined to that compartment. However, if the compartment is breached (e.g., by an open corridor door, or by collapse of the compartment wall), the increased internal

pressure will spread beyond the initial compartment quite rapidly. The next compartment may contain the higher pressure, or it too could be breached, thereby allowing the high internal pressure to continue to propagate.

Because of the great amount of air leakage that often occurs at large hangar doors, designers of hangars should consider utilizing the internal pressure coefficients for partially enclosed buildings in Table 6-7.

C6.5.12 Design Wind Loads on Buildings
This version of the standard states in the body of the Standard specific wind pressure equations for both the main wind force resisting systems and components and cladding.

In Eqs. 6-15, 6-17, and 6-19 a new velocity pressure term "q_i" appears that is defined as the "velocity pressure for internal pressure determination." The positive internal pressure is dictated by the positive exterior pressure on the windward face at the point where there is an opening. The positive exterior pressure at the opening is governed by the value of q at the level of the opening, not q_h. Therefore the old provision which used q_h as the velocity pressure is not in accord with the physics of the situation. For low buildings this does not make much difference, but for the example of a 300 ft tall building in Exposure B with a highest opening at 60 ft, the difference between q_{300} and q_{60} represents a 59% increase in internal pressure. This is unrealistic and represents an unnecessary degree of conservatism. Accordingly, $q_i = q_z$ for positive internal pressure evaluation in partially enclosed buildings where height z is defined as the level of the highest opening in the building that could affect the positive internal pressure. For buildings sited in wind borne debris regions, glazing that is not impact resistant or protected with an impact resistant covering, q_i should be treated as an opening. For positive internal pressure evaluation, q_i may conservatively be evaluated at height h ($q_i = q_h$).

C6.5.12.3 Full and Partial Loading
Tall buildings should be checked for torsional response induced by partial wind loading and by eccentricity of the elastic center with respect to the resultant wind load vector and the center of mass. The load combinations described in Fig. 6-9 reflect surface pressure patterns that have been observed on tall buildings in turbulent wind. Wind tunnel tests have demonstrated that even a 25% selective load reduction can underestimate the wind-induced torsion in buildings with a uniform rectangular cross-section

[C6-50]. In some structural systems, more severe effects are observed when the resultant wind load acts diagonally to the building or other structure. To account for this effect and the fact that many structures exhibit maximum response in the across-wind direction, a structure should be capable of resisting 75% of the design wind load applied simultaneously along the principal axes. Additional information on torsional response due to full and partial loading can be found in the literature [C6-2, C6-4, C6-17].

C6.6 METHOD 3—WIND-TUNNEL PROCEDURE

Wind tunnel testing is specified when a structure contains any of the characteristics defined in Section 6.5.2 or when the designer wishes to more accurately determine the wind loads. For some building shapes wind tunnel testing can reduce the conservatism due to enveloping of wind loads inherent in Methods 1 and 2. Also, wind tunnel testing accounts for shielding or channeling and can more accurately determine wind loads for a complex building shape than Methods 1 and 2. It is the intent of the standard that any building or other structure be allowed to use the wind tunnel testing method to determine wind loads. Requirements for proper testing are given in Section 6.6.2.

Wind-tunnel tests are recommended when the building or other structure under consideration satisfies one or more of the following conditions:

1. has a shape which differs significantly from a uniform rectangular prism or "box-like" shape,
2. is flexible with natural frequencies normally below 1 Hz,
3. is subject to buffeting by the wake of upwind buildings or other structures, or
4. is subject to accelerated flow caused by channeling or local topographic features.

It is common practice to resort to wind-tunnel tests when design data are required for the following wind-induced loads:

1. curtain wall pressures resulting from irregular geometry,
2. across-wind and/or torsional loads,
3. periodic loads caused by vortex shedding, and
4. loads resulting from instabilities such as flutter or galloping.

Boundary-layer wind tunnels capable of developing flows that meet the conditions stipulated in Section 6.4.3.1 typically have test-section dimensions in the following ranges; width of 6–12 ft (2–4 m), height of 6–10 ft (2–3 m), and length of 50–100 ft (15–30 m). Maximum wind speeds are ordinarily in the range of 25–100 mph (10–45 m/s). The wind tunnel may be either an open-circuit or closed-circuit type.

Three basic types of wind-tunnel test models are commonly used. These are designated as follows: (1) rigid pressure model (PM); (2) rigid high-frequency base balance model (H-FBBM); and (3) aeroelastic model (AM). One or more of the models may be employed to obtain design loads for a particular building or structure. The PM provides local peak pressures for design of elements such as cladding and mean pressures for the determination of overall mean loads. The H-FBBM measures overall fluctuating loads (aerodynamic admittance) for the determination of dynamic responses. When motion of a building or structure influences the wind loading, the AM is employed for direct measurement of overall loads, deflections and accelerations. Each of these models, together with a model of the surroundings (proximity model), can provide information other than wind loads such as snow loads on complex roofs, wind data to evaluate environmental impact on pedestrians, and concentrations of air-pollutant emissions for environmental impact determinations. Several references provide detailed information and guidance for the determination of wind loads and other types of design data by wind-tunnel tests [C6-4, C6-7, C6-8, C6-33].

Wind tunnel tests frequently measure wind loads which are significantly lower than required by Section 6.5 due to the shape of the building, shielding in excess of that implied by exposure categories, and necessary conservatism in enveloping load coefficients in Section 6.5. In some cases, adjacent structures may shield the structure sufficiently that removal of one or two structures could significantly increase wind loads. Additional wind tunnel testing without specific nearby buildings (or with additional buildings if they might cause increased loads through channeling or buffeting) is an effective method for determining the influence of adjacent buildings. It would be prudent for the designer to test any known conditions that change the test results and apply good engineering judgment in interpreting the test results. Discussion between the owner, designer and wind-tunnel laboratory can be an important part of this decision. However, it is impossible to anticipate all possible changes to the surrounding environment that could significantly impact pressure for the main wind force resisting system and for cladding pressures. Also, additional testing may not be cost effective. Suggestions, written in mandatory language for users (e.g., code writers) desiring to place a lower limit on the results of wind tunnel testing are shown below.

Lower limit on pressures for main wind force resisting system. Forces and pressures determined by wind tunnel testing shall be limited to not less than 80% of the design forces and pressures which would be obtained in Section 6.5 for the structure unless specific testing is performed to show that it is the aerodynamic coefficient of the building itself, rather than shielding from nearby structures, that is responsible for the lower values. The 80% limit may be adjusted by the ratio of the frame load at critical wind directions as determined from wind tunnel testing without specific adjacent buildings (but including appropriate upwind roughness), to that determined by Section 6.5.

Lower limit on pressures for components and cladding. The design pressures for components and cladding on walls or roofs shall be selected as the greater of the wind tunnel test results or 80% of the pressure obtained for Zone 4 for walls and Zone 1 for roofs as determined in Section 6.5, unless specific testing is performed to show that it is the aerodynamic coefficient of the building itself, rather than shielding from nearby structures, that is responsible for the lower values. Alternatively, limited tests at a few wind directions without specific adjacent buildings, but in the presence of an appropriate upwind roughness, may be used to demonstrate that the lower pressures are due to the shape of the building and not to shielding.

REFERENCES

[C6-1] Abbey, R.F. "Risk probabilities associated with tornado wind speeds." *Proc., Symposium on Tornadoes: Assessment of Knowledge and Implications for Man,* R.E. Peterson, Ed., Institute for Disaster Research, Texas Tech University, Lubbock, Tex., 1976.

[C6-2] Akins, R.E., and Cermak, J.E. Wind pressures on buildings. Technical Report CER 7677REAJEC15, Fluid Dynamics and Diffusion Lab, Colorado State University, Fort Collins, Colo., 1975.

[C6-3] ASCE Wind forces on structures. *Trans. ASCE,* 126(2), 1124–1198, 1961.

[C6-4] Wind tunnel model studies of buildings and structures. *Manuals and Reports on Engineering Practice,* No. 67, American Society of Civil Engineers, New York, N.Y., 1987.

[C6-5] Batts, M.E., Cordes, M.R., Russell, L.R., Shaver, J.R., and Simiu, E. Hurricane wind speeds in the United States. NBS Building Science Series 124, National Bureau of Standards, Washington, D.C., 1980.

[C6-6] Best, R.J., and Holmes, J.D. Model study of wind pressures on an isolated single-story house. James Cook University of North Queensland, Australia, Wind Engineering Rep. 3/78, 1978.

[C6-7] Boggs, D.W., and Peterka, J.A. "Aerodynamic model tests of tall buildings." *J. Engrg. Mech.,* ASCE, 115(3), New York, N.Y., 618–635, 1989.

[C6-8] Cermak, J.E. "Wind-tunnel testing of structures." *J. Engrg. Mech. Div.,* ASCE, 103(6), New York, N.Y., 1125–1140, 1977.

[C6-9] Cheung, J.C.J., and Melbourne, W.H. "Wind loadings on porous cladding." *Proc., 9th Australian Conference on Fluid Mechanics,* 308, 1986.

[C6-10] Davenport, A.G., Surry, D., and Stathopoulos, T. Wind loads on low-rise buildings. Final Report on Phases I and II. BLWT-SS8, University of Western Ontario, London, Ontario, Canada, 1977.

[C6-11] Davenport, A.G., Surry, D., and Stathopoulos, T. Wind loads on low-rise buildings. Final Report on Phase III, BLWT-SS4, University of Western Ontario, London, Ontario, Canada, 1978.

[C6-12] Interim guidelines for building occupants' protection from tornadoes and extreme winds. TR-83A, Defense Civil Preparedness Agency, Washington, D.C., 1975. [Available from Superintendent of Documents, U.S. Government Printing Office, Washington, D.C. 20402].

[C6-13] Durst, C.S. "Wind speeds over short periods of time." *Meteor. Mag.,* 89, 181–187, 1960.

[C6-14] Eaton, K.J., and Mayne, J.R. "The measurement of wind pressures on two-story houses at Aylesbury." *J. Industrial Aerodynamics,* 1(1), 67–109, 1975.

[C6-15] Georgiou, P.N., Davenport, A.G., and Vickery, B.J. "Design wind speeds in regions dominated by tropical cyclones." *J. Wind Engrg. and Industrial Aerodynamics,* 13, 139–152, 1983.

[C6-16] Haig, J.R. *Wind Loads on Tiles For USA*. Redland Technology Limited, Horsham, West Sussex, England, June, 1990.

[C6-17] Isyumov, N. "The aeroelastic modeling of tall buildings." *Proc., International Workshop on Wind Tunnel Modeling Criteria and Techniques in Civil Engineering Applications*, Cambridge University Press, Gaithersburg, Md., 373–407, 1982.

[C6-18] Jackson, P.S., and Hunt, J.C.R. "Turbulent wind flow over a low hill." *Quarterly J. of the Royal Meteorological Society*, 101, 929–955, 1975.

[C6-19] Kavanagh, K.T., Surry, D., Stathopoulos, T., and Davenport, A.G. Wind loads on low-rise buildings: Phase IV. BLWT-SS14, University of Western Ontario, London, Ontario, Canada, 1983.

[C6-20] Krayer, W.R., and Marshall, R.D. "Gust factors applied to hurricane winds." *Bulletin of the American Meteorological Society*, 73, 613–617, 1992.

[C6-21] Lemelin, D.R., Surry, D., and Davenport, A.G. "Simple approximations for wind speed-up over hills." *J. Wind Engrg. and Industrial Aerodynamics*, 28, 117–127, 1988.

[C6-22] Marshall, R.D. The measurement of wind loads on a full-scale mobile home. NBSIR 77-1289, National Bureau of Standards, U.S. Dept. of Commerce, Washington, D.C., 1977.

[C6-23] Macdonald, P.A., Kwok, K.C.S., and Holmes, J.H. Wind loads on isolated circular storage bins, silos and tanks: Point pressure measurements. Research Report No. R529, School of Civil and Mining Engineering, University of Sydney, Sydney, Australia, 1986.

[C6-24] McDonald, J.R. A methodology for tornado hazard probability assessment. NUREG/CR3058, U.S. Nuclear Regulatory Commission, Washington, D.C., 1983.

[C6-25] Mehta, K.C., Minor, J.E., and McDonald, J.R. "Wind speed analyses of April 3–4 Tornadoes." *J. Struct. Div.*, ASCE, 102(9), 1709–1724, 1976.

[C6-26] Minor, J.E. "Tornado technology and professional practice." *J. Struct. Div.*, ASCE, 108(11), 2411–2422, 1982.

[C6-27] Minor, J.E., and Behr, R.A. "Improving the performance of architectural glazing systems in hurricanes." *Proc., Hurricanes of 1992*, American Society of Civil Engineers (Dec. 1–3, 1993, Miami, Fla.), C1–11, 1993.

[C6-28] Minor, J.E., McDonald, J.R., and Mehta, K.C. The tornado: An engineering oriented perspective. TM ERL NSSL-82, National Oceanic and Atmospheric Administration, Environmental Research Laboratories, Boulder, Colo., 1977.

[C6-29] Peterka, J.A. "Improved extreme wind prediction for the United States." *J. Wind Engrg. and Industrial Aerodynamics*, 41, 533–541, 1992.

[C6-30] Peterka, J.A., and Shahid, S. "Extreme gust wind speeds in the U.S." *Proc., 7th U.S. National Conference on Wind Engineering*, UCLA, Los Angeles, Calif., 2, 503–512, 1993.

[C6-31] Peterka, J.A., and Cermak, J.E. "Wind pressures on buildings—Probability densities." *J. Struct. Div.*, ASCE, 101(6), 1255–1267, 1974.

[C6-32] Perry, D.C., Stubbs, N., and Graham, C.W. "Responsibility of architectural and engineering communities in reducing risks to life, property and economic loss from hurricanes." *Pro., ASCE Conference on Hurricanes of 1992*, Miami, Fla., 1993.

[C6-33] Reinhold, T.A. (Ed.). "Wind tunnel modeling for civil engineering applications." *Proc., International Workshop on Wind Tunnel Modeling Criteria and Techniques in Civil Engineering Applications*, Cambridge University Press, Gaithersburg, Md., 1982.

[C6-34] Australian Standard SAA Loading Code, Part 2: Wind Loads. Published by Standards Australia, Standards House, 80 Arthur St., North Sydney, NSW, Australia, 1989.

[C6-35] Saathoff, P., and Stathopoulos, T. "Wind loads on buildings with sawtooth roofs." *J. Struct. Engrg.*, ASCE, 118(2) 429–446, 1992.

[C6-36] *Normen fur die Belastungsannahmen, die Inbetriebnahme und die Uberwachung der Bauten*, SIA Technische Normen Nr 160, Zurich, Switzerland, 1956.

[C6-37] *Standard Building Code.* Southern Building Code Congress International, 1994.

[C6-38] Stathopoulos, T., Surry, D., and Davenport, A.G. "Wind-induced internal pressures in low buildings." *Proc., 5th International Conference on Wind Engineering*, Colorado State University, Fort Collins, Colo., 1979.

[C6-39] Stathopoulos, T., Surry, D., and Davenport, A.G. "A simplified model of wind pressure coefficients for low-rise buildings." *4th Colloquium on Industrial Aerodynamics*, Aachen, West Germany, June 18–20, 1980.

[C6-40] Stathopoulos, T. "Wind loads on eaves of low buildings." *J. Struct. Div.*, ASCE, 107(10), 1921–1934, 1981.

[C6-41] Stathopoulos, T., and Zhu, X. "Wind pressures on buildings with appurtenances." *J. Wind Engrg. and Industrial Aerodynamics*, 31, 265–281, 1988.

[C6-42] Stathopoulos, T., and Dumitrescu-Brulotte, M. "Design recommendations for wind loading on buildings of intermediate height." *Canadian J. Civ. Engrg*, 16(6), 910–916, 1989.

[C6-43] Stathopoulos, T., and Zhu, X. "Wind pressures on buildings with mullions." *J. Struct. Engrg.*, ASCE, 116(8), 2272–2291, 1990.

[C6-44] Stathopoulos, T., and Luchian, H.D. "Wind pressures on building configurations with stepped roofs." *Canadian J. Civ. Engrg.*, 17(4), 569–577, 1990.

[C6-45] Stathopoulos, T., and Luchian, H. "Wind-induced forces on eaves of low buildings." *Wind Engineering Society Inaugural Conference*, Cambridge, England, 1992.

[C6-46] Stathopoulos, T., and Mohammadian, A.R. "Wind loads on low buildings with mono-sloped roofs." *J. Wind Engrg. and Industrial Aerodynamics*, 23, 81–97, 1986.

[C6-47] Stathopoulos, T., and Saathoff, P. "Wind pressures on roofs of various geometries." *J. Wind Engrg. and Industrial Aerodynamics*, 38, 273–284, 1991.

[C6-48] Stubbs, N., and Boissonnade, A. "Damage simulation model or building contents in a hurricane environment." *Proc., 7th U.S. National Conference on Wind Engineering*, UCLA, Los Angeles, Calif., 2, 759–771, 1993.

[C6-49] Stubbs, N., and Perry, D.C. "Engineering of the building envelope." *Proc., ASCE Conference on Hurricanes of 1992*, Miami, Fla., 1993.

[C6-50] Surry, D., Kitchen, R.B., and Davenport, A.G. "Design effectiveness of wind tunnel studies for buildings of intermediate height." *Canadian J. Civ. Engrg.*, 4(1), 96–116, 1977.

[C6-51] Surry, D., and Stathopoulos, T. The wind loading of buildings with monosloped roofs.

Final Report, BLWT-SS38, University of Western Ontario, London, Ontario, Canada, 1988.

[C6-52] Taylor, T.J. "Wind pressures on a hemispherical dome." *J. Wind Engrg. and Industrial Aerodynamics*, 40(2), 199–213, 1992.

[C6-53] Templin, J.T., and Cermak, J.E. Wind pressures on buildings: Effect of mullions. Tech. Rep. CER76–77JTT-JEC24, Fluid Dynamics and Diffusion Lab, Colorado State University, Fort Collins, Colo., 1978.

[C6-54] Vickery, P.J., and Twisdale, L.A. "Wind field and filling models for hurricane wind speed predictions." *J Struct Engrg.*, ASCE, 121(11), 1700–1709, 1995.

[C6-55] Vickery, P.J., Davenport, A.G., and Surry, D. "Internal pressures on low-rise buildings." *4th Canadian Workshop on Wind Engineering*, Toronto, Ontario, 1984.

[C6-56] Walmsley, J.L., Taylor, P.A., and Keith, T. "A simple model of neutrally stratified boundary-layer flow over complex terrain with surface roughness modulations." *Boundary-Layer Meteorology*, 36, 157–186, 1986.

[C6-57] Wen, Y.K., and Chu, S.L. "Tornado risks and design wind speed." *J. Struct. Div.*, ASCE, 99(12), 2409–2421, 1973.

[C6-58] Womble, J.A., Yeatts, B.B., and Mehta, K.C. "Internal wind pressures in a full and small scale building." *Proc., 9th International Conference on Wind Engineering*, New Delhi, India: Wiley Eastern Ltd., 1995.

[C6-59] Yeatts, B.B., and Mehta, K.C. "Field study of internal pressures." *Proc., 7th U.S. National Conference on Wind Engineering*, UCLA, Los Angeles, Calif., 2, 889–897, 1993.

[C6-60] Gurley, K., and Kareem, A. "Gust Loading Factors for Tension Leg Platforms." *Applied Ocean Research*, 15(3), 1993.

[C6-61] Kareem, A. "A dynamic response of high-rise buildings to stochastic wind loads." *J. Wind Engrg. and Industrial Aerodynamics*, 41–44, 1992.

[C6-62] Kareem, A. "Lateral-torsional motion of tall buildings to wind loads." *J. Struct. Engrg.*, ASCE, 111(11), 1985.

[C6-63] Solari, G. "Gust buffeting I: Peak wind velocity and equivalent pressure." *J. Struct. Engrg.*, ASCE, 119(2), 1993.

[C6-64] Solari, G. "Gust buffeting II: Dynamic along-wind response." *J. Struct. Engrg.*, ASCE, 119(2), 1993.

[C6-65] Kareem, A., and Smith, C. "Performance of off-shore platforms in hurricane Andrew." *Proc., Hurricanes of 1992*, ASCE, Miami, Fla., Dec., 1993.

[C6-66] Ho, E. "Variability of low building wind lands." Doctoral Dissertation, University of Western Ontario, London, Ontario, Canada, 1992.

[C6-67] Simiu, E., and Scanlan, R.H. *Wind effects on structures.* Third Edition, John Wiley & Sons, New York, N.Y., 1996.

[C6-68] Cook N. *The designer's guide to wind loading of building structures, Part I,* Butterworths Publishers, 1985.

[C6-69] Isyumov, N., and Case, P. "*Evaluation of structural wind loads for low-rise buildings contained in ASCE standard 7-1995.*" BLWT-SS17-1995, Univ. of Western Ontario, London, Ontario, Canada.

[C6-70] Standard Test Method for Performance of Exterior Windows, Curtain Walls, Doors and Storm Shutters Impacted by Missile(s) and Exposed to Cyclic Pressure Differentials. ASTM E1886-97, ASTM Inc., West Conshohocken, Pa., 1997.

[C6-71] SBCCI Test Standard for Determining Impact Resistance from Windborne Debris. SSTD 12-97, Southern Building Code Congress, International, Birmingham, Ala., 1997.

[C6-72] Minor J.E. "Windborne debris and the building envelope." *J. Wind Engrg. and Industrial Aerodynamics*, 53, 207–227, 1994.

[C6-73] Peterka, J.A., Cermak, J.E., Cochran, L.S., Cochran, B.C., Hosoya, N., Derickson, R.G., Harper, C., Jones, J., and Metz, B. "Wind uplift model for asphalt shingles." *J. Arch. Engrg.*, December, 1997.

[C6-74] Peterka, J.A., and Shahid, S. "Design gust wind speeds for the United States." *J. Struct. Div.*, ASCE, February, 1998.

[C6-75] Lettau, H. "Note on aerodynamic roughness parameter estimation on the basis of roughness element description." *J. Applied Meteorology*, 8, 828–832, 1969.

[C6-76] Ellingwood, B.R., MacGregor, J.G., Galambos, T.V., and Cornell, C.A. "Probability-based load criteria: Load factors and load combinations." *J. Struct. Div.*, ASCE 108(5), 978–997, 1981.

[C6-77] Ellingwood, B. "Wind and snow load statistics for probability design." *J. Struct. Div.*, ASCE 197(7), 1345–1349, 1982.

[C6-78] Vickery, P.J., Skerlj, P.S., Steckley, A.C., and Twisdale, L.A. "Hurricane wind field and gust factor models for use in hurricane wind speed simulations." Submitted for publication.

[C6-79] Vickery, P.J., Skerlj, P.S., and Twisdale, L.A. "Simulation of hurricane risk in the United States using an empirical storm track modeling technique." Submitted for publication.

[C6-80] Twisdale, L.A., Vickery, P.J., and Steckley, A.C. "Analysis of hurricane windborne debris impact risk for residential structures." State Farm Mutual Automobile Insurance Companies, March, 1996.

[C6-81] Means, B., Reinhold, T.A., and Perry, D.C. "Wind loads for low-rise buildings on escarpments." *Proc., ASCE Structures Congress 14*, Chicago, Ill., April, 1996.

[C6-82] Beste, F., and Cermak, J.E. "Correlation of internal and area-averaged wind pressures on low-rise buildings." *3rd International Colloquium on Bluff Body Aerodynamics and Applications*, Blacksburg, Va., July 28–August 1, 1996.

[C6-83] Irwin, P.A. "Pressure model techniques for cladding loads." *J. Wind Engrg. and Industrial Aerodynamics*, 29, 69–78, 1987.

[C6-84] Vickery, P.J., and Bloxham, C. "Internal pressure dynamics with a dominant opening." *J. Wind Engrg. and Industrial Aerodynamics*, 41–44, 193–204, 1992.

[C6-85] Irwin, P.A., and Dunn, G.E. "Review of internal pressures on low-rise buildings." RWDI Report 93-270 for Canadian Sheet Building Institute, February 23, 1994.

[C6-86] Vickery, P.J., and Skerlj, P.S. On the Elimination of Exposure D along the Hurricane Coastline in ASCE-7. Report for Andersen Corporation by Applied Research Associates, ARA Project 4667, March, 1998.

[C6-87] Mehta, K.C., and Marshall, R.D. *Guide to the Use of the Wind Load Provisions of ASCE 7-95*, ASCE Press, 1998.

[C6-88] Vickery, P.J., and Twisdale, L.A. "Prediction of hurricane wind speeds in the United States." *J. Struct. Engrg.*, ASCE, 121(11), 1691–1699, 1995.

[C6-89] Georgiou, P.N. "Design wind speeds in tropical cyclone regions." Ph.D. Thesis, University of Western Ontario, London, Ontario, Canada, 1985.

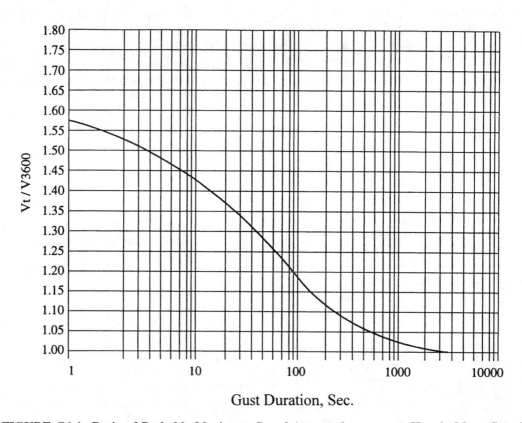

FIGURE C6-1. Ratio of Probable Maximum Speed Averaged over *t* s to Hourly Mean Speed

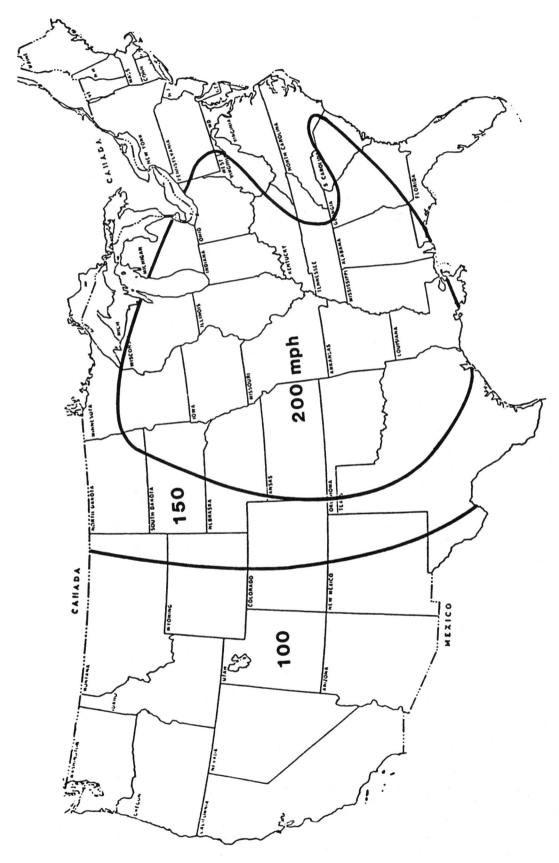

FIGURE C6-1A. Tornadic Gust Wind Speed Corresponding to Annual Probability of 10^{-5} (Mean Recurrence Interval of 100,000 Years) (from ANSI/ANS 1983)

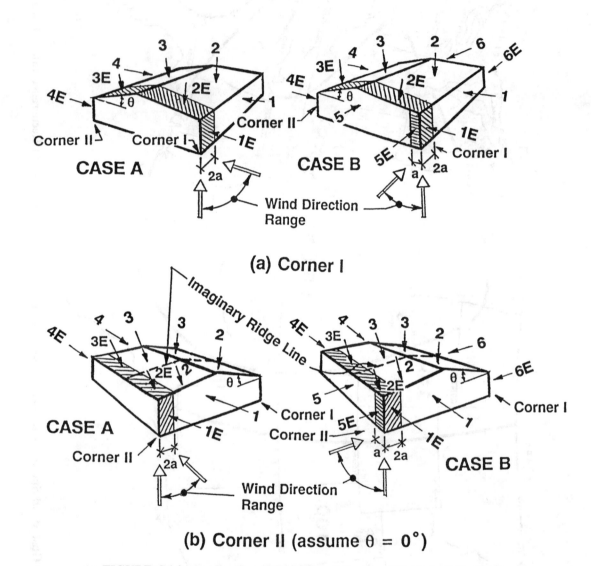

(a) Corner I

(b) Corner II (assume θ = 0°)

FIGURE C6-2. Application of Load Cases for Two Windward Corners

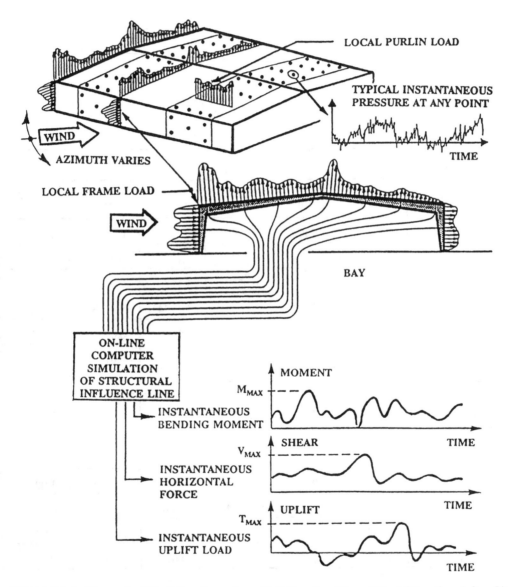

FIGURE C6-3. Unsteady Wind Loads on Low Building for Given Wind Direction (after [11])

Ambient Air Density Values for Various Altitudes

Table C6-1

Altitude		Ambient Air Density					
Feet	Meters	Minimum (lbm/ft^3)	Minimum (kg/m^3)	Average (lbm/ft^3)	Average (kg/m^3)	Maximum (lbm/ft^3)	Maximum (kg/m^3)
0	0	0.0712	1.1392	0.0765	1.2240	0.0822	1.3152
1000	305	0.0693	1.1088	0.0742	1.1872	0.0795	1.2720
2000	610	0.0675	1.0800	0.0720	1.1520	0.0768	1.2288
3000	914	0.0657	1.0512	0.0699	1.1184	0.0743	1.1888
3281	1000	0.0652	1.0432	0.0693	1.1088	0.0736	1.1776
4000	1219	0.0640	1.0240	0.0678	1.0848	0.0718	1.1488
5000	1524	0.0624	0.9984	0.0659	1.0544	0.0695	1.1120
6000	1829	0.0608	0.9728	0.0639	1.0224	0.0672	1.0752
6562	2000	0.0599	0.9584	0.0629	1.0064	0.0660	1.0560
7000	2134	0.0592	0.9472	0.0620	0.9920	0.0650	1.0400
8000	2438	0.0577	0.9232	0.0602	0.9632	0.0628	1.0048
9000	2743	0.0561	0.8976	0.0584	0.9344	0.0607	0.9712
9843	3000	0.0549	0.8784	0.0569	0.9104	0.0591	0.9456
10,000	3048	0.0547	0.8752	0.0567	0.9072	0.0588	0.9408

Probability of Exceeding Design Wind Speed During Reference Period

Table C6-2

Annual Probability P_a	Reference (Exposure) Period, n (years)					
	1	5	10	25	50	100
0.04 (1/25)	0.04	0.18	0.34	0.64	0.87	0.98
0.02 (1/50)	0.02	0.10	0.18	0.40	0.64	0.87
0.01 (1/100)	0.01	0.05	0.10	0.22	0.40	0.64
0.005 (1/200)	0.005	0.02	0.05	0.10	0.22	0.39

Conversion Factors for Other Mean Recurrence Intervals

Table C6-3

MRI (years)	Peak gust wind speed, V (mph) m/s		Alaska
	Continental U.S.		
	V = 85-100 (38-45 m/s)	V > 100 (hurricane) (44.7 m/s)	
500	1.23	1.23	1.18
200	1.14	1.14	1.12
100	1.07	1.07	1.06
50	1.00	1.00	1.00
25	0.93	0.88	0.94
10	0.84	0.74 (76 mph min.) (33.9 m/s)	0.87
5	0.78	0.66 (70 mph min.) (31.3 m/s)	0.81

Note: Conversion factors for the column "V > 100 (hurricane)" are approximate. For the MRI = 50 as shown, the actual return period, as represented by the design wind speed map in Fig. 6-1, varies from 50 to approximately 90 years. For an MRI = 500, the conversion factor is theoretically "exact" as shown.

Gust Effect Factor - Example

Table C6-4

Values Obtained From Table 6-4	
z_{min}	60 ft
$\bar{\in}$	0.5
c	0.45
$\bar{b}$	0.3
$\bar{\alpha}$	0.33
$\hat{b}$	0.64
$\hat{\alpha}$	0.2
ℓ	180
C_{fx}	1.3
ξ	1
Height (h)	600 ft
Base (B)	100 ft
Depth (L)	100 ft

Gust Effect Factor - Example

Table C6-5

Calculated Values	
V	132 ft/s
$\bar{z}$	360 ft
$I_{\bar{z}}$	0.302
$L_{\bar{z}}$	594.52 ft
Q^2	0.589
$\bar{V}_{\bar{z}}$	87.83 ft/s
$\hat{V}_{\bar{z}}$	136.24 ft/s
N_1	1.354
R_n	0.111
η	1.047
R_B	0.555
η	6.285
R_h	0.146
η	3.507
R_L	0.245
R^2	0.580
G	1.055
K	0.502
m_1	745,400 slugs
g_R	3.787

Along Wind Response - Example

Table C6-6

z (ft)	$\phi(z)$	X_{max} (z)	Rms.Acc. (ft/sec^2)	Rms. Acc. (milli-g)	Max. Acc. (ft/sec^2)	Max. Acc. (milli-g)
0	0.00	0.00	0.00	0.00	0.00	0.00
60	0.10	0.07	0.02	0.6	0.07	2.2
120	0.20	0.15	0.04	1.2	0.14	4.5
180	0.30	0.22	0.06	1.8	0.22	6.7
240	0.40	0.30	0.08	2.4	0.29	8.9
300	0.50	0.37	0.10	3.0	0.36	11.2
360	0.60	0.45	0.11	3.5	0.43	13.4
420	0.70	0.52	0.13	4.1	0.50	15.7
480	0.80	0.60	0.15	4.7	0.58	17.9
540	0.90	0.67	0.17	5.3	0.65	20.1
600	1.00	0.75	0.19	5.9	0.72	22.4

C7.0 SNOW LOADS

Methodology. The procedure established for determining design snow loads is as follows:

1. Determine the ground snow load for the geographic location (Section 7.2 and the Commentary's Section C7.2).
2. Generate a flat roof snow load from the ground load with consideration given to: (a) roof exposure (Section 7.3.1 and the Commentary's Sections C7.3 and C7.3.1); (b) roof thermal condition (Section 7.3.2 and the Commentary's Sections C7.3 and C7.3.2); and (c) occupancy and function of structure (Section 7.3.3 and the Commentary's Section C7.3.3).
3. Consider roof slope (Sections 7.4 through 7.4.5 and the Commentary's Section C7.4).
4. Consider partial loading (Section 7.5 and the Commentary's Section C7.5).
5. Consider unbalanced loads (Sections 7.6 through 7.6.4 and the Commentary's Section C7.6).
6. Consider snow drifts: (a) on lower roofs (Sections 7.7 through 7.7.2 and the Commentary's Section C7.7); and (b) from projections (Section 7.8 and the Commentary's Section C7.8).
7. Consider sliding snow (Section 7.9 and the Commentary's Section C7.9).
8. Consider extra loads from rain on snow (Section 7.10 and the Commentary's Section C7.10).
9. Consider ponding loads (Section 7.11 and the Commentary's Section C7.11).
10. Consider existing roofs (Section 7.12 and the Commentary's Section C7.12).
11. Consider other roofs and sites (the Commentary's Section C7.13).
12. Consider the consequences of loads in excess of the design value (immediately following).

Loads in Excess of the Design Value. The philosophy of the probabilistic approach used in this standard is to establish a design value that reduces the risk of a snow load-induced failure to an acceptably low level. Since snow loads in excess of the design value may occur, the implications of such "excess" loads should be considered. For example, if a roof is deflected at the design snow load so that slope to drain is eliminated, "excess" snow load might cause ponding (as discussed in the Commentary's Section C7.11) and perhaps progressive failure.

The snow load/dead load ratio of a roof structure is an important consideration when assessing the implications of "excess" loads. If the design snow load is exceeded, the percentage increase in total load would be greater for a lightweight structure (that is, one with a high snow load/dead load ratio) than for a heavy structure (that is, one with a low snow load/dead load ratio). For example, if a 40-lb/ft^2 (1.92 kN/m^2) roof snow load is exceeded by 20 lb/ft^2 (0.96 kN/m^2) for a roof having a 25-lb/ft^2 (1.19 kN/m^2) dead load, the total load increases by 31% from 65 to 85 lb/ft^2 (3.11 to 4.07 kN/m^2). If the roof had a 60-lb/ft^2 (2.87 kN/m^2) dead load, the total load would increase only by 20% from 100 to 120 lb/ft^2 (4.79 to 5.75 kN/m^2).

C7.2 GROUND SNOW LOADS

The snow load provisions were developed from an extreme-value statistical analysis of weather records of snow on the ground [C7-1]. The log normal distribution was selected to estimate ground snow loads, which have a 2% annual probability of being exceeded (50-year mean recurrence interval).

Maximum measured ground snow loads and ground snow loads with a 2% annual probability of being exceeded are presented in Table C7-1 for 204 National Weather Service (NWS) "first-order" stations at which ground snow loads have been measured for at least 11 years during the period 1952–1992.

Concurrent records of the depth and load of snow on the ground at the 204 locations in Table C7-1 were used to estimate the ground snow load and the ground snow depth having a 2% annual probability of being exceeded for each of these locations. The period of record for these 204 locations, where both snow depth and snow load have been measured, averages 33 years up through the winter of 1991–1992. A mathematical relationship was developed between the 2% depths and the 2% loads. The nonlinear best-fit relationship between these extreme values was used to estimate 2% (50-year mean recurrence interval) ground snow loads at about 9,200 other locations at which only snow depths were measured. These loads, as well as the extreme-value loads developed directly from snow load measurements at 204 first-order locations, were used to construct the maps.

In general, loads from these two sources were in agreement. In areas where there were differences, loads from the 204 first-order locations were considered to be more valuable when the map was constructed. This procedure ensures that the map is referenced to the NWS observed loads and contains spatial detail provided by snow-depth measurements at about 9,200 other locations.

The maps were generated from data current through the 1991–1992 winter. Where statistical studies using more recent information are available, they may be used to produce improved design guidance.

However, adding a big snow year to data developed from periods of record exceeding 20 years will usually not change 50-year values much. As examples, the data bases for Boston and Chattanooga were updated to include the winters of 1992–1993 and 1993–1994 since record snows occurred there during that period. In Boston, 50-year loads based on water equivalent measurements only increased from 34 to 35 psf (1.63 to 1.68 kN/m^2) and loads generated from snow depth measurements remained at 25 psf (1.20 kN/m^2). In Chattanooga, loads generated from water equivalent measurements increased from 6 to 7 psf (0.29 to 0.34 kN/m^2) and loads generated from snow depth measurements remained at 6 psf (0.29 kN/m^2).

The following additional information was also considered when establishing the snow load zones on the map of the United States (Fig. 7-1).

1. The number of years of record available at each location.
2. Additional meterological information available from NWS, Soil Conservation Service (SCS) snow surveys and other sources.
3. Maximum snow loads observed there.
4. Regional topography.
5. The elevation of each location.

The map is an updated version of those in the 1993 edition of this Standard and is the same as that in the 1995 edition.

In much of the south, infrequent but severe snowstorms disrupted life in the area to the point that meteorological observations were missed. In these and similar circumstances more value was given to the statistical values for stations with complete records. Year-by-year checks were made to verify the significance of data gaps.

The mapped snow loads cannot be expected to represent all the local differences that may occur within each zone. Because local differences exist, each zone has been positioned so as to encompass essentially all the statistical values associated with normal sites in that zone. Although the zones represent statistical values, not maximum observed values,

the maximum observed values were helpful in establishing the position of each zone.

For sites not covered in Fig. 7-1 design values should be established from meteorological information, with consideration given to the orientation, elevation, and records available at each location. The same method can also be used to improve upon the values presented in Fig. 7-1. Detailed study of a specific site may generate a design value lower than that indicated by the generalized national map. It is appropriate in such a situation to use the lower value established by the detailed study. Occasionally a detailed study may indicate that a higher design value should be used than the national map indicates. Again, results of the detailed study should be followed.

The area covered by a site-specific case study will vary depending on local climate and topography. In some places, a single case study will suffice for an entire community, but in others, varying local conditions limit a "site" to a much smaller area. The area of applicability usually becomes clear as information in the vicinity is examined for the case study.

As suggested by the footnote, it is not appropriate to use only the site-specific information in Table C7-1 for design purposes. It lacks an appreciation for surrounding station information and, in a few cases, is based on rather short periods of record. The map or a site-specific case study provide more valuable information.

The importance of conducting detailed studies for locations not covered in Fig. 7-1 is shown in Table C7-2.

For some locations within the CS areas of the northeast (Fig. 7-1), ground snow loads exceed 100 lb/ft^2 (4.79 kN/m^2). Even in the southern portion of the Appalachian Mountains, not far from sites where a 15-lb/ft^2 (0.72 kN/m^2) ground snow load is appropriate, ground loads exceeding 50 lb/ft^2 (2.39 kN/m^2) may be required. Lake-effect storms create requirements for ground loads in excess of 75 lb/ft^2 (3.59 kN/m^2) along portions of the Great Lakes. In some areas of the Rocky Mountains, ground snow loads exceed 200 lb/ft^2 (9.58 kN/m^2).

Local records and experience should also be considered when establishing design values.

The values in Table 7-1 are for specific Alaskan locations only and generally do not represent appropriate design values for other nearby locations. They are presented to illustrate the extreme variability of snow loads within Alaska. This variability precludes statewide mapping of ground snow loads there.

Valuable information on snow loads for the Rocky Mountain states is contained in references [C7-2–C7-12].

Most of these references for the Rocky Mountain states use annual probabilities of being exceeded that are different from the 2% value (50-year mean recurrence interval) used in this standard. Reasonable, but not exact, factors for converting from other annual probabilities of being exceeded to the value herein are presented in Table C7-3.

For example, a ground snow load based on a 3.3% annual probability of being exceeded (30-year mean recurrence interval) should be multiplied by 1.18 to generate a value of p_g for use in Eq. 7-1.

The snow load provisions of several editions of the National Building Code of Canada served as a guide in preparing the snow load provisions in this standard. However, there are some important differences between the Canadian and the United States data bases. They include:

1. The Canadian ground snow loads are based on a 3.3% annual probability of being exceeded (30-year mean recurrence interval) generated by using the extreme-value, Type-I (Gumbel) distribution, while the normal-risk values in this standard are based on a 2% annual probability of being exceeded (50-year mean recurrence interval) generated by a log-normal distribution.
2. The Canadian loads are based on measured depths and regionalized densities based on 4 or less measurements per month. Because of the infrequency of density measurements, an additional weight of rain is added [C7-13]. In this Standard the weight of the snow is based on many years of frequently-measured weights obtained at 204 locations across the United States. Those measurements contain many rain-on-snow events and thus a separate rain-on-snow surcharge load is not needed except for some roofs with a slope less than 1/2 in./ft (2.38°).

C7.3 FLAT-ROOF SNOW LOADS, p_f

The live load reductions in Section 4.8 should not be applied to snow loads. The minimum allowable values of p_f presented in Section 7.3 acknowledge that in some areas a single major storm can generate loads that exceed those developed from an analysis of weather records and snow load case studies.

The factors in this standard that account for the thermal, aerodynamic, and geometric characteristics

of the structure in its particular setting were developed using the National Building Code of Canada as a point of reference. The case study reports in references [C7-14–C7-22] were examined in detail.

In addition to these published references, an extensive program of snow load case studies was conducted by eight universities in the United States, by the Corps of Engineers' Alaska District, and by the U.S. Army Cold Regions Research and Engineering Laboratory (CRREL) for the Corps of Engineers. The results of this program were used to modify the Canadian methodology to better fit United States conditions. Measurements obtained during the severe winters of 1976–1977 and 1977–1978 are included. A statistical analysis of some of that information is presented in [C7-23]. The experience and perspective of many design professionals, including several with expertise in building failure analysis, have also been incorporated.

The minimum values of p_f account for a number of situations that develop on roofs. They are particularly important considerations where p_g is 20 lb/ft^2 (0.96 kN/m^2) or less. In such areas, single storms result in loadings for which Eq. 7-1 and the C_e and C_t values in Tables 7-2 and 7-3, respectively, underestimate loads.

C7.3.1 Exposure Factor, C_e

Except in areas of "aerodynamic shade," where loads are often increased by snow drifting, less snow is present on most roofs than on the ground. Loads in unobstructed areas of conventional flat roofs average less than 50% of ground loads in some parts of the country. The values in this standard are above-average values, chosen to reduce the risk of snow load-induced failures to an acceptably low level. Because of the variability of wind action, a conservative approach has been taken when considering load reductions by wind.

The effects of exposure are handled on two scales. First Eq. 7-1 contains a basic exposure factor of 0.7. Second, the type of terrain and the exposure of the roof are handled by exposure factor C_e. This two-step procedure generates ground-to-roof load reductions as a function of exposure that range from 0.49 to 0.91.

Table 7-2 has been changed from what appeared in a prior (1988) version of this Standard to separate regional wind issues associated with terrain from local wind issues associated with roof exposure. This was done to better define categories without significantly changing the values of C_e.

Although there is a single "regional" terrain category for a specific site, different roofs of a structure may have different exposure factors due to obstruction provided by higher portions of the structure or by objects on the roof. For example in terrain Category C, an upper level roof could be fully exposed (C_e = 0.9) while a lower level roof would be partially exposed (C_e = 1.0) due to the presence of the upper level roof, as shown in Example 3 in the Commentary.

The adjective "windswept" is used in the "mountainous areas" terrain category to preclude use of this category in those high mountain valleys that receive little wind.

The normal, combined exposure reduction in this standard is 0.70 as compared to a normal value of 0.80 for the ground-to-roof conversion factor in the 1990 National Building Code of Canada. The decrease from 0.80 to 0.70 does not represent decreased safety but arises due to increased choices of exposure and thermal classification of roofs (that is, six terrain categories, three roof exposure categories and four thermal categories in this standard versus three exposure categories and no thermal distinctions in the Canadian code).

It is virtually impossible to establish exposure definitions that clearly encompass all possible exposures that exist across the country. Because individuals may interpret exposure categories somewhat differently, the range in exposure has been divided into several categories rather than just two or three. A difference of opinion of one category results in about a 10% "error" using these several categories and an "error" of 25% or more if only three categories are used.

C7.3.2 Thermal Factor, C_t

Usually, more snow will be present on cold roofs than on warm roofs. An exception to this is discussed below. The thermal condition selected from Table 7-3 should represent that which is likely to exist during the life of the structure. Although it is possible that a brief power interruption will cause temporary cooling of a heated structure, the joint probability of this event and a simultaneous peak snow load event is very small. Brief power interruptions and loss of heat are acknowledged in C_t = 1.0 category. Although it is possible that a heated structure will subsequently be used as an unheated structure, the probability of this is rather low. Consequently, heated structures need not be designed for this unlikely event.

Some dwellings are not used during the winter. Although their thermal factor may increase to 1.2 at that time, they are unoccupied, so their importance factor reduces to 0.8. The net effect is to require the same design load as for a heated, occupied dwelling.

Discontinuous heating of structures may cause thawing of snow on the roof and subsequent refreezing in lower areas. Drainage systems of such roofs have become clogged with ice, and extra loads associated with layers of ice several inches thick have built up in these undrained lower areas. The possibility of similar occurrences should be investigated for any intermittently heated structure.

Similar icings may build up on cold roofs subjected to meltwater from warmer roofs above. Exhaust fans and other mechanical equipment on roofs may also generate meltwater and icings.

Icicles and ice dams are a common occurrence on cold eaves of sloped roofs. They introduce problems related to leakage and to loads. Large ice dams that can prevent snow from sliding off roofs are generally produced by heat losses from within buildings. Icings associated with solar melting of snow during the day and refreezing along eaves at night are often small and transient. Although icings can occur on cold or warm roofs, roofs that are well insulated and ventilated are not commonly subjected to serious icings at their eaves. Because ice dams can prevent load reductions by sliding on some warm (C_t = 1.0) roofs, the "unobstructed slippery surface" curve in Fig. 7-2a now only applies to unventilated roofs with a thermal resistance equal to or greater than 30 ft^2·h·F°/Btu (5.3 K·m^2/W) and to ventilated roofs with a thermal resistance equal to or greater than 20 ft^2·h·F°/Btu (3.5 K·m^2/W). Roofs that are well insulated and ventilated have been given a C_t = 1.1 in Table 7-3. This increases their flat roof snow load p_f. These provisions have been changed from the 1995 version of this Standard to cause a partial increase in loads when C_t = 1.1 instead of the "full" increase at both C_t = 1.1 and 1.2. Methods of minimizing eave icings are discussed in [C7-24–C7-29].

Glass, plastic, and fabric roofs of continuously heated structures are seldom subjected to much snow load because their high heat losses cause snow melt and sliding. For such specialty roofs, knowledgeable manufacturers and designers should be consulted. The National Greenhouse Manufacturers Association [C7-30] recommends use of C_t = 0.83 for continuously heated greenhouses and C_t = 1.00 for unheated or intermittently heated greenhouses. They suggest a value of I = 1.0 for retail greenhouses and I = 0.8 for all other greenhouses. To qualify as a continuously heated greenhouse, a production or retail greenhouse must have a constantly maintained temperature of 50°F (10°C) or higher during winter months. In addition it must also have a maintenance attendant on duty at all times or an adequate temperature alarm system to provide warning in the event of a heating system failure. Finally, the greenhouse roof material must have a thermal resistance, R-value, less than 2 ft$^2 \cdot$h$\cdot$F°/Btu (0.4 K$\cdot$m^2/W). In this Standard, the C_t factor for such continuously heated greenhouses is set at 0.85. An unheated or intermittently heated greenhouse is any greenhouse that does not meet the requirements of a continuously heated single or double glazed greenhouse. Greenhouses should be designed so that the structural supporting members are stronger than the glazing. If this approach is used, any failure caused by heavy snow loads will be localized and in the glazing. This should avert progressive collapse of the structural frame. Higher design values should be used where drifting or sliding snow is expected.

Little snow accumulates on warm air-supported fabric roofs because of their geometry and slippery surface. However, the snow that does accumulate is a significant load for such structures and should be considered. Design methods for snow loads on air structures are discussed in [C7-31, C7-32].

The combined consideration of exposure and thermal conditions generates ground-to-roof factors that range from a low of 0.49 to a high of 1.09. The equivalent ground-to-roof factors in the 1990 National Building Code of Canada are 0.8 for sheltered roofs, 0.6 for exposed roofs and 0.4 for exposed roofs in exposed areas north of the tree line, all regardless of their thermal condition.

Reference [C7-33] indicates that loads exceeding those calculated using this Standard can occur on roofs that receive little heat from below. Limited case histories for freezer buildings suggest that the C_t factor may be larger than 1.2.

C7.3.3 Importance Factor, I

The importance Factor I has been included to account for the need to relate design loads to the consequences of failure. Roofs of most structures having normal occupancies and functions are designed with an importance factor of 1.0, which corresponds to unmodified use of the statistically determined ground snow load for a 2% annual probability of being exceeded (50-year mean recurrence interval).

A study of the 204 locations in Table C7-1 showed that the ratio of the values for 4% and 2%

annual probabilities of being exceeded (the ratio of the 25-year to 50-year mean recurrence interval values) averaged 0.80 and had a standard deviation of 0.06. The ratio of the values for 1% and 2% annual probabilities of being exceeded (the ratio of the 100-year to 50-year mean recurrence interval values) averaged 1.22 and had a standard deviation of 0.08. On the basis of the nationwide consistency of these values it was decided that only one snow load map need be prepared for design purposes and that values for lower and higher risk situations could be generated using that map and constant factors.

Lower and higher risk situations are established using the importance factors for snow loads in Table 7-4. These factors range from 0.8 to 1.2. The factor 0.8 bases the average design value for that situation on an annual probability of being exceeded of about 4% (about a 25-year mean recurrence interval). The factor 1.2 is nearly that for a 1% annual probability of being exceeded (about a 100-year mean recurrence interval).

C7.3.4 Minimum Allowable Values of p_f for Low Slope Roofs

These minimums account for a number of situations that develop on low slope roofs. They are particularly important considerations where p_g is 20 lb/ft^2 (0.96 kN/m^2) or less. In such areas, single storms can result in loading for which the basic exposure factor of 0.7 as well as the C_e and C_t factors do not apply.

C7.4 SLOPED-ROOF SNOW LOADS, p_s

Snow loads decrease as the slopes of roofs increase. Generally, less snow accumulates on a sloped roof because of wind action. Also, such roofs may shed some of the snow that accumulates on them by sliding and improved drainage of meltwater. The ability of a sloped roof to shed snow load by sliding is related to the absence of obstructions not only on the roof but also below it, the temperature of the roof, and the slipperiness of its surface. It is difficult to define "slippery" in quantitative terms. For that reason a list of roof surfaces that qualify as slippery and others that do not, are presented in the Standard itself. Most common roof surfaces are on that list. The slipperiness of other surfaces is best determined by comparisons with those surfaces. Some tile roofs contain built-in protrusions or have a rough surface which prevents snow from sliding. However, snow will slide off other smooth-surfaced tile roofs. When a surface may or may not be slippery the implications of treating it either as a slippery or non-slippery surface should be determined.

Discontinuous heating of a building may reduce the ability of a sloped roof to shed snow by sliding, since meltwater created during heated periods may refreeze on the roof's surface during periods when the building is not heated, thereby "locking" the snow to the roof.

All these factors are considered in the slope reduction factors presented in Fig. 7-2 which are supported by references [C7-33–C7-36]. The thermal resistance requirements have been added to the "unobstructed slippery surfaces" curve in Fig. 7-2a to prevent its use for roofs on which ice dams often form since ice dams prevent snow from sliding. Mathematically the information in Fig. 7-2 can be represented as follows:

1. Warm roofs ($C_t = 1.0$ or less):
   (a) Unobstructed slippery surfaces with $R \geq 30$ ft$^2 \cdot$h$\cdot$F°/Btu (5.3 K$\cdot$m^2/W) if unventilated and $R \geq 20$ ft$^2 \cdot$h$\cdot$F°/Btu (3.5 K$\cdot$m^2/W) if ventilated:

   0–5° slope $C_s = 1.0$
   5–70° slope $C_s = 1.0 - ($slope$ - 5°)/65°$
   >70° slope $C_s = 0$

   (b) All other surfaces:

   0–30° slope $C_s = 1.0$
   30–70° slope $C_s = 1.0 - ($slope$ - 30°)/40°$
   >70° slope $C_s = 0$

2. Cold Roofs ($C_t = 1.2$):
   (a) Unobstructed slippery surfaces:

   0–15° slope $C_s = 1.0$
   15–70° slope $C_s = 1 - ($slope$ - 15°)/55°$
   >70° slope $C_s = 0$

   (b) All other surfaces:

   0–45° slope $C_s = 1.0$
   45–70° slope $C_s = 1.0 - ($slope$ - 45°)/25°$
   >70° slope $C_s = 0$

If the ground (or another roof of less slope) exists near the eave of a sloped roof, snow may not be able to slide completely off the sloped roof. This may result in the elimination of snow loads on upper portions of the roof and their concentration on lower portions. Steep A-frame roofs that nearly reach the ground are subject to such conditions. Lateral as well as vertical loads induced by such snow should be considered for such roofs.

C7.4.3 Roof Slope Factor for Curved Roofs

These provisions have been changed from those in the 1993 edition of this Standard to cause the load to diminish along the roof as the slope increases.

C7.4.4 Roof Slope Factor for Multiple Folded Plate, Sawtooth, and Barrel Vault Roofs

Because these types of roofs collect extra snow in their valleys by wind drifting and snow creep and sliding, no reduction in snow load should be applied because of slope.

C7.4.5 Ice Dams and Icicles Along Eaves

The intent is to consider heavy loads from ice that forms along eaves only for structures where such loads are likely to form. It is also not considered necessary to analyze the entire structure for such loads, just the eaves themselves.

C7.5 UNLOADED PORTIONS

In many situations a reduction in snow load on a portion of a roof by wind scour, melting, or snow-removal operations will simply reduce the stresses in the supporting members. However, in some cases a reduction in snow load from an area will induce heavier stresses in the roof structure than occur when the entire roof is loaded. Cantilevered roof joists are a good example; removing half the snow load from the cantilevered portion will increase the bending stress and deflection of the adjacent continuous span. In other situations adverse stress reversals may result.

The intent is not to require consideration of multiple "checkerboard" loadings.

Separate, simplified provisions have been added for continuous beams to provide specific partial loading requirements for that common structural system.

C7.6 UNBALANCED ROOF SNOW LOADS

Unbalanced snow loads may develop on sloped roofs because of sunlight and wind. Winds tend to reduce snow loads on windward portions and increase snow loads on leeward portions. Since it is not possible to define wind direction with assurance, winds from all directions should generally be considered when establishing unbalanced roof loads.

C7.6.1 Unbalanced Snow Loads on Hip and Gable Roofs

The unbalanced uniform snow load on the leeward side was $1.5p_s/C_e$ and $1.3p_s/C_e$ in the 1993 and 1995 editions of this Standard. In this edition, a 1993 approach is prescribed for roofs with small eave to ridge distances and the 15° cutoff is eliminated. For moderate to large roof slopes (i.e., roof slopes between $275\beta p_f/\gamma W$ and 70°), the unbalanced snow load on the leeward side varies between $1.5p_s/C_e$ and $1.8p_s/C_e$ as a function of β. The gable roof drift parameter, β, quantifies the likelihood of wind induced drifting across the ridge line. That is, for small values of L/W one expects significant leeward drifts only for wind essentially perpendicular to the ridge line, while for larger values of L/W one gets significant drifts for a larger range of wind directions. For these moderate to large roof slopes, the maximum leeward side drift load tends to occur between the ridge and the eave. For these cases, a uniform load intensity of $1.2(1 + \beta/2)p_s/C_e$ is prescribed.

For smaller roof slopes the load at the eave is limited to the "balanced" portion, $1.2p_f/C_e$, plus γh_e where h_e is the elevation difference between the ridge and eave. The balanced portion, $1.2p_f/C_e$, nominally corresponds to the ground load if the structure is unheated (i.e., $C_t = 1.2$). For very small roof slopes the aerodynamic shade region fills, resulting in the top of the leeward snow being nominally horizontal.

The angle, $275\beta p_f/W$, corresponds to the case where the maximum leeward side snow load occurs at the midpoint between the ridge and the eave. As sketched in Fig. C7-1, it is calculated by equating the area of snow transported from the windward side

$$A_w = \frac{\beta}{2}\left(1.2\,\frac{p_f}{\gamma}\right)W \qquad \text{(Eq. C7-1)}$$

with the area of a triangular surcharge drift on the leeward side.

$$A_d = \frac{1}{2}\left(\frac{W}{2}\right)\left(\frac{W}{2}\tan\theta\right) \qquad \text{(Eq. C7-2)}$$

Using the small angle approximation (i.e., $\theta \cong \tan\theta$) for angles in radians and converting from radians to degrees,

$$\theta = 275\beta p_f/\gamma W \qquad \text{(Eq. C7-3)}$$

The gable roof drift parameter β is based upon an analysis of case histories presented in [C7-56].

The design snow load on the windward side for the unbalanced case, $0.3p_f$ (or $0.3p_s$) is based upon case histories presented in [C7-21 and C7-56]. The lower limit of $\theta = 70/W + 0.5$ is intended to exclude low slope roofs, such as membrane roofs, on which significant unbalanced loads have not been observed [C7-57].

The provisions of the Standard and this Commentary correspond to the case where the gable or hip roof, in plan, is symmetric about the ridge line, specifically roofs for which the eave to ridgeline distance, W, is the same for both sides. For asymmetric roofs, the unbalanced load for the side with the smaller W should be increased to account for the larger snow source area on the opposite side, following the procedures discussed in [C7-56].

C7.6.2 Unbalanced Snow Loads for Curved Roofs

The method of determining roof slope is the same as the 1995 edition of this Standard. C_s is based on the actual slope, not an equivalent slope. These provisions do not apply to roofs that are concave upward. For such roofs, see Section 7.13.

C7.6.3 Unbalanced Snow Loads for Multiple Folded Plate, Sawtooth, and Barrel Vault Roofs

A minimum slope of 3/8 in./ft (1.79°) has been established to preclude the need to determine unbalanced loads for most internally-drained, membrane roofs which slope to internal drains. Case studies indicate that significant unbalanced loads can occur when the slope of multiple gable roofs is as low as 1/2 in./ft (2.38°).

The unbalanced snow load in the valley is $2p_f/C_e$ to create a total unbalanced load that does not exceed a uniformly distributed ground snow load in most situations.

Sawtooth roofs and other "up-and-down" roofs with significant slopes tend to be vulnerable in areas of heavy snowfall for the following reasons:

1. They accumulate heavy snow loads and are therefore expensive to build.
2. Windows and ventilation features on the steeply sloped faces of such roofs may become blocked with drifting snow and be rendered useless.
3. Meltwater infiltration is likely through gaps in the steeply sloped faces if they are built as walls,

since slush may accumulate in the valley during warm weather. This can promote progressive deterioration of the structure.

4. Lateral pressure from snow drifted against clerestory windows may break the glass.

5. The requirement that snow above the valley not be at an elevation higher than the snow above the ridge may limit the unbalanced load to less than $2p_f/C_e$.

C7.6.4 Unbalanced Snow Loads for Dome Roofs

This provision is based on a similar provision in the 1990 National Building Code of Canada.

C7.7 DRIFTS ON LOWER ROOFS (AERODYNAMIC SHADE)

When a rash of snow-load failures occurs during a particularly severe winter, there is a natural tendency for concerned parties to initiate across-the-board increases in design snow loads. This is generally a technically ineffective and expensive way of attempting to solve such problems, since most failures associated with snow loads on roofs are caused not by moderate overloads on every square foot (square meter) of the roof but rather by localized significant overloads caused by drifted snow.

It is extremely important to consider localized drift loads in designing roofs. Drifts will accumulate on roofs (even on sloped roofs) in the wind shadow of higher roofs or terrain features. Parapets have the same effect. The affected roof may be influenced by a higher portion of the same structure or by another structure or terrain feature nearby if the separation is 20 ft (6.1 m) or less. When a new structure is built within 20 ft (6.1 m) of an existing structure, drifting possibilities should also be investigated for the existing structure. The snow that forms drifts may come from the roof on which the drift forms, from higher or lower roofs or, on occasion, from the ground.

The leeward drift load provisions are based on studies of snow drifts on roofs [C7-37–C7-40]. Drift size is related to the amount of driftable snow as quantified by the upwind roof length and the ground snow load. Drift loads are considered for ground snow loads as low as 5 lb/ft^2 (0.24 kN/m^2). Case studies show that, in regions with low ground snow loads, drifts 3 to 4 ft (0.9 to 1.2 m) high can be caused by a single storm accompanied by high winds.

A change from a prior (1988) edition of this Standard involves the width w when the drift height h_d from Fig. 7-9, exceeds the clear height h_c. In this situation the width of the drift is taken as $4h_d^2/h_c$ with a maximum value of $8h_c$. This drift width relation is based upon equating the cross-sectional area of this drift (i.e., $1/2h_c \times w$) with the cross-sectional area of a triangular drift where the drift height is not limited by h_c (i.e., $1/2h_d \times 4h_d$). The upper limit of drift width is based on studies by Finney [C7-41] and Tabler [C7-42] which suggest that a ''full'' drift has a rise-to-run of about 1:6.5, and case studies [C7-43] which show observed drifts with a rise-to-run greater than 1:10.

The drift height relationship in Fig. 7-9 is based on snow blowing off a high roof upwind of a lower roof. The change in elevation where the drift forms is called a ''leeward step.'' Drifts can also form at ''windward steps.'' An example is the drift that forms at the downwind end of a roof that abuts a higher structure there. Fig. 7-7 shows ''windward step'' and ''leeward step'' drifts.

For situations having the same amount of available snow (i.e., upper and lower roofs of the same length) the drifts which form in leeward steps are larger than those which form in windward steps. In the previous version of the Standard, the windward drifts height was given as $1/2h_d$ from Fig. 7-9 using the length of the lower roof for l_u. Based upon recent case history experience in combination with a prior study [C7-45], a value of 3/4 is now prescribed.

Depending on wind direction, any change in elevation between roofs can be either a windward or leeward step. Thus the height of a drift is determined for each wind direction as shown in Example 3, and the larger of the two heights is used as the design drift.

The drift load provisions cover most, but not all, situations. References [C7-41 and C7-46] document a larger drift than would have been expected based on the length of the upper roof. The larger drift was caused when snow on a somewhat lower roof, upwind of the upper roof, formed a drift between those two roofs allowing snow from the upwind lower roof to be carried up onto the upper roof then into the drift on its downwind side. It was suggested that the sum of the lengths of both roofs could be used to calculate the size of the leewind drift.

In another situation [C7-47] a long ''spike'' drift was created at the end of a long skylight with the wind about 30° off the long axis of the skylight. The skylight acted as a guide or deflector which concentrated drifting snow. This caused a large drift to accumulate in the lee of the skylight. This drift was replicated in a wind tunnel.

As shown in Fig. 7-8, the clear height, h_c, is determined based on the assumption that the upper roof is blown clear of snow in the vicinity of the drift. This is a reasonable assumption when the upper roof is nearly flat. However, sloped roofs often accumulate snow at eaves as illustrated in Figs. 7-3 and 7-5. For such roofs, it is appropriate to assume that snow at the upper roof edge effectively increases the height difference between adjacent roofs. Using half the depth of the unbalanced snow load in the calculation of h_c produces more realistic estimates of drift loads.

Tests in wind tunnels [C7-48 and C7-49] and flumes [C7-44] have proven quite valuable in determining patterns of snow drifting and drift loads. For roofs of unusual shape or configuration, wind tunnel or water-flume tests may be needed to help define drift loads.

C7.8 ROOF PROJECTIONS

Drifts around penthouses, roof obstructions and parapet walls are also of the "windward step" type since the length of the upper roof is small or no upper roof exists. Solar panels, mechanical equipment, parapet walls, and penthouses are examples of roof projections that may cause "windward" drifts on the roof around them. The drift-load provisions in Sections 7.7 and 7.8 cover most of these situations adequately, but flat-plate solar collectors may warrant some additional attention. Roofs equipped with several rows of them are subjected to additional snow loads. Before the collectors were installed, these roofs may have sustained minimal snow loads, especially if they were windswept. Since a roof with collectors is apt to be somewhat "sheltered" by the collectors, it seems appropriate to assume the roof is partially exposed and calculate a uniform snow load for the entire area as though the collectors did not exist. Second, the extra snow that might fall on the collectors and then slide onto the roof should be computed using the "cold roofs—all other surfaces" curve in Fig. 7-2b. This value should be applied as a uniform load on the roof at the base of each collector over an area about 2 ft (0.6 m) wide along the length of the collector. The uniform load combined with the load at the base of each collector probably represents a reasonable design load for such situations, except in very windy areas where extensive snow drifting is

to be expected among the collectors. By elevating collectors several feet (a meter or more) above the roof on an open system of structural supports, the potential for drifting will be diminished significantly. Finally, the collectors themselves should be designed to sustain a load calculated by using the "unobstructed slippery surfaces" curve in Fig. 7-2a. This last load should not be used in the design of the roof itself, since the heavier load of sliding snow from the collectors has already been considered. The influence of solar collectors on snow accumulation is discussed in [C7-50 and C7-51].

C7.9 SLIDING SNOW

Situations that permit snow to slide onto lower roofs should be avoided [C7-52]. Where this is not possible, the extra load of the sliding snow should be considered. Roofs with little slope have been observed to shed snow loads by sliding. Consequently, it is prudent to assume that any upper roof sloped to an unobstructed eave is a potential source of sliding snow.

The dashed lines in Figs. 7-2a and 7-2b should not be used to determine the total load of sliding snow available from an upper roof, since those lines assume that unobstructed slippery surfaces will have somewhat less snow on them than other surfaces because they tend to shed snow by sliding. To determine the total sliding load available from the upper roof, it is appropriate to use the solid lines in Figs. 7-2a and 7-2b. The final resting place of any snow that slides off a higher roof onto a lower roof will depend on the size, position, and orientation of each roof [C7-35]. Distribution of sliding loads might vary from a uniform load 5 ft (1.5 m) wide, if a significant vertical offset exists between the two roofs, to a 20-ft (6.1 m)-wide uniform load, where a low-slope upper roof slides its load onto a second roof that is only a few feet (about a meter) lower or where snow drifts on the lower roof create a sloped surface that promotes lateral movement of the sliding snow.

In some instances a portion of the sliding snow may be expected to slide clear of the lower roof. Nevertheless, it is prudent to design the lower roof for a substantial portion of the sliding load in order to account for any dynamic effects that might be associated with sliding snow.

Snow guards are needed on some roofs to prevent roof damage and eliminate hazards associated with sliding snow [C7-53]. When snow guards are added to a sloping roof, snow loads on the roof can be expected to increase. Thus, it may be necessary to strengthen a roof before adding snow guards. When designing a roof that will likely need snow guards in the future, it may be appropriate to use the "all other surfaces" curves in Fig. 7-2 not the "unobstructed slippery surfaces" curves.

C7.10 RAIN ON SNOW SURCHARGE LOAD

The ground snow-load measurements on which this standard is based contain the load effects of light rain on snow. However, since heavy rains percolate down through snowpacks and may drain away, they might not be included in measured values. Where p_g is greater than 20 psf (0.96 kN/m²), it is assumed that the full rain-on-snow effect has been measured and a separate rain-on-snow surcharge is not needed. The temporary roof load contributed by a heavy rain may be significant. Its magnitude will depend on the duration and intensity of the design rainstorm, the drainage characteristics of the snow on the roof, the geometry of the roof, and the type of drainage provided. Loads associated with rain on snow are discussed in [C7-54] and [C7-55].

Water tends to remain in snow much longer on relatively flat roofs than on sloped roofs. Therefore, slope is quite beneficial, since it decreases opportunities for drain blockages and for freezing of water in the snow.

For a roof with a 1/4-in./ft (1.19°) slope, where p_g = 20 lb/ft² (0.96 kN/m²), p_f = 18 lb/ft² (0.86 kN/m²), and the minimum allowable value of p_f is 20 lb/ft² (0.96 kN/m²), the rain-on-snow surcharge of 5 lb/ft² (0.24 kN/m²) would be added to the 18-lb/ft² (0.86 kN/m²) flat roof snow load to generate a design load of 23 lb/ft² (1.10 kN/m²).

C7.11 PONDING INSTABILITY

Where adequate slope to drain does not exist, or where drains are blocked by ice, snow meltwater and rain may pond in low areas. Intermittently heated structures in very cold regions are particularly susceptible to blockages of drains by ice. A roof designed without slope or one sloped with only 1/8 in./ft (0.6°) to internal drains probably contains low spots away from drains by the time it is constructed. When a heavy snow load is added to such a roof, it is even more likely that undrained low spots exist. As

rainwater or snow meltwater flows to such low areas, these areas tend to deflect increasingly, allowing a deeper pond to form. If the structure does not possess enough stiffness to resist this progression, failure by localized overloading can result. This mechanism has been responsible for several roof failures under combined rain and snow loads.

It is very important to consider roof deflections caused by snow loads when determining the likelihood of ponding instability from rain-on-snow or snow meltwater.

Internally drained roofs should have a slope of at least 1/4 in./ft (1.19°) to provide positive drainage and to minimize the chance of ponding. Slopes of 1/4 in./ft (1.19°) or more are also effective in reducing peak loads generated by heavy spring rain on snow. Further incentive to build positive drainage into roofs is provided by significant improvements in the performance of waterproofing membranes when they are sloped to drain.

Rain loads and ponding instability are discussed in detail in Section 8 of this Standard.

C7.12 EXISTING ROOFS

Numerous existing roofs have failed when additions or new buildings nearby caused snow loads to increase on the existing roof. A prior (1988) edition of this Standard mentioned this issue only in its Commentary where it was not a mandatory provision. The 1995 edition moved this issue to the Standard.

The addition of a gable roof alongside an existing gable roof as shown in Fig. C7-2 most likely explains why some such metal buildings failed in the South during the winter of 1992–1993. The change from a simple gable roof to a multiple folded plate roof increased loads on the original roof as would be expected from Section 7.6.3. Unfortunately, the original roofs were not strengthened to account for these extra loads and they collapsed.

If the eaves of the new roof in Fig. C7-2 had been somewhat higher than the eaves of the existing roof, the exposure factor C_e, for the original roof may have increased thereby increasing snow loads on it. In addition, drift loads and loads from sliding snow would also have to be considered.

C7.13 OTHER ROOFS AND SITES

Wind tunnel model studies, similar tests employing fluids other than air, for example water flumes, and other special experimental and computational methods have been used with success to establish design snow loads for other roof geometries and complicated sites [C7-44, C7-48, C7-49]. To be reliable, such methods must reproduce the mean and turbulent characteristics of the wind and the manner in which snow particles are deposited on roofs then redistributed by wind action. Reliability should be demonstrated through comparisons with situations for which full-scale experience is available.

EXAMPLES

The following three examples illustrate the method used to establish design snow loads for most of the situations discussed in this Standard.

Example 1: Determine balanced and unbalanced design snow loads for an apartment complex in a suburb of Boston, Massachusetts. Each unit has an 8-on-12 slope unventilated gable roof. The building length is 100 ft (30.5 m) and the eave to ridge distance, W, is 30 ft (9.1 m). Composition shingles clad the roofs. Trees will be planted among the buildings.

Flat-roof snow load:

$$p_f = 0.7C_eC_tIp_g$$

where

$p_g = 30$ lb/ft^2 (1.44 kN/m^2) (from Fig. 7-1);
$C_e = 1.0$ (from Table 7-2 for Terrain Category B and a partially exposed roof); and
$C_t = 1.0$ (from Table 7-3); and $I = 1.0$ (from Table 7-4).

Thus:

$p_f = (0.7)(1.0)(1.0)(1.0)(30)$

= 21 lb/ft^2 (balanced load)

[in SI: $p_f = (0.7)(1.0)(1.0)(1.0)(1.44) = 1.01$ kN/m^2]

Since $p_g = 30$ psf (1.44 kN/m^2) and $I = 1.0$, the minimum value of $p_f = 20(1.0) = 20$ psf (0.96 kN/m^2) and hence does not control, see Section 7.3.

Sloped-roof snow load:

$p_s = C_sp_f$ where $C_s = 0.91$

[from solid line, Fig. 7-2a].

Thus:

$p_s = 0.91(21) = 19$ lb/ft^2

[in SI: $p_s = 0.91(1.01) = 0.92$ kN/m^2]

Unbalanced snow load:

Since the roof slope is greater than $70/W + 0.5$ = $70/30 + 0.5 = 2.38°$, unbalanced loads must be considered. The gable roof length to width (eave to ridge) ratio $L/W = 100/30 = 3.33$ and $\beta = 0.89$ as calculated using Eq. 7-3. For $p_g = 30$ psf (1.44 kN/m^2), the snow density $\gamma = 17.9$ pcf (2.81 kN/m^3) as calculated using Eq. 7-4. The roof slope (8 on 12) of 33.6° is between $275\beta p_f/\gamma W = 9.6°$ and 70°, hence from Fig. 7-5 the windward load is $0.3p_s = 6$ psf (0.29 kN/m^2) while the leeward unbalanced load is $1.2(1 + \beta/2)p_s/C_e = 33$ psf (1.6 kN/m^2).

Rain on snow surcharge:

A rain-on-snow surcharge load need not be considered, since the slope is greater than 1/2 in./ft (2.38°) (see Section 7.10). See Fig. C7-3 for both loading conditions.

Example 2: Determine the roof snow load for a vaulted theater which can seat 450 people, planned for a suburb of Chicago, Illinois. The building is the tallest structure in a recreation-shopping complex surrounded by a parking lot. Two large deciduous trees are located in an area near the entrance. The building has an 80-ft (24.4-m) span and 15-ft (4.6-m) rise circular arc structural concrete roof covered with insulation and aggregate surfaced built-up roofing. The unventilated roofing system has a thermal resistance of 20 ft$^2\cdot$h$\cdot$F°/Btu (3.5 K$\cdot$m^2/W). It is expected that the structure will be exposed to winds during its useful life.

Flat-roof snow load:

$$p_f = 0.7C_eC_tIp_g$$

where

$p_g = 25$ lb/ft^2 (1.20 kN/m^2) (from Fig. 7-1);
$C_e = 0.9$ (from Table 7-2 for Terrain Category B and a fully exposed roof);
$C_t = 1.0$ (from Table 7-3); and
$I = 1.1$ (from Table 7-4).

Thus:

$$p_f = (0.7)(0.9)(1.0)(1.1)(25) = 17 \text{ lb/ft}^2$$

[in SI: $p_f = (0.7)(0.9)(1.0)(1.1)(1.19) = 0.83 \text{ kN/m}^2$]

Tangent of vertical angle from eaves to crown = 15/40 = 0.375 Angle = 21°

Since the vertical angle exceeds 10°, the minimum allowable values of p_f do not apply. Use $p_f = 17 \text{ lb/ft}^2$ (0.83 kN/m²), see Section 7.3.4.

Sloped-roof snow load:

$$p_s = C_s p_f$$

From Fig. 7-2a, $C_s = 1.0$ until slope exceeds 30° which (by geometry) is 30 ft (9.1 m) from the centerline. In this area $p_s = 17(1) = 17 \text{ lb/ft}^2$ [in SI: $p_s = 0.83(1) = 0.83 \text{ kN/m}^2$]. At the eaves, where the slope is (by geometry) 41°, $C_s = 0.72$ and $p_s = 17(0.72) = 12 \text{ lb/ft}^2$ [in SI: $p_s = 0.83(0.72) = 0.60 \text{ kN/m}^2$]. Since slope at eaves is 41°, Case II loading applies.

Unbalanced snow load:

Since the vertical angle from the eaves to the crown is greater than 10° and less than 60°, unbalanced snow loads must be considered.

Unbalanced load at crown

$$= 0.5 p_f = 0.5(17) = 9 \text{ lb/ft}^2$$

[in SI: $= 0.5(0.83) = 0.41 \text{ kN/m}^2$]

Unbalanced load at 30-degree point

$$= 2 \, p_f C_s / C_e = 2(17)(1.0)/0.9 = 38 \text{ lb/ft}^2$$

[in SI: $= 2(0.83)(1.0)/0.9 = 1.84 \text{ kN/m}^2$]

Unbalanced load at eaves

$$= 2(17)(0.72)/0.9 = 27 \text{ lb/ft}^2$$

in SI: $= 2(0.83)(0.72)/0.9 = 1.33 \text{ kN/m}^2$

Rain on snow surcharge:

A rain-on-snow surcharge load need not be considered, since the slope is greater than 1/2 in./ft (2.38°) (see Section 7.10). See Fig. C7-4 for both loading conditions.

Example 3: Determine design snow loads for the upper and lower flat roofs of a building located where $p_g = 40$ psf (1.92 kN/m²). The elevation difference between the roofs is 10 ft (3 m). The 100-ft by 100-ft (30.5 m by 30.5 m) unventilated high portion is heated and the 170 ft wide (51.8 m), 100-ft long (30.5 m) long low portion is an unheated storage area. The building is in an industrial park in flat open country with no trees or other structures offering shelter.

High roof:

$$p_f = 0.7 C_e C_t I p_g$$

where

$p_g = 40 \text{ lb/ft}^2$ (1.92 kN/m²) (given);
$C_e = 0.9$ (from Table 7-2);
$C_t = 1.0$ (from Table 7-3); and
$I = 1.0$ (from Table 7-4).

Thus:

$$p_f = 0.7(0.9)(1.0)(1.0)(40) = 25 \text{ lb/ft}^2$$

[in SI: $p_f = 0.7(0.9)(1.0)(1.0)(1.92) = 1.21 \text{ kN/m}^2$]

Since $p_g = 40$ psf (1.92 kN/m²) and $I = 1.0$, the minimum value of $p_f = 20 \, (1.0) = 20$ psf (0.96 kN/m²) and hence does not control, see Section 7.3.

Low roof:

$$p_f = 0.7 C_e C_t I p_g$$

where

$p_g = 40 \text{ lb/ft}^2$ (1.92 kN/m²) (given);
$C_e = 1.0$ (from Table 7-2) partially exposed due to presence of high roof;
$C_t = 1.2$ (from Table 7-3); and
$I = 0.8$ (from Table 7-4).

Thus:

$$p_f = 0.7(1.0) \, (1.2) \, (0.8) \, (40) = 27 \text{ lb/ft}^2$$

[in SI: $p_f = 0.7(1.0)(1.2)(0.8)(1.92) = 1.29 \text{ kN/m}^2$]

Since $p_g = 40$ psf (1.92 kN/m²) and $I = 0.8$, the minimum value of $p_f = 20 \, (0.8) = 16$ psf (0.77 kN/m²) and hence does not control, see Section 7.3.

Drift load calculation:

$$\gamma = 0.13(40) + 14 = 19 \text{ lb/ft}^3 \text{ (Eq. 7-3)}$$

$$[\text{in SI: } \gamma = 0.426(1.92) + 2.2 = 3.02 \text{ kN/m}^3]$$

$$h_b = p_f/19 = 27/19 = 1.4 \text{ ft}$$

$$[\text{in SI: } h_b = 1.29/3.02 = 0.43 \text{ m}]$$

$$h_c = 10 - 1.4 = 8.6 \text{ ft}$$

$$[\text{in SI: } h_c = 3.05 - 0.43 = 2.62 \text{ m}]$$

$$h_c/h_b = 8.6/1.4 = 6.1$$

$$[\text{in SI: } h_c/h_b = 2.62/0.43 = 6.1]$$

Since $h_c/h_b \geq 0.2$ drift loads must be considered (see Section 7.7.1).

h_d (leeward step) = 3.8 ft (1.16 m) [Fig. 7-9 with p_g = 40 lb/ft^2 (1.92 kN/m^2) and l_u = 100 ft (30.5 m)]

h_d (windward step) = 3/4 × 4.8 ft (1.5 m) = 3.6 ft (1.1 m) [4.8 ft (1.5 m) from Fig. 7-9 with p_g = 40 lb/ft^2 (1.92 kN/m^2) and l_u = length of lower roof = 170 ft (52 m)]

Leeward drift governs, use h_d = 3.8 ft (1.16 m)

Since $h_d < h_c$,

$$h_d = 3.8 \text{ ft (1.16 m)}$$

$$w = 4\,h_d = 15.2 \text{ ft (4.64 m), say 15 ft (4.6 m)}$$

$$p_d = h_d\gamma = 3.8(19) = 72 \text{ lb/ft}^2$$

$$\text{in SI: } p_d = 1.16(3.02) = 3.50 \text{ kN/m}^2)$$

Rain on snow surcharge:

A rain-on-snow surcharge load need not be considered even though the slope is less than 1/2 in./ft (2.38°), since p_g is greater than 20 lb/ft^2 (0.96 kN/m^2). See Fig. C7-5 for snow loads on both roofs.

REFERENCES

[C7-1] Ellingwood, B., and Redfield, R. "Ground snow loads for structural design." *J. Struct. Engrg.*, ASCE, 109(4), 950–964, 1983.

[C7-2] MacKinlay, I., and Willis, W.E. *Snow country design.* National Endowment for the Arts, Washington, D.C., 1965.

[C7-3] Sack, R.L., and Sheikh-Taheri, A. *Ground and roof snow loads for Idaho.* Dept. of Civil Engineering, University of Idaho, Moscow, Idaho, ISBN 0-89301-114-2, 1986.

[C7-4] Structural Engineers Association of Arizona. *Snow load data for Arizona.* Univ. of Arizona, Tempe, Ariz., 1973.

[C7-5] Structural Engineers Association of Colorado. *Snow load design data for Colorado.* Denver, Colo., 1971. [Available from: Structural Engineers Association of Colorado, Denver, Colo.]

[C7-6] Structural Engineers Association of Oregon. *Snow load analysis for Oregon.* Oregon Dept. of Commerce, Building Codes Division, Salem, Or., 1971.

[C7-7] Structural Engineers Association of Washington. *Snow loads analysis for Washington.* SEAW, Seattle, Wash., 1981.

[C7-8] USDA Soil Conservation Service. *Lake Tahoe basin snow load zones.* U.S. Dept. of Agriculture, Soil Conservation Service, Reno, Nev., 1970.

[C7-9] Videon, F.V., and Stenberg, P. *Recommended snow loads for Montana structures.* Montana State Univ., Bozeman, Mont., 1978.

[C7-10] Structural Engineers Association of Northern California. *Snow load design data for the Lake Tahoe area.* San Francisco, Calif., 1964.

[C7-11] Placer County Building Division. *Snow Load Design, Placer County Code.* Chapter 4, Section 4.20(V), Auburn, Calif., 1985.

[C7-12] Brown, J. "An approach to snow load evaluation." *Proc., 38th Western Snow Conference*, 1970.

[C7-13] Newark, M. "A new look at ground snow loads in Canada." *Proc., 41st Eastern Snow Conference*, Washington, D.C., 37–48, 1984.

[C7-14] Elliott, M. "Snow load criteria for western United States, case histories and state-of-the-art." *Proc., 1st Western States Conference of Structural Engineer Associations.* Sun River, Or., June, 1975.

[C7-15] Lorenzen, R.T. "Observations of snow and wind loads precipitant to building failures in New York State, 1969–1970." American Society of Agricultural Engineers North Atlantic Region meeting, Paper NA 70-305,

Newark, Del., August, 1970. [Available from: American Society of Agricultural Engineers, St. Joseph, Missouri.]

[C7-16] Lutes, D.A., and Schriever, W.R. "Snow accumulation in Canada: Case histories: II." National Research Council of Canada. DBR Tech. Paper 339, NRCC 11915, Ottawa, Ontario, Canada, March, 1971.

[C7-17] Meehan, J.F. "Snow loads and roof failures." *Proc., 1979 Structural Engineers Association of California Convention.* [Available from Structural Engineers Association of California, San Francisco, Calif.]

[C7-18] Mitchell, G.R. "Snow loads on roofs—An interim report on a survey." *Wind and snow loading*, The Construction Press Ltd., Lancaster, England, 177–190, 1978.

[C7-19] Peter, B.G.W., Dalgliesh, W.A., and Schriever, W.R. "Variations of snow loads on roofs." *Trans. Engrg. Inst. Can*, 6(A-1), 8 p., April, 1963.

[C7-20] Schriever, W.R., Faucher, Y., and Lutes, D.A. "Snow accumulation in Canada: Case histories: I." National Research Council of Canada, Division of Building Research. NRCC 9287, Ottawa, Ontario, Canada, January, 1967.

[C7-21] Taylor, D.A. "A survey of snow loads on roofs of arena-type buildings in Canada." *Can. J. Civil Engrg.*, 6(1), 85–96, March, 1979.

[C7-22] Taylor, D.A. "Roof snow loads in Canada." *Can. J. Civil Engrg.* 7(1), 1–18, March, 1980.

[C7-23] O'Rourke, M., Koch, P., and Redfield, R. "Analysis of roof snow load case studies: Uniform loads," U.S. Dept. of the Army, Cold Regions Research and Engineering Laboratory, CRREL Report 83-1, Hanover, NH, 1983.

[C7-24] Grange, H.L., and Hendricks, L.T. "Roof-snow behavior and ice-dam prevention in residential housing." Univ. of Minnesota, Agricultural Extension Service, *Extension Bull. 399*, St. Paul, Minn, 1976.

[C7-25] Klinge, A.F. "Ice dams." *Popular Science*, 119–120, Nov., 1978.

[C7-26] Mackinlay, I. "Architectural design in regions of snow and ice." *Proc., 1st International Conference on Snow Engineering*, Santa Barbara, Calif., 441–455, July, 1988.

[C7-27] Tobiasson, W. "Roof design in cold regions." *Proc., 1st International Conference on Snow Engineering*, Santa Barbara, Calif., 462–482, July, 1988.

[C7-28] de Marne, H. "Field experience in control and prevention of leaking from ice dams in New England." *Proc., 1st International Conference on Snow Engineering*, Santa Barbara, Calif., 473–482, July, 1988.

[C7-29] Tobiasson, W., and Buska J. "Standing seam metal roofs in cold regions." *Proc., 10th Conference on Roofing Technology*, Gaithersburg, Md., 34–44, April, 1993.

[C7-30] National Greenhouse Manufacturers Association, *Design loads in greenhouse structures*, Taylors, South Carolina, 1988.

[C7-31] Air Structures Institute. *Design and standards manual. ASI-77.* Available from the Industrial Fabrics Assn. International, St. Paul, Minn., 1977.

[C7-32] American Society of Civil Engineers, *Air supported structures*, New York, 1994.

[C7-33] Sack, R.L. "Snow loads on sloped roofs." *J. Struct. Engrg.*, ASCE, 114(3), 501–517, 1988.

[C7-34] Sack, R., Arnholtz, D., and Haldeman, J. "Sloped roof snow loads using simulation." *J. Struct. Engrg.*, ASCE, 113(8), 1820–1833, 1987.

[C7-35] Taylor, D. "Sliding snow on sloping roofs." *Canadian Building Digest 228*. National Research Council of Canada, Ottawa, Ontario, Canada, Nov., 1983.

[C7-36] Taylor, D. "Snow loads on sloping roofs: Two pilot studies in the Ottawa area." Division of Building Research Paper 1282, *Can. J. Civil Engrg.*, (2), 334–343, June, 1985.

[C7-37] O'Rourke, M., Tobiasson, W., and Wood, E. "Proposed code provisions for drifted snow loads." *J. Struct. Engrg.*, ASCE, 112(9), 2080–2092, 1986.

[C7-38] O'Rourke, M., Speck, R., and Stiefel, U. "Drift snow loads on multilevel roofs." *J. Struct. Engrg.*, ASCE, 111(2), 290–306, 1985.

[C7-39] Speck, R., Jr. "Analysis of snow loads due to drifting on multilevel roofs." Thesis, presented to the Department of Civil Engineering at Rensselaer Polytechnic Institute, Troy, N.Y., in partial fulfillment of the requirements for the degree of Master of Science.

[C7-40] Taylor, D.A. "Snow loads on two-level flat roofs." *Proc., Eastern Snow Conference, 29,*

41st Annual Meeting, Washington, D.C., June 7–8, 1984.

[C7-41] Finney, E. "Snow drift control by highway design." *Bulletin 86*, Michigan State College Engineering Station, Lansing, Mich., 1939.

[C7-42] Tabler, R. "Predicting profiles of snow drifts in topographic catchments." *Proc., Western Snow Conference*, Coronado, Calif., 1975.

[C7-43] Zallen, R. "Roof collapse under snow drift loading and snow drift design criteria." *J. Perf. Constr. Fac.*, ASCE, 2(2), 80–98, 1988.

[C7-44] O'Rourke, M., and Weitman, N. "Laboratory studies of snow drifts on multilevel roofs." *Proc., 2nd International Conference on Snow Engrg.*, Santa Barbara, Calif., June, 1992.

[C7-45] O'Rourke, M., and El Hamadi, K. "Roof snow loads: Drifting against a higher wall." *Proc., 55th Western Snow Conference*, Vancouver, B.C., 124–132, April, 1987.

[C7-46] O'Rourke, M. "Discussion of 'Roof collapse under snow drift loading and snow drift design criteria.'" *J. Perf. Constr. Fac.*, ASCE, 3(11), 266–268, 1989.

[C7-47] Kennedy, D., Isyumov, M., and Mikitiuk, M. "The Effectiveness of code provisions for snow accumulations on stepped roofs." *Proc., 2nd International Conference on Snow Engrg.*, Santa Barbara, Calif., June, 1992.

[C7-48] Isyumou, N., and Mikitiuk, M. "Wind tunnel modeling of snow accumulation on large roofs." *Proc., 2nd International Conference on Snow Engrg.*, Santa Barbara, Calif., June, 1992.

[C7-49] Irwin, P., William, C., Gamle, S., and Retziaff, R. "Snow prediction in Toronto and the Andes Mountains: FAE simulation capabilities." *Proc., 2nd International Conference on Snow Engrg.*, Santa Barbara, Calif., June, 1992.

[C7-50] O'Rourke, M.J. Snow and ice accumulation around solar collector installations. U.S. Dept. of Commerce, National Bureau of Standards, NBS-GCR-79 180, Washington, D.C., Aug., 1979.

[C7-51] Corotis, R.B., Dowding, C.H., and Rossow, E.C. "Snow and ice accumulation at solar collector installations in the Chicago metropolitan area." U.S. Dept. of Commerce, National Bureau of Standards, NBS-GCR-79 181, Washington, D.C., Aug, 1979.

[C7-52] Paine, J.C. "Building design for heavy snow areas." *Proc., 1st International Conference on Snow Engineering*, Santa Barbara, Calif., 483–492, July, 1988.

[C7-53] Tobiasson, W., Buska, J., and Greatorex, A. "Snow Guards for Metal Roofs." *Proc., 8th Conference on Cold Regions Engineering*, ASCE, NY, Aug., 1996.

[C7-54] Colbeck, S.C. "Snow loads resulting from rain-on-snow." U.S. Dept. of the Army, Cold Regions Research and Engineering Laboratory. CRREL Rep. 77-12, Hanover, N.H., 1977.

[C7-55] Colbeck, S.C. "Roof loads resulting from rain-on-snow: Results of a physical model." *Can. J. Civil Engrg.*, 4, 482–490, 1977.

[C7-56] O'Rourke, M., and Auren, M. "Snow loads on gable roofs." *J. Struct. Engrg.*, ASCE, 123(12), 1645–1651, 1997.

[C7-57] Tobiasson, W. "Discussion of 'Snow loads on gable roofs.'" *J. Struct. Engrg.*, ASCE, 125(4), 470–471, 1999.

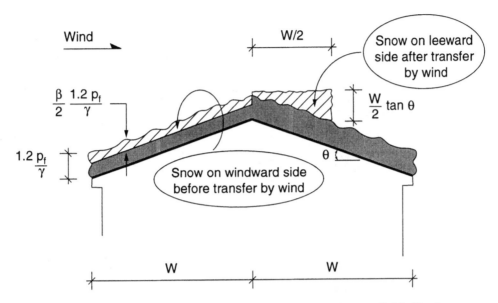

FIGURE C7-1. Formation of a Unbalanced Load on a Gable Roof

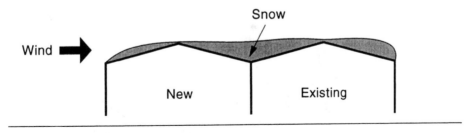

FIGURE C7-2. Valley in Which Snow Will Drift Is Created When New Gable Roof Is Added Alongside Existing Gable Roof

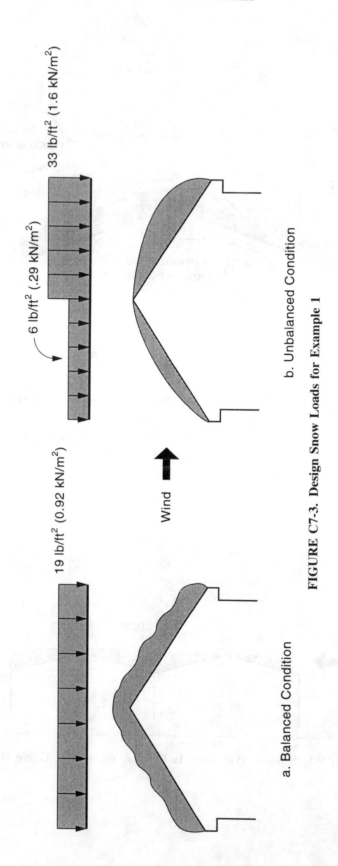

33 lb/ft² (1.6 kN/m²)

6 lb/ft² (.29 kN/m²)

19 lb/ft² (0.92 kN/m²)

Wind

a. Balanced Condition

b. Unbalanced Condition

FIGURE C7-3. Design Snow Loads for Example 1

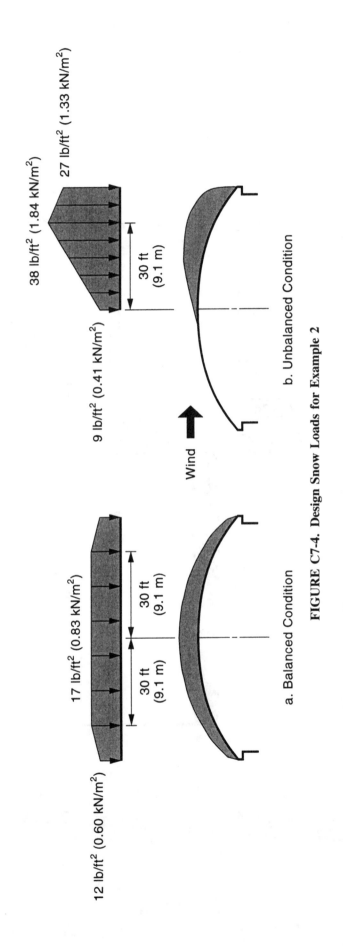

FIGURE C7-4. Design Snow Loads for Example 2

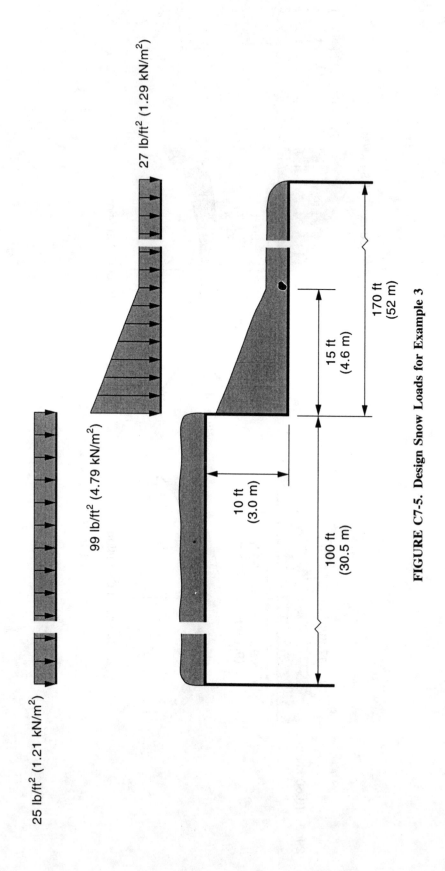

27 lb/ft² (1.29 kN/m²)

99 lb/ft² (4.79 kN/m²)

25 lb/ft² (1.21 kN/m²)

170 ft (52 m)

15 ft (4.6 m)

10 ft (3.0 m)

100 ft (30.5 m)

FIGURE C7-5. Design Snow Loads for Example 3

TABLE C7-1. Ground Snow Loads at 204 National Weather Service Locations at Which Load Measurements Are Made

(Note: To convert lb/ft² to kN/m², multiply by 0.0479)

Location	Ground Snow Load (lb/ft²)			Location	Ground Snow Load (lb/ft²)		
	Years of Record	Maximum Observed	2% Annual Probability*		Years of Record	Maximum Observed	2% Annual Probability*
ALABAMA				IOWA			
Birmingham	40	4	3	Burlington	11	15	17
Huntsville	33	7	5	Des Moines	40	22	22
Mobile	40	1	1	Dubuque	39	34	32
ARIZONA				Sioux City	38	28	28
Flagstaff	38	88	48	Waterloo	33	25	32
Tucson	40	3	3	KANSAS			
Winslow	39	12	7	Concordia	30	12	17
ARKANSAS				Dodge City	40	10	14
Fort Smith	37	6	5	Goodland	39	12	15
Little Rock	24	6	6	Topeka	40	18	17
CALIFORNIA				Wichita	40	10	14
Bishop	31	6	8	KENTUCKY			
Blue Canyon	26	213	242	Covington	40	22	13
Mt. Shasta	32	62	62	Jackson	11	12	18
Red Bluff	34	3	3	Lexington	40	15	13
COLORADO				Louisville	39	11	12
Alamosa	40	14	14	LOUISIANA			
Colorado Springs	39	16	14	Alexandria	17	2	2
Denver	40	22	18	Shreveport	40	4	3
Grand Junction	40	18	16	MAINE			
Pueblo	33	7	7	Caribou	34	68	95
CONNECTICUT				Portland	39	51	60
Bridgeport	39	21	24	MARYLAND			
Hartford	40	23	33	Baltimore	40	20	22
New Haven	17	11	15	MASSACHUSETTS			
DELAWARE				Boston	39	25	34
Wilmington	39	12	16	Nantucket	16	14	24
GEORGIA				Worcester	33	29	44
Athens	40	6	5	MICHIGAN			
Atlanta	39	4	3	Alpena	31	34	48
Augusta	40	8	7	Detroit City	14	6	10
Columbus	39	1	1	Detroit Airport	34	27	18
Macon	40	8	7	Detroit-Willow	12	11	22
Rome	28	3	3	Flint	37	20	24
IDAHO				Grand Rapids	40	32	36
Boise	38	8	9	Houghton Lake	28	33	48
Lewiston	37	6	9	Lansing	35	34	36
Pocatello	40	12	10	Marquette	16	44	53
ILLINOIS				Muskegon	40	40	51
Chicago—O'Hare	32	25	17	Sault Ste. Marie	40	68	77
Chicago	26	37	22	MINNESOTA			
Moline	39	21	19	Duluth	40	55	63
Peoria	39	27	15	International Falls	40	43	44
Rockford	26	31	19	Minneapolis-St. Paul	40	34	51
Springfield	40	20	21	Rochester	40	30	47
INDIANA				St. Cloud	40	40	53
Evansville	40	12	17	MISSISSIPPI			
Fort Wayne	40	23	20	Jackson	40	3	3
Indianapolis	40	19	22	Meridian	39	2	2
South Bend	39	58	41				

297

TABLE C7-1. Ground Snow Loads at 204 National Weather Service Locations at Which Load Measurements are Made (*Continued*)

(Note: To convert lb/ft^2 to kN/m^2, multiply by 0.0479)

Location	Ground Snow Load (lb/ft^2)			Location	Ground Snow Load (lb/ft^2)		
	Years of Record	Maximum Observed	2% Annual Probability*		Years of Record	Maximum Observed	2% Annual Probability*
MISSOURI				Wilmington	39	14	7
Columbia	39	19	20	Winston-Salem	12	14	20
Kansas City	40	18	18	NORTH DAKOTA			
St. Louis	37	28	21	Bismark	40	27	27
Springfield	39	14	14	Fargo	39	27	41
MONTANA				Williston	40	28	27
Billings	40	21	15	OHIO			
Glasgow	40	18	19	Akron-Canton	40	16	14
Great Falls	40	22	15	Cleveland	40	27	19
Havre	26	22	24	Columbus	40	11	11
Helena	40	15	17	Dayton	40	18	11
Kalispell	29	27	45	Mansfield	30	31	17
Missoula	40	24	22	Toledo Express	36	10	10
NEBRASKA				Youngstown	40	14	10
Grand Island	40	24	23	OKLAHOMA			
Lincoln	20	15	22	Oklahoma City	40	10	8
Norfolk	40	28	25	Tulsa	40	5	8
North Platte	39	16	13	OREGON			
Omaha	25	23	20	Astoria	26	2	3
Scottsbluff	40	10	12	Burns City	39	21	23
Valentine	26	26	22	Eugene	37	22	10
NEVADA				Medford	40	6	6
Elko	12	12	20	Pendleton	40	9	13
Ely	40	10	9	Portland	39	10	8
Las Vegas	39	3	3	Salem	39	5	7
Reno	39	12	11	Sexton Summit	14	48	64
Winnemucca	39	7	7	PENNSYLVANIA			
NEW HAMPSHIRE				Allentown	40	16	23
Concord	40	43	63	Erie	32	20	18
NEW JERSEY				Harrisburg	19	21	23
Atlantic City	35	12	15	Philadelphia	39	13	14
Newark	39	18	15	Pittsburgh	40	27	20
NEW MEXICO				Scranton	37	13	18
Albuquerque	40	6	4	Williamsport	40	18	21
Clayton	34	8	10	RHODE ISLAND			
Roswell	22	6	8	Providence	39	22	23
NEW YORK				SOUTH CAROLINA			
Albany	40	26	27	Charleston	39	2	2
Binghamton	40	30	35	Columbia	38	9	8
Buffalo	40	41	39	Florence	23	3	3
NYC—Kennedy	18	8	15	Greenville-Spartanburg	24	6	7
NYC—LaGuardia	40	23	16	SOUTH DAKOTA			
Rochester	40	33	38	Aberdeen	27	23	43
Syracuse	40	32	32	Huron	40	41	46
NORTH CAROLINA				Rapid City	40	14	15
Asheville	28	7	14	Sioux Falls	39	40	40
Cape Hatteras	34	5	5	TENNESSEE			
Charlotte	40	8	11	Bristol	40	7	9
Greensboro	40	14	11	Chattanooga	40	6	6
Raleigh-Durham	36	13	14				

TABLE C7-1. Ground Snow Loads at 204 National Weather Service Locations at Which Load Measurements are Made (*Continued*)

(Note: To convert lb/ft² to kN/m², multiply by 0.0479)

Location	Ground Snow Load (lb/ft²)			Location	Ground Snow Load (lb/ft²)		
	Years of Record	Maximum Observed	2% Annual Probability*		Years of Record	Maximum Observed	2% Annual Probability*
Knoxville	40	10	9	Norfolk	38	9	10
Memphis	40	7	6	Richmond	40	11	16
Nashville	40	6	9	Roanoke	40	14	20
TEXAS				WASHINGTON			
Abilene	40	6	6	Olympia	40	23	22
Amarillo	39	15	10	Quillayute	25	21	15
Austin	39	2	2	Seattle-Tacoma	40	15	18
Dallas	23	3	3	Spokane	40	36	42
El Paso	38	8	8	Stampede Pass	36	483	516
Fort Worth	39	5	4	Yakima	39	19	30
Lubbock	40	9	11	WEST VIRGINIA			
Midland	38	4	4	Beckley	20	20	30
San Angelo	40	3	3	Charleston	38	21	18
San Antonio	40	9	4	Elkins	32	22	18
Waco	40	3	2	Huntington	30	15	19
Wichita Falls	40	4	5	WISCONSIN			
UTAH				Green Bay	40	37	36
Milford	23	23	14	La Crosse	16	23	32
Salt Lake City	40	11	11	Madison	40	32	35
Wendover	13	2	3	Milwaukee	40	34	29
VERMONT				WYOMING			
Burlington	40	43	36	Casper	40	9	10
VIRGINIA				Cheyenne	40	18	18
Dulles Airport	29	15	23	Lander	39	26	24
Lynchburg	40	13	18	Sheridan	40	20	23
National Airport	40	16	22				

*It is not appropriate to use only the site specific information in this table for design purposes. Reasons are given in Commentary Section 7.2.

TABLE C7-2. Comparison of Some Site-Specific Values and Zoned Values in Fig. 7-1

State	Location	Elevation, ft (m)	Zoned value lb/ft^2 (kN/m^2)	Case Study Value* psf (kN/m^2)
California	Mount Hamilton	4210 (1283)	0 to 2400' (732 m)	30 (1.44)
			0 to 3500' (1067 m)	
Arizona	Palisade Ranger Station	7950 (2423)	5 to 4600' (0.24 to 1402 m)	120 (5.75)
			10 to 5000' (0.48 to 1524 m)	
Tennessee	Monteagle	1940 (591)	10 to 1800' (0.48 to 549 m)	15 (0.72)
Maine	Sunday River Ski Area	900 (274)	90 to 700' (4.31 to 213 m)	100 (4.79)

*Based on a detailed study of information in the vicinity of each location.

TABLE C7-3. Factors for Converting from Other Annual Probabilities of Being Exceeded and Other Mean Recurrence Intervals, to that Used in This Standard

Annual Probability of Being Exceeded (%)	Mean Recurrence Interval (years)	Multiplication Factor
10	10	1.82
4	25	1.20
3.3	30	1.15
1	100	0.82

C8.0 RAIN LOADS

C8.1 SYMBOLS AND NOTATION

A = roof area serviced by a single drainage system, in square feet (square meters);

 i = design rainfall intensity as specified by the code having jurisdiction, in inches per hour (millimeters per hour); and

Q = flow rate out of a single drainage system, in gallons per minute (cubic meters per second).

C8.2 ROOF DRAINAGE

Roof drainage systems are designed to handle all the flow associated with intense, short-duration rainfall events. (For example, the 1993 BOCA National Plumbing Code [C8-1], and Factory Mutual Loss Prevention Data 1-54, "Roof Loads for New Construction" [C8-2] use a 1-h duration event with a 100-year return period; the 1994 Standard Plumbing Code [C8-3] uses 1-h and 15-min duration events with 100-year return periods for the primary and secondary drainage systems, respectively, and the 1990 National Building Code [C8-4] of Canada uses a 15-min event with a 10-year return period. A very severe local storm or thunderstorm may produce a deluge of such intensity and duration that properly designed primary drainage systems are temporarily overloaded. Such temporary loads are adequately covered in design when blocked drains (see Section 8.3) and ponding instability (see Section 8.4) are considered.

Roof drainage is a structural, architectural and mechanical (plumbing) issue. The type and location of secondary drains and the hydraulic head above their inlets at the design flow must be known in order to determine rain loads. Design team coordination is particularly important when establishing rain loads.

C8.3 DESIGN RAIN LOADS

The amount of water that could accumulate on a roof from blockage of the primary drainage system is determined and the roof is designed to withstand the load created by that water plus the uniform load caused by water that rises above the inlet of the secondary drainage systems at its design flow. If parapet walls, cant strips, expansion joints, and other features create the potential for deep water in an area, it may

be advisable to install in that area secondary (overflow) drains with separate drain lines rather than overflow scuppers to reduce the magnitude of the design rain load. Where geometry permits, free discharge is the preferred form of emergency drainage.

When determining these water loads, it is assumed that the roof does not deflect. This eliminates complexities associated with determining the distribution of water loads within deflection depressions. However, it is quite important to consider this water when assessing ponding instability in Section 8.4.

The depth of water, d_h, above the inlet of the secondary drainage system (i.e., the hydraulic head) is a function of the rainfall intensity at the site, the area of roof serviced by that drainage system and the size of the drainage system.

The flow rate through a single drainage system is as follows:

$$Q = 0.0104 \, Ai \quad [\text{in SI: } Q = 0.278 \times 10^{-6} \, Ai]$$
$$(\text{Eq. C8-1})$$

The hydraulic head, d_h, is related to flow rate, Q, for various drainage systems in Table C8-1. That table indicates that d_h can vary considerably depending on the type and size of each drainage system and the flow rate it must handle. For this reason the single value of 1 in. (25 mm) [i.e., 5 lb/ft^2 (0.24 kN/m^2)] used in ASCE 7-93 has been eliminated.

The hydraulic head, d_h, is zero when the secondary drainage system is simply overflow all along a roof edge.

C8.4 PONDING INSTABILITY

Water may accumulate as ponds on relatively flat roofs. As additional water flows to such areas, the roof tends to deflect more, allowing a deeper pond to form there. If the structure does not possess enough stiffness to resist this progression, failure by localized overloading may result. References [C8-1–C8-16] contain information on ponding and its importance in the design of flexible roofs. Rational design methods to preclude instability from ponding are presented in references [C8-5–C8-6].

By providing roofs with a slope of 1/4 in./ft (1.19°) or more, ponding instability can be avoided. If the slope is less than 1/4 in./ft (1.19°), the roof structure must be checked for ponding instability because construction tolerances and long-term deflections under dead load can result in flat portions susceptible to ponding.

C8.5 CONTROLLED DRAINAGE

In some areas of the country, ordinances are in effect that limit the rate of rainwater flow from roofs into storm drains. Controlled-flow drains are often used on such roofs. Those roofs must be capable of sustaining the storm water temporarily stored on them. Many roofs designed with controlled-flow drains have a design rain load of 30 lb/ft^2 (1.44 kN/m^2) and are equipped with a secondary drainage system (for example, scuppers) that prevents water depths ($d_s + d_h$) greater than 5-3/4 in. (145 mm) on the roof.

Examples

The following two examples illustrate the method used to establish design rain loads based on Section 8 of this Standard.

Example 1: Determine the design rain load, R, at the secondary drainage for the roof plan shown in Fig. C8-1, located at a site in Birmingham, Ala. The design rainfall intensity, i, specified by the plumbing code for a 100-year, 1-h rainfall is 3.75 in./h (95 mm/h). The inlet of the 4 in. diameter (102 mm) secondary roof drains are set 2 in. (51 mm) above the roof surface.

Flow rate, Q, for the secondary drainage 4 in. diameter (102 mm) roof drain:

$$Q = 0.0104\ Ai \qquad \text{(Eq. C8-1)}$$

$$Q = 0.0104\ (2{,}500)(3.75)$$

$$= 97.5 \text{ gal./min } (0.0062 \text{ m}^3/\text{s})$$

Hydraulic head, d_h:

Using Table C8-1, for a 4 in. diameter (102 mm) roof drain with a flow rate of 97.5 gal./min (0.0062 m^3/s) interpolate between a hydraulic head of 1 and 2 in. (25 and 51 mm) as follows:

$$d_h = 1 + [(97.5 - 80) \div (170 - 80)]$$

$$= 1.19 \text{ in. } (30.2 \text{ mm})$$

Static head $d_s = 2$ in. (51 mm); the water depth from drain inlet to the roof surface.

Design rain load, R, adjacent to the drains:

$$R = 5.2\ (d_s + d_h) \qquad \text{(Eq. 8-1)}$$

$$R = 5.2\ (2 + 1.19) = 16.6 \text{ psf } (0.80 \text{ kN/m}^2)$$

Example 2: Determine the design rain load, R, at the secondary drainage for the roof plan shown in Fig. C8-2, located at a site in Los Angeles, Calif. The design rainfall intensity, i, specified by the plumbing code for a 100-year, 1-h rainfall is 1.5 in./h (38 mm/h). The inlet of the 12 in. (305 mm) secondary roof scuppers are set 2 in. (51 mm) above the roof surface.

Flow rate, Q, for the secondary drainage, 12 in. (305 mm) wide channel scupper:

$$Q = 0.0104\ A_i \qquad \text{(Eq. C8-1)}$$

$$Q = 0.0104\ (11{,}500)(1.5)$$

$$= 179 \text{ gal./min } (0.0113 \text{ m}^3/\text{s})$$

Hydraulic head, d_h:

Using Table C8-1, by interpolation, the flow rate for a 12 in. (305 mm) wide channel scupper is twice that of a 6 in. (152 mm) wide channel scupper. Using Table C8-1, the hydraulic head, d_h, for one-half the flow rate, Q, or 90 gal./min (0.0057 m^3/s), through a 6 in. (152 mm) wide channel scupper is 3 in. (76 mm).

$d_h = 3$ in. (76 mm) for a 12 in. wide (305 mm) channel scupper with a flow rate, Q, of 179 gal./min (0.0113 m^3/s).

Static head, $d_s = 2$ in. (51 mm); depth of water from the scupper inlet to the roof surface.

Design rain load, R, adjacent to the scuppers:

$$R = 5.2(d_h + d_s) \qquad \text{(Eq. 8-1)}$$

$$R = 5.2\ (2 + 3) = 26 \text{ psf } (1.2 \text{ kN/m}^2)$$

REFERENCES

[C8-1] Building Officials and Code Administrators International. "The BOCA National Plumbing Code/1993,'' BOCA Inc., Country Club Hills, Illinois, Jan., 1993.

[C8-2] Factory Mutual Engineering Corp. "Loss Prevention Data 1-54, Roof Loads for New Construction." Norwood, Mass., Aug., 1991.

[C8-3] Southern Building Code Congress International. "Standard Plumbing Code, 1991 Edition." SBCCI Inc., Birmingham, Alabama, 1991.

[C8-4] Associate Committee on the National Building Code. "National Building Code of Canada 1990," National Research Council of Canada, Ottawa, Ontario, Jan., 1990.

[C8-5] American Institute of Steel Construction. "Specification for structural steel for buildings, allowable stress design and plastic design." AISC, New York, June, 1989.

[C8-6] American Institute of Steel Construction. "Load and resistance factor design specification for structural steel buildings." AISC, New York, Sept., 1986.

[C8-7] American Institute of Timber Construction. "Roof slope and drainage for flat or nearly flat roofs." AITC, Tech. Note No. 5, Englewood, Colo., Dec., 1978.

[C8-8] Burgett, L.B. "Fast check for ponding." *Engrg. J. Am. Inst. Steel Construction*, 10(1), 26–28, 1973.

[C8-9] Chinn, J., Mansouri, A.H., and Adams, S.F. "Ponding of liquids on flat roofs." *J. Struct. Div*., ASCE, 95(5), 797–808, 1969.

[C8-10] Chinn, J. "Failure of simply-supported flat roofs by ponding of rain." *Engrg. J. Am. Inst. Steel Construction*, 3(2), 38–41, 1965.

[C8-11] Haussler, R.W. "Roof deflection caused by rainwater pools." *Civil Eng*. 32, 58–59, Oct., 1962.

[C8-12] Heinzerling, J.E. "Structural design of steel joist roofs to resist ponding loads." Steel Joist Institute, Tech. Digest No. 3, Arlington, Va., May, 1971.

[C8-13] Marino, F.J. "Ponding of two-way roof systems." *Engrg. J. Am. Inst. Steel Construction*, 3(3), 93–100, 1966.

[C8-14] Salama, A.E., and Moody, M.L. "Analysis of beams and plates for ponding loads." *J. Struct. Div*., ASCE. 93(1), 109–126, 1967.

[C8-15] Sawyer, D.A. "Ponding of rainwater on flexible roof systems." *J. Struct. Div*., ASCE, 93(1), 127–148, 1967.

[C8-16] Sawyer, D.A. "Roof-structural roof-drainage interactions." *J. Struct. Div*., ASCE, 94(1), 175–198, 1969.

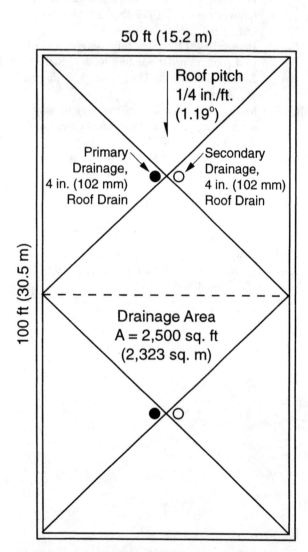

FIGURE C8-1. Example 1 Roof Plan (Dashed lines indicate the boundary between separate drainage areas)

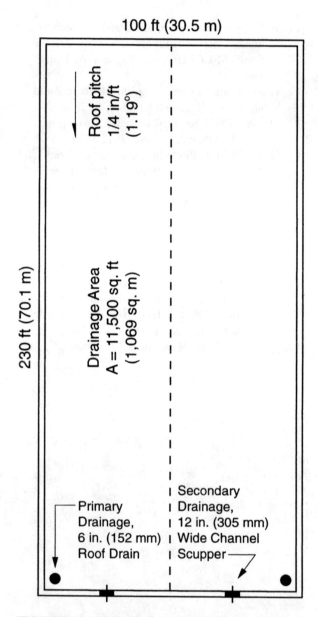

FIGURE C8-2. Example 2 Roof Plan (Dashed lines indicate the boundary between separate drainage areas)

TABLE C8-1. Flow Rate, Q, in Gallons per Minute of Various Drainage Systems at Various Hydraulic Heads, d_h in Inches [2]

Drainage	Hydraulic Head, d_h, inches									
	1	2	2.5	3	3.5	4	4.5	5	7	8
4 in. diameter drain	80	170	180							
6 in. diameter drain	100	190	270	380	540					
8 in. diameter drain	125	230	340	560	850	1,100	1,170			
6 in. wide, channel scupper[2]	18	50	—[1]	90	—[1]	140	—[1]	194	321	393
24 in. wide, channel scupper	72	200	—[1]	360	—[1]	560	—[1]	776	1,284	1,572
6 in. wide, 4 in. high, closed scupper[2]	18	50	—[1]	90	—[1]	140	—[1]	177	231	253
24 in. wide, 4 in. high, closed scupper	72	200	—[1]	360	—[1]	560	—[1]	708	924	1,012
6 in. wide, 6 in. high, closed scupper	18	50	—[1]	90	—[1]	140	—[1]	194	303	343
24 in. wide, 6 in. high, closed scupper	72	200	—[1]	360	—[1]	560	—[1]	776	1,212	1,372

[1] Interpolation is appropriate, including between widths of each scupper.
[2] Channel scuppers are open-topped (i.e., 3-sided). Closed scuppers are 4-sided.

In SI, Flow Rate, Q, in Cubic Meters per Second of Various Drainage Systems at Various Hydraulic Heads, d_h in Millimeters [2]

Drainage System	Hydraulic Head d_h, mm									
	25	51	64	76	89	102	114	127	178	203
102 mm diameter drain	.0051	.0107	.0114							
152 mm diameter drain	.0063	.0120	.0170	.0240	.0341					
203 mm diameter drain	.0079	.0145	.0214	.0353	.0536	.0694	.0738			
152 mm wide, channel scupper[2]	.0011	.0032	—[1]	.0057	—[1]	.0088	—[1]	.0122	.0202	.0248
610 mm wide, channel scupper	.0045	.0126	—[1]	.0227	—[1]	.0353	—[1]	.0490	.0810	.0992
152 mm wide, 102 mm high, closed scupper[2]	.0011	.0032	—[1]	.0057	—[1]	.0088	—[1]	.0112	.0146	.0160
610 mm wide, 102 mm high, closed scupper	.0045	.0126	—[1]	.0227	—[1]	.0353	—[1]	.0447	.0583	.0638
152 mm wide, 152 mm high, closed scupper	.0011	.0032	—[1]	.0057	—[1]	.0088	—[1]	.0122	.0191	.0216
610 mm wide, 152 mm high, closed scupper	.0045	.0126	—[1]	.0227	—[1]	.0353	—[1]	.0490	.0765	.0866

[1] Interpolation is appropriate, including between widths of each scupper.
[2] Channel scuppers are open-topped (i.e., 3-sided). Closed scuppers are 4-sided.

C9.0 EARTHQUAKE LOADS

The 1998 edition incorporates several new topics and includes several revisions of the 1995 edition. Major revisions include the adoption of new spectral response seismic design maps that reflect seismic hazards on the basis of contours. These maps were completed by the U.S. Geological Survey. The maps can be accessed via the internet at the following sites: the ASCE 7 portion of the ASCE/Structural Engineering Institute's website at *www.seinstitute.org/codediv.html*; or the U.S. Geological Survey at *www.usgs.gov*; or the National Institute of Building Sciences at *www.nibs.org*. Care must be exercised to access the correct maps at the various websites. New design procedures based on these maps were developed by the BSSC for use in the 1997 NEHRP Provisions and are adapted for use in this document. Improvements in the design of procedures for high-seismic, near-source areas are also of major significance.

The 1998 provisions are based on the *NEHRP Recommended Provisions for the Development of Seismic Regulations for New Buildings and Other Structures* (1), which is prepared by the Building Seismic Safety Council (BSSC) under sponsorship of the Federal Emergency Management Agency (FEMA). NEHRP stands for the National Earthquake Hazard Reduction Program, which is managed by FEMA. These provisions are a direct descendent of the *Tentative Provisions for the Development of Seismic Regulations for Buildings* (2), developed by the Applied Technology Council (ATC) under sponsorship of the National Science Foundation and the National Bureau of Standard (now the National Institute for Standards and Technology).

The ATC and BSSC efforts had their origin in the 1971 San Fernando Valley Earthquake, which demonstrated that design rules of that time for seismic resistance had some serious shortcomings. Each subsequent major earthquake has taught new lessons. ATC, BSSC, and ASCE have endeavored to work individually and collectively to improve each succeeding document so that it would be based upon the best earthquake engineering research as applicable to design and construction, and that they would have nationwide applicability. The natural evolution of this process has led BSSC's NEHRP Provisions into ASCE 7, some current model building codes, and was utilized to develop seismic provisions for The International Building Code (IBC 2000). The IBC 2000 provisions have also influenced the ASCE 7-98 provisions. There was considerable support in the ASCE-7 committee to review and adopt many changes IBC 2000 made to the original NEHRP 1997 provisions.

Content of Commentary

This commentary does not attempt to explain the earthquake loading provisions in great detail. The reader is referred to two excellent resources:

Part 2, Commentary, of the *NEHRP Recommended Provisions for the Development of Seismic Regulations for New Buildings and Other Structures*, Building Seismic Safety Council, Federal Emergency Management Agency, 1997 edition.

Recommended Lateral Force Requirements and Commentary, Seismology Committee, Structural Engineers Association of California, 1996.

Section 9 is organized such that the BSSC Commentary is easily used; simply remove the first "9" or ("A.9") from the section number in this standard to find the corresponding section in the BSSC Commentary. Most of this commentary is primarily devoted to noting and explaining the differences of major substance between Section 9 of ASCE 7-98 and the 1997 edition of the *NEHRP Recommended Provisions*.

Nature of Earthquake "Loads"

The 1988 edition of ASCE 7 and the 1982 edition of ANSI A58.1 contained seismic provisions based upon those in the *Uniform Building Code* (UBC) of 1985 and earlier. The UBC provisions for seismic safety have been based upon recommendations of the Structural Engineers Association of California (SEAOC) and predecessor organizations. Until 1988, the UBC and SEAOC provisions had not yet been fully influenced by the ATC and BSSC efforts. The 1972 and 1955 editions of A58.1 contained seismic provisions based upon much earlier versions of SEAOC and UBC recommendations.

The two most far reaching differences between the 1993, 1995, and 1998 editions and these prior editions are that the newer editions are based upon a strength level limit state rather than an equivalent loading for use with allowable stress design and that it contains a much larger set of provisions that are not directly statements of loading. The intent is to provide a more reliable and consistent level of seismic safety in new building construction.

Earthquakes "load" structures indirectly. As the ground displaces, a building will follow and vibrate.

The vibration produces deformations with associated strains and stresses in the structure. Computation of dynamic response to earthquake ground shaking is complex. As a simplification, this Standard is based upon the concept of a response spectrum. A response spectrum for a specific earthquake ground motion does not reflect the total time history of response, but only approximates the maximum value of response for simple structures to that ground motion. The design response spectrum is a smoothed and normalized approximation for many different ground motions, adjusted at the extremes for characteristics of larger structures. The BSSC *NEHRP Commentary*, Chapters 4 and 5, contains a much fuller description of the development of the design response spectrum and the maps that provide the background information for various levels of seismic hazard and various ground conditions.

The provisions of Section 9 are stated in terms of forces and loads; however, the user should always bear in mind that there are no external forces applied to the above-ground portion of a structure during an earthquake. The design forces are intended only as approximations to produce the same deformations, when multiplied by the Deflection Amplification factor C_d, as would occur in the same structure should an earthquake ground motion at the design level occur.

The design limit state for resistance to an earthquake is unlike that for any other load within the scope of ASCE 7. The earthquake limit state is based upon system performance, not member performance, and considerable energy dissipation through repeated cycles of inelastic straining is assumed. The reason is the large demand exerted by the earthquake and the associated high cost of providing enough strength to maintain linear elastic response in ordinary buildings. This unusual limit state means that several conveniences of elastic behavior, such as the principle of superposition, are not applicable, and makes it difficult to separate design provisions for loads from those for resistance. This is the reason the *NEHRP Provisions* contain so many provisions that modify customary requirements for proportioning and detailing structural members and systems. It is also the reason for the construction quality assurance requirements. All these "non-load" provisions are presented in the seismic safety appendix to Section 9.

Use of Allowable Stress Design Standards

The conventional design of nearly all masonry structures and many wood and steel structures has been accomplished using allowable stress design (ASD) standards. Although the fundamental basis for the earthquake loads in Section 9 is a strength limit state beyond first yield of the structure, the provisions are written such that the conventional ASD standards can be used by the design engineer. Conventional ASD standards may be used in one of two fashions:

1. the earthquake load as defined in Section 9 may be used directly in allowable stress load combinations of Section 2.4 and the resulting stresses compared directly with conventional allowable stresses, or
2. the earthquake load may be used in strength design load combinations and resulting stresses compared with amplified allowable stresses (for those materials for which the design standard gives the amplified allowable stresses, e.g. masonry).

Method 1 is changed somewhat since the 1995 edition of the standard. The factor on E in the ASD combinations has been reduced to 0.7 from 1.0. This change was accomplished simultaneously with reducing the factor on D in the combination where dead load resists the effects of earthquake loads from 1.0 to 0.6.

The factor 0.7 was selected as somewhat of a compromise among the various materials for which ASD may still be used. The basic premise suggested herein is that for earthquake loadings ASD is an alternative to strength based design, and that ASD should generally result in a member or cross section with at least as much true capacity as would result in strength based design. As this commentary will explain, this is not always precisely the case.

There are two general load combinations, one where the effects of earthquake load and gravity load add, and a second where they counteract. In the second, the gravity load is part of the resistance, and therefore only dead load is considered. These combinations can be expressed as follows, where α is the factor on E in the ASD combination, calibrated to meet the premise of the previous paragraph. Using the combinations from Sections 2.3 and 2.4:

1. Additive combinations:
   Strength: $1.2D + 0.5L + 0.2S + 1.0E \leq \varphi * $ Strength
   ASD: $1.0D + 0.75 * (1.0L + 1.0S + \alpha E) \leq$ Allowable Stress
   $1.0D + \alpha E \leq$ Allowable Stress

2. Counteracting combinations:
   Strength: $0.9D + 1.0E \leq \varphi * \text{Strength}$
   ASD: $0.6D + \alpha E \leq \text{Allowable Stress}$

For any given material and limit state, the factor α depends on the central factor of safety between the strength and the allowable stress, the resistance factor φ, and ratios of the effects of the various loads. Table C9.1 summarizes several common cases of interest, including those where the designer opts to use the 1/3 increase in allowable stress permitted in various reference standards.

The bold entries indicate circumstances in which the 0.7 factor in the ASD equations will result in a structural capacity less than required by strength design. Given the current basis of the earthquake load provisions, such situations should be carefully considered in design. This review suggests that the one-third increase in allowable stress given in some reference standards is not appropriate in some situations, such as the design of certain steel members. The review also suggest that a load factor on E of 0.8 would be more appropriate for the design of wood structures.

The amplification for Method 2 is accomplished by the introduction of two sets of factors to amplify conventional allowable stresses to approximate the equivalent yield strength: one is a stress increase factor (1.7 for steel, 2.16 for wood, and 2.5 for masonry) and the second is a resistance or strength reduction factor (less than or equal to 1.0) that varies depending on the type of stress resultant and component. The 2.16 factor is selected for conformance with the new design standard for wood (*Load and Resistance Factor Standard for Engineered Wood Construction*, ASCE 16-95) and with an existing ASTM standard; it should not be taken to imply an accuracy level for earthquake engineering.

Although the modification factors just described accomplish a transformation of allowable stresses to the earthquake strength limit state, it is not conservative to ignore the provisions in the standard as well as the supplementary provisions in the appendix that deal with design or construction issues that do not appear directly related to computation of equivalent loads, because **the specified loads are derived assuming certain levels of damping and ductile behavior**. In many instances this behavior is not necessarily delivered by designs conforming to conventional standards, which is why there are so many seemingly "non-load" provisions in this standard and appendix.

Past design practices (the SEAOC and UBC requirements prior to 1997) for earthquake loads produce loads intended for use with allowable stress design methods. Such procedures generally appear very similar to this standard, but a coefficient R_w is used in place of the response modification factor R. R_w is always larger than R, generally by a factor of about 1.5, thus the loads produced are smaller, much as allowable stresses are smaller than nominal strengths. However, the other procedures contain as many, if not more, seemingly "non-load" provisions for seismic design to assure the assumed performance.

Story Above Grade. Fig. C9.1 illustrates this definition.

Occupancy Importance Factor

The NEHRP 1997 Provisions introduce the Occupancy Importance Factor, I. It is a new factor in NEHRP provisions but not for ASCE 7 or UBC provisions. Editions of this standard prior to 1995 as well as other current design procedures for earthquake loads make use of an occupancy importance factor, I, in the computation of the total seismic force. This factor was removed from the 1995 edition of the standard when it introduced the provisions consistent with the 1994 edition of NEHRP provisions. The 1995 edition did include a classification of buildings by occupancy, but this classification did not affect the total seismic force.

The NEHRP provisions in the opening Section 1.1 Purpose, identify two purposes of the provisions, one of which specifically is to "improve the capability of essential facilities and structures containing substantial quantities of hazardous materials to function during and after design earthquakes." This is achieved by introducing the occupancy importance factor of 1.25 for Seismic Use Group II structures and 1.5 for Seismic Use Group III structures. The NEHRP Commentary Section 1.4, 5.2, and 5.2.8 explain that the factor is intended to reduce the ductility demands and result in less damage. When combined with the more stringent drift limits for such essential or hazardous facilities the result is improved performance of such facilities.

FEDERAL GOVERNMENT CONSTRUCTION

The Interagency Committee on Seismic Safety in Construction has prepared an order executed by the President, Executive Order 12699, that all federally owned or leased building construction, as well as federally regulated and assisted construction, should be constructed to mitigate seismic hazards and that the *NEHRP Provisions* are deemed to be the suitable standard. The ICSSC has also issued a report that currently only the 1997 UBC and ASCE 7-95 have been deemed to be in "substantial compliance" with the executive order. It is expected that this standard would also be deemed equivalent, but the reader should bear in mind that there are certain differences, which are summarized in this Commentary.

C9.4.1.1 Maximum Considered Earthquake Ground Motions

This edition replaces the use in the 1995 edition of seismic ground acceleration maps to define ground motion with a general procedure for determining maximum considered earthquake and design spectral response accelerations based on new maps included in NEHRP 1997 Provisions. The reasons and procedures for the development of these new maps are explained in Appendix B of NEHRP 1997 Recommended Provisions and Commentary documents.

C9.5.2.4 Redundancy

This standard introduces a new reliability factor for structures in Seismic Design Categories D, E, and F to quantify redundancy. The value of this factor varies from 1.0 to 1.5. This factor has an effect of reducing the R factor for less redundant structures thereby increasing the seismic demand. The factor is introduced in recognition of the need to address the issue of redundancy in the design. The NEHRP Commentary Section 5.2.4 explains that this new requirement is "intended to quantify the importance of redundancy." The Commentary points out that "many non-redundant structures have been designed in the past using values of R that were intended for use in designing structures with higher levels of redundancy." In other words, the use of R factor in the design has led to slant in design in the wrong direction. The NEHRP Commentary cites the source of this factor as the work of SEAOC for inclusion of this factor in UBC 1997.

C9.5.2.5.1 Seismic design Category A

This edition includes a new provision of minimum lateral force for Seismic Design Category A structures. The minimum load is a structural integrity issue related to the load path. It is intended to specify design forces in excess of wind loads in heavy low-rise construction. The design calculation is simple

and easily done to ascertain if it governs or the wind load governs. This provision requires a minimum lateral force of 1% of the total gravity load assigned to a story to assure general structural integrity.

C9.14 NONBUILDING STRUCTURES

The NEHRP Provisions contain additional design requirements for nonbuilding structures in an Appendix to Chapter 14. The NEHRP Commentary contains, in addition to its Chapter 14 for nonbuilding structures, additional guidance in a separate chapter titled Appendix to Chapter 14. These additional resources should be referred to in designing nonbuilding structures for seismic loads.

CA.9.6.6 Bolts and Headed Stud Anchors in Concrete

The provisions contained in this section were developed by the American Concrete Institute (ACI) and are copyrighted by ACI. ASCE has received permission from ACI to reprint this material. These provisions are not an official ACI document but rather are part of a document being processed as an ACI consensus standard. Upon the adoption of these provisions into an ACI consensus standard it is the intent of ASCE to remove these provisions from ASCE 7 and adopt them by reference. These provisions are identified within ACI as CR 30-A.

CA.9.11 MASONRY

The appendix for masonry design is primarily necessary because the reference standard itself refers to the 1993 edition of ASCE 7. Since that time significant changes have been made in terminology to distinguish among various types of masonry walls based upon the method of engineering design (as reinforced or unreinforced) and on the minimum amount of reinforcement. This change in terminology was mainly needed to accommodate the change in philosophy that the detailing rules be driven by the R factor rather than the old Seismic Performance Category.

Most of the serious recent development in improved seismic resistant design for masonry structures has been expressed in terms of strength design methodology, whereas the reference standard is written solely in terms of allowable stress design. This is not to say that the committee responsible for the reference standard has not considered strength design; the committee simply has not reached consensus on an overall strength design standard. Even though this consensus has not been reached, a few of the concepts are necessary when using the highest R factors for design of masonry systems. One very significant rule from the strength design provisions of the NEHRP Recommended Provisions and the proposed new International Building Code has been incorporated into this appendix for those masonry walls with relatively generous R factors: the upper limit on reinforcement in shear walls. This requirement intends to deliver ductile performance by requiring substantial yield of tension steel before crushing of masonry at the compressive end of the wall (Method A) or by providing a boundary member to control the spread of crushing (Method B).

CA.9.12 WOOD

Tables of strengths for wood diaphragms and shear walls are not provided because a referenceable form exists as a supplement to ASCE 16. The comparable tables in NEHRP 97 result in a reduction of 10% for wood sheathed shear walls, which was incorporated shortly after the Northridge earthquake. It is now felt that this reduction was unnecessary. The 10% reduction is not incorporated in the IBC, nor in the referenced tables. The reader should note that the referenced tables do not include staples, and where staple construction is desired it will be necessary to use other sources and to petition the authority having jurisdiction for their approval.

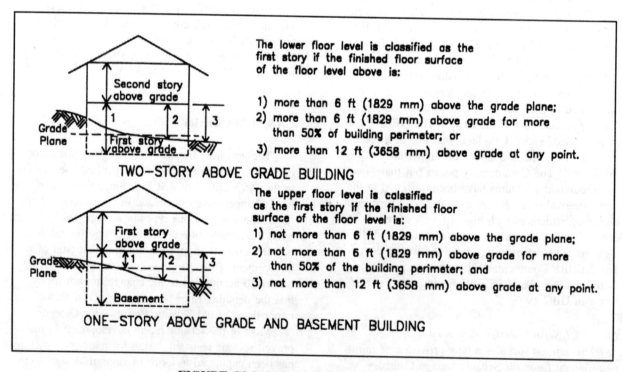

The lower floor level is classified as the first story if the finished floor surface of the floor level above is:

1) more than 6 ft (1829 mm) above the grade plane;
2) more than 6 ft (1829 mm) above grade for more than 50% of building perimeter; or
3) more than 12 ft (3658 mm) above grade at any point.

TWO-STORY ABOVE GRADE BUILDING

The upper floor level is calssified as the first story if the finished floor surface of the floor level is:

1) not more than 6 ft (1829 mm) above the grade plane;
2) not more than 6 ft (1829 mm) above grade for more than 50% of the building perimeter; and
3) not more than 12 ft (3658 mm) above grade at any point.

ONE-STORY ABOVE GRADE AND BASEMENT BUILDING

FIGURE C9.1. Definition of Story above Grade

TABLE C9.1.

Structural Element and Limit State	ASD Rules	Ratio of Load Effects (Moment, Axial Load, etc.) in Load Combination Being Considered		Equivalency Factor α
		D/E	L/E	
Steel girder; bending	reduction per 2.4.3	0	0	0.67
		0.5	0.25	0.65
		1	0.5	0.48
	1/3 increase per reference standard	0	0	**0.89**
		0.5	0.25	**0.79**
		1	0.5	0.71
Steel brace, tension	reduction per 2.4.3	0	0	0.67
		1	0	0.67
	1/3 increase per reference standard	0	0	**0.89**
		0.5	0	**0.80**
Steel brace, compression	reduction per 2.4.3	0.25	0.25	0.66
		0.5	0.5	0.45
	1/3 increase per reference standard	0.25	0.25	0.67
Masonry wall, reinforcement for in-plane bending	reduction per 2.4.3	0	0	0.50
		1.11	0	0.64
	1/3 increase per reference standard	0	0	0.67
		1.11	0	0.67
Masonry wall, in-plane shear (reinforced, with M/Vd $\geq$ 1)	reduction per 2.4.3	0	0	0.47
	1/3 increase per reference standard	0	0	0.63
Wood shear panel	reduction per 2.4.3 (reference standard	0	0	**0.77**
Wood collector, tension	does not have explicit increase,	0	0	**0.77**
	although duration factors are not the same in LRFD and ASD)	1	0	**0.85**
Bolts in wood		0	0	**0.75**
		0.25	0.5	**0.76**

C10.0 ICE LOADS—ATMOSPHERIC ICING

C10.1 DEFINITIONS

Accreted ice can be classified by its process of formation or its physical characteristics. Freezing rain or drizzle is the most common icing mechanism in the United States. The glaze ice that forms in these conditions is usually clear but may appear opaque because of included air bubbles. Severe ice loads may also result from snow that can accrete to large radial thicknesses by adhering to wires, cables, conductors and guys. In-cloud icing is caused by the collision of supercooled cloud or fog droplets with a structure. The resulting ice ranges from hard, clear glaze ice to softer, lower density, white rime ice containing entrapped air. Hoarfrost is a deposit of ice crystals formed by the deposition of water vapor. It does not usually impose significant loads on structures because the amount of ice accreted by vapor deposition is small.

In a general sense, the meteorological parameters that influence the mass, density and shape of an ice accretion are well known. Wind speed and liquid water content or precipitation intensity control the amount of water available for ice formation. The density and shape of an ice accretion are influenced by the air temperature, wind speed, droplet size, and liquid water content or precipitation intensity. The density of snow accretions depends on the wetness of the snow. Because it is important to understand the meteorological conditions that result in their formation, what follows is a more detailed description of the various types of ice accretions.

Freezing Rain. Freezing rain (or drizzle) is a common icing mechanism. Freezing rain occurs when warm moist air is forced over a layer of subfreezing air at the earth's surface. The precipitation usually begins as snow that melts as it falls through the layer of warm air aloft. The drops cool as they fall through the cold surface air layer and freeze on contact with structures or the ground. Upper air data indicates that the cold surface air layer is typically between 1,000 ft (300 m) and 3,900 ft (1,200 m) thick (Young, 1978), averaging 1,600 ft (500 m) (Bocchieri, 1980). The warm air layer aloft averages 5,000 ft (1,500 m) thick in freezing rain, but in freezing drizzle the entire temperature profile may be below 32°F (0°C) (Bocchieri, 1980). As freezing rain is often associated with frontal systems, it usually does not last more

than a day or two. Precipitation associated with slowly moving frontal systems can alternate between snow and freezing rain to form a composite snow-glaze accretion on structures. The density of glaze is usually assumed to be 57 pcf (910 kg/m^3).

In freezing rain the water impingement rate is usually greater than the freezing rate. The excess water drips off and may freeze as icicles, resulting in a variety of accretion shapes from a smooth cylindrical sheath, through a crescent on the windward side with icicles hanging on the bottom, to large irregular protuberances. The shape of a glaze accretion depends on a combination of varying meteorological factors and the cross-sectional shape of the structural member, its spatial orientation and flexibility.

Snow. The snow that falls on a round cross-sectional structural member, component or appurtenance (such as a wire, cable, conductor or guy) may deform and/or slide around it, as a result of either its own weight or aerodynamic lift. Because of the shear and tensile strength of the snow resulting from capillary forces, interparticle freezing (Colbeck and Ackley, 1982) and/or sintering (Kuroiwa, 1962), the accreting snow may not fall off the structural member during this process. Ultimately the snow forms a cylindrical sleeve, even around bundled conductors and wires. The formation of the snow sleeve is enhanced by torsional rotation of flexible structural members, components or appurtenances because of the eccentric weight of the snow. The density of accreted snow ranges from below 5 up to 50 pcf (80 up to 800 kg/m^3) and may be much higher than the density of the same snowfall on the ground.

Damaging snow accretions have been observed at surface air temperatures ranging from the low 20s up to about 36°F (−5° to 2°C). Snow with a high moisture content appears to stick more readily than drier snow. Snow falling at a surface air temperature above 32°F (0°C) may accrete even at wind speeds above 25 mph (10 m/s), producing dense (37- to 50-pcf [600- to 800-kg/m^3]) accretions. Snow with a lower moisture content is not as sticky, blowing off the structure in high winds. These accreted snow densities are typically between 12 and 25 pcf (200 and 400 kg/m^3) (Kuroiwa, 1965). Even apparently dry snow can accrete on structures (Gland and Admirat, 1986). The cohesive strength of the dry snow is initially supplied by the interlocking of the flakes and ultimately by sintering, as molecular diffusion increases the bond area between adjacent snowflakes. These dry snow accretions appear to form only in very low winds and have densities estimated at between 5 and 10 pcf (80 and 150 kg/m³) (Sakamoto et al., 1990; Peabody, 1993).

In-Cloud Icing. This icing condition occurs when supercooled cloud or fog water droplets, 100 μm or less in diameter, collide with a structure. It occurs in mountainous areas where adiabatic cooling causes saturation of the atmosphere to occur at temperatures below freezing, in free air in supercooled clouds, and in supercooled fogs produced by a stable air mass with a strong temperature inversion. Because these conditions can persist for days or weeks, significant accumulations of ice can result. Large concentrations of supercooled droplets are not common at air temperatures below about 0°F (−18°C).

In-cloud icing forms rime or glaze ice with a density between about 10 and 57 pcf (150 and 910 kg/m³). If the heat of fusion that is released by the freezing droplets is removed by convective and evaporative cooling faster than it is released, the droplets freeze on impact. The degree to which the droplets spread as they collide with the structure and freeze governs how much air is incorporated in the accretion and thus its density. If the cooling rate is relatively low, not all the colliding droplets freeze. The resulting accretion will be clear or opaque ice, possible with attached icicles.

The collision efficiency of a structure is defined as the fraction of cloud droplets in the volume swept out by the structure that actually collide with it. The basic theory of the collision efficiency of smooth circular cylinders perpendicular to the flow of droplets carried by a constant wind was developed by Langmuir and Blodgett (1946). Collision efficiency increases with wind speed and droplet diameter and decreases as the diameter of the cylinder increases. For a given wind speed and droplet size, the theory defines a critical cylinder diameter beyond which accretion will not occur. This concept of a critical diameter has been confirmed by observation.

The amount of ice accreted during in-cloud icing depends on the duration of the icing condition and the wind speed. If, as often occurs, wind speed increases and air temperature decreases with height above ground, larger amounts of ice will accrete on higher structures. The accretion shape depends on the flexibility of the structural member, component or appurtenance. If it is free to rotate, such as a long guy or a long span of a single conductor or wire, the ice accretes with a roughly circular cross section. On more rigid structural members, components and ap-

purtenances, the ice forms in irregular pennant shapes extending into the wind.

Hoarfrost. Hoarfrost is an accumulation of ice crystals formed by direct deposition of water vapor from the air onto a structure. Because it forms when air with a dew point below freezing is brought to saturation by cooling, hoarfrost is often found early in the morning after a clear, cold night. It is feathery in appearance and typically accretes up to about 1 in. (25 mm) in thickness with very little weight. Normally, hoarfrost does not constitute a significant loading problem; however, it is a very good collector of supercooled fog or cloud droplets and in light winds may accrete rime of significant volume and weight (MEP, 1984).

These four ice accretion categories cover the spectrum of the atmospheric icing phenomenon. The distinctions made in defining each category are not often clearly identifiable in practice. There may be an overlap of more than one type of icing condition, such as snow and freezing rain.

Ice-sensitive structures. Ice-sensitive structures are structures for which the load effects from atmospheric icing control the design of part or all of the structural system. Many open structures are efficient ice collectors, so ice accretions can have a significant load effect. The sensitivity of an open structure to ice loads depends on the size and number of structural members, components and appurtenances and also on the other loads for which the structure is designed. For example, the additional weight of ice which may accrete on a heavy wide-flange member will be smaller in proportion to the dead load than the same ice thickness on a light angle member. Also, the increase in projected area for wind loads will be smaller for the wide-flange member than for the angle member. For some open structures other design loads, for example, snow loads and live loads on a catwalk floor, may be larger than the design ice load.

C10.2 GENERAL

Ice accretions can impose substantial vertical loads on the structural system, present a greater projected area exposed to the wind and affect the lift and drag coefficients. Ice loads on wires, cables, conductors and guys increase both the tension in that component and the load on the rest of the structural system.

Meteorological data suggest that atmospheric ice loads and the concurrent wind loads should be included in the design of ice-sensitive structures throughout most of the United States. This Section provides general guidance on the selection of design ice loads. In the following paragraphs the atmospheric icing risk in each of six regions in the contiguous 48 states is discussed, with particular emphasis on freezing rain. In any region which regularly experiences clouds with below-freezing temperatures, in-cloud icing will occur on tall structures and in mountainous terrain. Snow accretions can occur anywhere that snow falls, even in localities that may experience only one or two snow events a year. The following information was provided by the six regional climate centers of the National Oceanographic and Atmospheric Administration (NOAA).

Northeast Region
(ME, NH, VT, NY, MA, RI, CT, NJ, PA, DE, MD, WV). Freezing rain storms can occur anywhere in the 12-state area comprising the Northeastern United States. There is a somewhat greater risk for these storms to occur in the cooler, northern portions of the region than in the Northeast's southernmost states. Also, due to the moderating effect of the Atlantic Ocean, coastal regions are not at as great a risk as interior areas, more than about 50 miles (80 km) from the coast. The portions of this region most likely to experience freezing rain storms are Pennsylvania (except the southeast) and interior portions of New York and New England. The remainder of the region, although not at as great a risk, is still at a higher risk for freezing rain than many other parts of the country.

Southeast Region
(VA, NC, SC, GA, AL, FL). In the southeastern United States freezing rain occurs relatively frequently. A large V-shaped wedge of high event frequency is observed over inland/mountain Virginia and North Carolina. The highest average is over six events per year, and occurs at Roanoke and Lynchburg, Virginia; Greensboro, North Carolina; and Washington, D.C. The freezing rain event frequency decreases in the northern coastal regions (except in Washington, D.C.) and in the southern Piedmont and lowlands. Thus the general decreasing gradients of freezing rain events are north to south and mountain to coastal. No freezing rain events have been recorded in Tampa Bay, Key West and Palm Beach, Florida.

Freezing rain event duration is highest in the northern regions. Roanoke, Greensboro and Dulles Airport have the longest average durations, with the typical winter event lasting over 12 h. Most of upland Virginia and North Carolina experience events of this length. Asheville and Charlotte have recorded the longest single events of well over 100 h. The same north-south and mountain-coastal gradients are seen for duration. Northern Florida is not exempt from longer duration freezing rain events. Although few events have been recorded, some have lasted longer than 20 h.

Midwest Region
(IA, IL, IN, KY, MI, MN, MO, OH, WI). Almost all of the Midwest is located in the "glaze belt" where freezing rain events are common. This area experiences frequent interactions between cold polar air from the north and moist tropical air from the Gulf of Mexico which lead to favorable conditions for icing during the colder months of the year.

Southern Region
(AR, LA, MS, OK, TN, TX). Icing conditions due to freezing rain in the Southern Region demonstrate a north-to-south gradient in terms of frequency and severity. This latitudinal trend is further enhanced by the regional geography, with a continental climate dominant across the northern sections of the Southern Region, while the Gulf of Mexico produces a moderating maritime influence to the south. Data from the National Weather Service first-order stations indicate that freezing rain storms have been reported at every location across the region in recent years. Even Brownsville, Texas, has reported a handful of events. Freezing rain storms capable of creating significant damage, such as damaging trees, downing power lines and creating transportation hazards, can occur virtually anywhere within this region. These severe events tend to be somewhat rare in southern sections, particularly along the coastal margins.

Throughout these six states freezing rain has been observed everywhere during the months of December, January and February. Based on an examination of all January precipitation-hours from National Weather Service stations (i.e., hours with any form of precipitation reported), freezing rain is reported during about 20% of these hours in the plains of Oklahoma and Texas, while being reported during less than 10% of these hours in some of the southern and eastern sections of the Southern Region.

The panhandle regions of Texas and Oklahoma have the highest frequency of freezing rain events relative to precipitation-hours during the winter months with a January maximum of 20%. The percentage of rain-hours with freezing rain is substantially lower in Arkansas and Tennessee; 10 to 13% for most of Arkansas, falling to well under 10% across most of Tennessee. However, this frequency increases in eastern Tennessee. The ridge and valley complexes of eastern Tennessee may promote the occurrence of freezing rain. Warm-air advection over the ridges can create an environment where liquid precipitation falls along the slopes, while freezing rain occurs in the valley bottoms where cold air drainage has created subfreezing surface temperatures.

Further south the frequency of freezing rain events in central Texas, northern Louisiana and central Mississippi is lower. Proximity to the Gulf of Mexico becomes more important, particularly in Louisiana and Mississippi. In central Texas freezing rain events account for 10% or less of all winter rain-hours. In the drier sections of Texas, these precipitation events are more uncommon. Coastal stations throughout the entire Southern Region reported only a few events in twenty years.

High Plains Region

(ND, SD, NE, KS, CO, WY). The High Plains Region is characterized by a diverse climate with ice loads that are sometimes severe. The west-northwestern portions are the driest, have the greatest elevations, and experience the coldest winter seasons of the High Plains. This area also most frequently experiences a dry continental air mass of polar origin. The southeastern portion of the region is more humid than the north and west, has much longer warm seasons, and is frequently in contact with the warm maritime air mass. Winter snow occurrences and accumulations are common in the northern and western portions, while less common in southeastern portions of the High Plains.

Generally, freezing rain or drizzle is associated with the intrusion of warm moist air aloft from the south together with subfreezing air at the surface intruding from the north. The portions of the High Plains where this situation frequently occurs are in eastern and southeastern Nebraska, in southeastern South Dakota and in central and eastern Kansas. The northern and western portions of the High Plains experience freezing rain less frequently.

Western Region

(MT, ID, WA, OR, CA, NV, UT, NM, AZ). In the nine westernmost states, the area most likely to be subject to freezing rain is the Pacific Northwest. Freezing rain occurs most often when warm, moist Pacific air moves onshore and is forced to rise over domes of trapped cold air lying in valleys and basins. Because the layers of cold air are relatively shallow, the resulting precipitation maintains its liquid character until it falls to the ground and freezes. Usually the warm air aloft is slowly eroding away the cold air from the top, eventually reaching the surface at which point the freezing rain changes to rain. This is in contrast to situations near frontal boundaries, where cold air north of a front may be replenished at a rate comparable to that at which the warm air is eroding it away from the top. The duration of a freezing rain event can also be extended when a large pool of cold arctic air centered east of the Cascade Mountains spills westward into river valleys, such as down the Columbia River Gorge and into the Willamette Valley, the Fraser River Valley in southern British Columbia, the Santiam, MaKenzie, Umpqua, and Rogue passes in Oregon, and a few other rivers draining the west slopes of the Cascades. Freezing rain with accompanying elevated surface wind speeds can continue in these areas for quite some time as long as enough moisture and a sufficiently strong east-to-west pressure gradient is maintained.

Freezing rain is more likely to occur at lower elevations, from sea level up to 2,000–3,000 ft (600–900 m), and rarely above 4,000 ft (1,200 m). Places such as Walla Walla, Pendleton, and Yakima average about 10 h of freezing rain in January, and 20–40 h per year. Mountain valleys in central and northern Idaho and the lower elevation Pacific drainages of western Montana, where cold air can remain trapped, experience about 5 h of freezing rain in January on average. Missoula, Montana, averages about 10 h/year, and Kalispell about 20–30 h. Freezing rain also occurs farther to the south in the northern and central parts of the Sacramento Valley, but less frequently than in Oregon.

Rime occurs very frequently on exposed ridges and slopes in the mountains. Above the mean freezing level, heavy deposits (many inches or even many feet [100s of mm or even meters] thick) can form during the numerous storms that strike the region in winter. Steep cliff faces and any exposed structures or obstacles to the wind can become covered with deep coats of rime. Riming occurs on nearly all mountaintops in the West. While rime ice is not commonly found below elevations of about 3,000 ft (915 m), it does occasionally occur when freezing fog fills the basin regions of eastern Washington and Oregon during periods of strong wintertime temperature inversions.

C10.3 DESIGN FOR ICE LOADS

Ice sensitive structures should be designed for the appropriate type and amount of ice with concurrent wind. Of the three types of icing, freezing rain, in-cloud icing and snow; sufficient information to make recommendations for design is currently available only for freezing rain, and only in the Pacific northwest, midwest and eastern United States.

Ice Loads due to Freezing Rain and Drizzle. Figs. C10-1 and C10-2 show uniform radial thicknesses t_{33} of glaze ice with concurrent 3-s gust and fastest-mile wind speeds due to freezing rain and drizzle at 33 ft (10 m) above the ground for a 50-year mean recurrence interval. Multipliers to convert 50-year mean recurrence interval uniform ice thicknesses and concurrent 3-s gust wind speeds to 25- and 100-year recurrence intervals are presented in Table C10-1. Figs. C10-1 and C10-2 are based on wind exposure C, but also should be used for wind exposures A, B and D. Design thicknesses of ice t_z for other heights z can be obtained by using Eq. C10-1.

$$t_z = t_{33} \left(\frac{z}{33}\right)^{0.12} \quad \text{for } t_{33} \leq t_z \leq 1.5t_{33} \quad \text{(Eq. C10-1)}$$

Ice loads on a ridge, hill or escarpment will be higher than those in level terrain because of wind speed-up effects. The topographic factor for the ice thickness is $K_{zt}^{0.42}$, where K_{zt} is obtained from Eq. 6-1.

The values shown in the unshaded areas of Fig. C10-1 in the midwestern and eastern United States, and the values in Fig. C10-2 for the Pacific Northwest may be used for design. In the shaded areas of the maps, other sources of information about freezing rain and drizzle must be consulted to determine the design ice and wind-on-ice loads; see Section C10.2 and "Current Industry Practice" and "Other Sources of Information" below.

In the unshaded portion of Fig. C10-1 east of longitude 105° west, historical weather data from 230 National Weather Service (NWS), military, and Federal Aviation Administration (FAA) weather stations were used with an ice load model for freezing rain

(Jones, 1996a,b) to estimate glaze ice loads in past freezing rain storms. The station locations are shown in Fig. C10-4. The period of record of the meteorological data is typically 20–45 years. The model uses hourly precipitation and wind speed data to simulate the accretion of ice on a horizontal circular cylinder 33 ft (10 m) above ground in freezing rain storms. The uniform radial ice thickness determined by the model is independent of cylinder diameter. The ice is assumed to remain on the cylinder until after freezing rain ceases and the air temperature increases to at least 33°F (0.6°C). The maximum ice load and the maximum wind-on-ice load were determined for each storm. Severe storms with significant ice or wind-on-ice loads at one or more weather stations were researched in *Storm Data*, newspapers and utility reports to obtain corroborating qualitative information on the extent of and damage from the storm. The 50-year recurrence interval ice loads were determined through an extreme value analysis of the data using the maximum ice load from each storm. Extreme ice loads were determined using the peaks-over-threshold method (Abild et al., 1992; Hoskings and Wallis, 1987; and Wang, 1991). To reduce sampling error, weather stations were grouped into superstations (Peterka, 1992) based on the incidence of severe storms, the frequency of freezing rain storms, latitude, proximity to large bodies of water, elevation, and terrain. Periods of record for the superstations range from 40 to 200 years. A few stations that were judged to have unique freezing rain climatologies were not incorporated in superstations. Extreme wind-on-ice loads were also determined, and concurrent wind-on-ice speeds were calculated from them using the extreme ice loads assuming a uniform radial ice thickness. The actual shape of glaze ice accretions varies from the compact shape represented by a uniform radial accretion to very elongated, heavily icicled shapes, which may have larger projected areas.

This portion of the map represents the most consistent and best available nationwide map for design ice loads. The icing model used to produce the map has not, however, been verified with a large set of co-located measurements of meteorological data and ice loads. Furthermore, the weather stations used to develop this map are almost all at airports. Structures in more exposed locations at higher elevations, or in valleys or gorges, for example, Signal and Lookout Mountains in Tennessee, the Ponatock Ridge and the edge of the Yazoo Basin in Mississippi, and Shenandoah Valley in Virginia, may be subject to larger ice loads and higher concurrent wind speeds.

The freezing rain map for the Pacific Northwest shown in Fig. C10-2 is based on ice load maps produced by Meteorology Research Inc. (MRI) for the Bonneville Power Administration (BPA) (Richmond et al., 1977). BPA commissioned the study in recognition of the need for more complete and detailed information regarding transmission line icing throughout its service area. Data were gathered from a variety of sources, including BPA's in-house records of observed ice accumulations, observed icing events and/or outage reports from 93 power companies and 30 telephone companies, historical weather data for 42 weather stations from National Climatic Data Center (NCDC), and personal interviews of pilots and forecasters to learn of local climatological effects. These data were processed and analyzed using a mathematical model developed by MRI. Weighting factors were developed to compensate for localized effects on icing of terrain features such as valleys, ridges, lakes, elevation, and forest cover. The study produced 144 maps showing isopleths of equivalent solid radial ice thicknesses for 25-, 50-, and 100-year mean recurrence intervals. The ice thicknesses shown for the Pacific Northwest may not be consistent with those shown for the rest of the United States because of differences in the methodologies used.

Within the areas where ice thicknesses and wind speeds are mapped, there are "special icing regions" identified. As described above, freezing rain occurs only under special conditions with a cold relatively thin, surface air layer, and a layer of warm moist air aloft. Thus, severe freezing rain storms at high elevations in mountainous terrain will typically not occur in the same weather systems that cause severe freezing rain storms at the nearest airport with a weather station. Furthermore, in these regions ice and wind-on-ice loads may vary significantly over short distances because of the large variations in elevation, topography and exposure. In these mountainous regions, the values given in Figs. C10-1 and C10-2 should be adjusted, based on local historical records and experience, to account for possibly higher ice loads both from freezing rain and from in-cloud icing.

Figs. C10-1 and C10-2 represent ice loads appropriate for a single structure of small areal extent. Higher ice loads may be required to achieve the same reliability for a structure or system of structures that cover a larger area; for example, a long electric transmission line or a system of communication towers.

Loads due to Clouds and Fog (In-Cloud Icing) and due to Snow Accretions. Information to produce maps similar to Figs. C10-1 and C10-2 for in-cloud icing and snow accretions is not currently available.

In-cloud icing may cause significant loadings on ice-sensitive structures in mountainous regions and for very tall structures in other areas. Mulherin (1996) reports that of 120 communications tower failures in the United States due to atmospheric icing, 38 were due to in-cloud icing and an additional 26 were caused by in-cloud icing combined with freezing rain. A span of a transmission line in the Wasatch Mountains of Utah, which descends from a high plateau into a valley, was observed with rime ice bridging two conductors 18 in. (0.5 m) apart. This line was designed with input from the state meteorologist for 4 radial in. (102 mm) of 10-pcf rime ice (150 kg/m^3) with 40 mph (18 m/s) winds and also for 2 radial in. (51 mm) of 57-pcf (910-kg/m^3) glaze ice. In-cloud icing accretions are very sensitive to exposure related to terrain and the direction of the flows of moisture laden clouds. Large differences in accretion size can occur over a few hundred feet and cause severe load unbalances in overhead wire systems. Advice from a meteorologist familiar with the area is particularly valuable in these circumstances. In Arizona, New Mexico and the panhandles of Texas and Oklahoma, the U.S. Forest Service specifies ice loads due to in-cloud icing for towers constructed at specific mountaintop sites (USFS, 1994); see also Section C10.3.3 Partial Loading.

Snow accretions also can result in severe structural loads. A heavy wet snow storm on March 29, 1976 caused $15 million damage to the electric transmission and distribution system of Nebraska Public Power District (NPPD, 1976). Mozer and West (1983) report a transmission line failure on December 2, 1974 near Lonaconing, Maryland, due to heavy wet snow of 5-in. (127-mm) radial thickness on the wires with an estimated density of 19 pcf (304 kg/m^3). Goodwin et al. (1983) report measurements of snow accretions on wires in Pennsylvania with an approximate radial thickness of 4 in. (102 mm). The micrometeorological conditions along a transmission line that failed under vertical load in the Front Range of Colorado were analyzed after the failure. The study indicated that the failure was caused by a 1.7-radial-in. (43-mm), 30-pcf (480-kg/m^3) wet snow accretion with a 42-mph (19-m/s) wind. The mean recurrence interval of this event was estimated at 25 years (McCormick and Pohlman, 1993). Golden

317

Valley Electric Association in Fairbanks, Alaska made 27 field measurements of radial thickness and density in the winters of 1994–95 and 1996–97 of dry snow accretions. Densities ranged from 1.4 to 8 pcf (22 to 128 kg/m³) and radial thicknesses up to 4.4 in. (112 mm). The heaviest were equivalent in weight to 1 in. (25.4 mm) uniform radial thickness of glaze ice (GVEA, 1997).

For additional guidance in determining loads due to in-cloud icing and snow accretions, refer to C10.2 and "Current Industry Practice" and "Other Sources of Information" below.

Current Industry Practice. The telecommunication and broadcast industries use ANSI/EIA/TIA Standard 222 (1996) for the design of towers. It assumes 75% of the 50-year mean recurrence interval wind load on the ice-covered structure, unless a specific wind speed is specified simultaneously with ice such as is shown in Figs. C10-1 and C10-2. It requires that ice be considered when it is known to occur, but no ice thickness is specified. Typically 0.5- or 1-in. (12- or 25-mm) radial glaze ice thicknesses have been applied uniformly over the tower. Thicker ice has been used for towers designed for areas of severe icing such as near large bodies of water or at higher altitudes. This has been generally satisfactory. Problems have occurred, however, when ice loads have not been considered in regions where freezing rain occurs or when insufficient ice has been used in the design of tall towers, which, even at low altitudes, are subject to in-cloud icing. The 75% factor applied to the wind load on an ice covered structure is appropriate for the EIA/TIA standard due to its other less conservative requirements. This wind and ice combination is generally conservative compared to the combinations that are specified by other standards, for example, CSA S37 (CSA, 1994).

A 1979 survey of design practice for transmission line loadings (ASCE, 1982) obtained responses from 130 utilities operating 290,000 miles (470,000 km) of high-voltage transmission lines. Fifty-eight of these utilities specifically indicated "heavy icing areas" as one reason for special loadings in excess of code requirements. Design ice loads on conductors ranged from no ice, primarily in portions of the southern United States, up to a 2- or 2.25-in. (50- or 57-mm radial thickness of glaze ice in some states. Radial glaze ice thicknesses between 1.25 and 1.75 in. (32 and 45 mm) are commonly used. Most of the responding utilities design for heavy ice on the wire with no wind and less ice with wind. Few utilities

consider ice on the supporting structures in design. Ice loads on transmission lines are discussed in ASCE 74, *Guidelines for Electric Transmission Line Structural Loading* (1991).

Other Sources of Information. Bennett (1959) presents the geographical distribution of the occurrence of ice on utility wires from data compiled by various railroad, electric power and telephone associations covering the 9-year period from the winter of 1928–29 to the winter of 1936–37. The data includes measurements of all forms of ice accretion on wire including glaze ice, rime ice and accreted snow but does not differentiate between them. Ice thicknesses were measured on wires of various diameters, heights above ground and exposures. No standardized technique was used in measuring the thickness. The maximum ice thickness observed during the 9-year period in each of 975 squares, 60 miles (97 km) on a side, in a grid covering the contiguous United States, is reported. In every state except Florida, thickness measurements of accretions with unknown densities of approximately 1 radial in. were reported. Referring to the NWS regions in Section C10.2, the Northeast, Southeast and Southern Regions all had measurements as high as 2 radial in., the Midwest Region 1.75 radial in., the High Plains 2.4 radial in. and the Western Region 3.0 radial in. Information on the geographical distribution of the number of storms in this 9-year period with ice accretions greater than specified thicknesses is also included.

Tattelman and Gringorten (1973) reviewed ice load data, storm descriptions and damage estimates in several meteorological publications to estimate maximum ice thicknesses with a 50-year mean recurrence interval in each of seven regions in the United States.

Storm Data (NOAA, 1959–present) is a monthly publication that describes damage from storms of all sorts throughout the United States. The storms are sorted alphabetically by state within each month. The compilation of this qualitative information on storms causing damaging ice accretions in a particular region can be used to estimate the severity of ice and wind-on-ice loads. The Electric Power Research Institute has compiled a database of freezing rain events from the reports in *Storm Data*. Damage severity maps were also prepared (Shan and Marr 1996).

Robbins and Cortinas (1996) and Bernstein and Brown (1997) provide information on freezing rain climatology for the 48 contiguous states based on recent meteorological data.

For Alaska, what information is available indicates that moderate to severe ice loads of all types can be expected. The measurements made by Golden Valley Electric Association referred to above are consistent in magnitude with visual observations across a broad area of central Alaska (Peabody, 1993). Several meteorological studies using an ice load model to predict ice loads have been performed for high voltage transmission lines in Alaska (Richmond, 1985, 1991, and 1992; Gouze and Richmond, 1982a,b; Peterka, et al., 1996). Predicted 50-year mean recurrence interval glaze ice accretions range from 0.25 to 1.5 radial in. (6 to 38 mm), snow from 1.0 to 5.5 radial in. (25 to 140 mm), and rime from 0.5 to 6.0 radial in. (12 to 150 mm). The assumed accretion densities were glaze 57 pcf (910 kg/m³), snow 5 to 31 pcf (80 to 500 kg/m³) and rime 25 pcf (400 kg/m³). The loads are valid only for the particular regions studied and are highly dependent on the elevation and local terrain features. Large accretions of snow have been observed in most areas of Alaska that have overhead lines.

In areas where little information on ice loads is available, it is recommended that a meteorologist familiar with atmospheric icing be consulted. Factors to be kept in mind include that taller structures may accrete additional ice because of higher winds and colder temperatures aloft and that the influences of elevation, complex relief, proximity to water, potential for partial loading and structure size and shape are highly significant.

C10.3.2 Wind on Ice-Covered Structures

Figs. C10-1 and C10-2 include 3-s gust and fastest-mile wind speeds that are coincident with the ice loads. The 3-s gust should be applied to ice coated structures of small areal extent in accordance with Section 6. Loads on long spans of wire may be calculated by converting 3-s gust speeds to fastest mile speeds for use in the formulas in ASCE 74. Adjustments to the mapped wind speeds should be made for exposure, topography and height in accordance with Section 6.

C10.3.3 Partial Loading

Accretions from freezing rain rarely exceed a thickness of 2 in. (50 mm) because the associated meteorological conditions typically last no more than a day or two. Horizontal spatial variations over distances of about 1,000 ft (300 m) during a freezing rain storm are small. Therefore, partial icing of a structure from freezing rain is usually not significant (Cluts and Angelos, 1977).

In-cloud icing conditions can persist for several days, resulting in rime accretion thicknesses of 1 ft (300 mm) or more. Because the rime density and thickness increase with increasing wind speed, significant differences in ice loads over the structure are associated with differences in the exposure of the various structural members, components and appurtenances to the wind. The exposure is affected by shielding by other parts of the structure and by the upwind terrain. Partial loading due to differences in exposure to in-cloud icing may be significant. Partial loading associated with ice shedding will usually be small.

REFERENCES

Abild, J., Andersen, E. Y., and Rosbjerg, L. (1992). "The climate of extreme winds at the Great Belt, Denmark." *J. Wind Engineering and Industrial Aerodynamics*, 41-44: 521–532.

ANSI/EIA/TIA (1996). *Structural standards for steel towers and antenna supporting structures*. EIA/TIA-222F, Electronics Industries Association, Washington, D.C., 107 pp.

ASCE (1982). "Loadings for electrical transmission structures by the committee on electrical transmission structures." *J. Struct. Div.*, ASCE, 108(5): 1088–1105.

ASCE (1991). *Guidelines for electrical transmission line structural loading*. ASCE Manuals and Reports on Engineering Practice No. 74, American Society of Civil Engineers, New York, 139 pp.

Bennett, I. (1959). *Glaze: Its meteorology and climatology, geographical distribution and economic effects*. Quartermaster Research and Engineering Center, Environmental Protection Research Division Technical Report EP-105, 217 pp.

Bernstein, B. C., and Brown, B. G. (1997). "A climatology of supercooled large drop conditions based upon surface observations and pilot reports of icing." *Proc., 7th Conf. on Aviation, Range and Aerospace Meteorology*, Long Beach, CA., Feb. 2–7.

Bocchieri, J. R. (1980). "The objective use of upper air soundings to specify precipitation type." *Monthly Weather Review*, 108:596–603.

CSA (1994). "Antennas, towers and antenna-supporting structures." CAS-S37-94, Canadian Standards Association, Rexdale, Ontario.

Colbeck, S. C., and Ackley, S. F. (1982). "Mechanisms for ice bonding in wet snow accretions on power lines." *Proc., 1st Int. Workshop on Atmospheric Icing of Structures*, U.S. Army CRREL Special Report 83-17, Hanover, New Hampshire, 25–30.

Cluts, S., and Angelos, A. (1977). "Unbalanced forces on tangent transmission structures." *IEEE Winter Power Meeting*, Paper No. A77-220-7.

Gland, H., and Admirat, P. (1986). "Meteorological conditions for wet snow occurrence in France—Calculated and measured results in a recent case study on 5 March 1985." *Proc., 3rd Int. Workshop on Atmospheric Icing of Structures*, published by Canadian Climate Program in 1991, Vancouver, Canada, 91–96.

Goodwin, E. J., Mozer, J. D., DiGioia, A. M. Jr., and Power, B. A. (1983). "Predicting ice and snow loads for transmission line design." *Proc., 3rd Int. Workshop on Atmospheric Icing of Structures*, published by Canadian Climate Program in 1991, Vancouver, Canada, 267–275.

Gouze, S. C., and Richmond, M. C. (1982a). "Meteorological evaluation of the proposed Alaska transmission line routes." Meteorology Research, Inc., Altadena, CA.

Gouze, S. C., and Richmond, M. C. (1982b). "Meteorological evaluation of the proposed Palmer to Glennallen transmission line route." Meteorology Research, Inc., Altadena, CA.

GVEA (1997). Unpublished data of Golden Valley Electric Association, Fairbanks, Alaska.

Hoskings, J. R. M., and Wallis, J. R. (1987). "Parameter and quantile estimation for the generalized Pareto distribution." *Technometrics*, 29(3): 339–349.

IEC (1990). *Loading and strength of overhead transmission lines international standard 826*. International Electrotechnical Commission, Technical Committee 11, Geneva, Switzerland, Second Edition.

Jones, K. F. (1996a). "A simple model for freezing rain loads." *Proc., 7th Int. Workshop on Atmospheric Icing of Structures*, Chicoutimi, Quebec, Canada, June 3–7, 412–416.

Jones, K. F. (1996b). *Ice accretion in freezing rain*. CRREL Report 96-2, Cold Regions Research and Engineering Laboratory, Hanover, New Hampshire.

Kuroiwa, D. (1962). "A study of ice sintering." Research Report 86, U.S. Army CRREL, Hanover, New Hampshire, 8 pp.

Kuroiwa, D. (1965). "Icing and snow accretion on electric wires." Research Paper 123, U.S. Army CRREL, Hanover, New Hampshire, 10 pp.

Langmuir, I., and Blodgett, K. (1946). "Mathematical investigation of water droplet trajectories." *The Collected Works of Irving Langmuir*, Pergamon Press, Elmsford, New York, 335–393.

McCormick, T., and Pohlman, J. C. (1993). "Study of compact 220 kV line system indicates need for micro-scale meteorological information." *Proc., 6th Int. Workshop on Atmospheric Icing of Structures*, Budapest, Hungary.

Mozer, J. D., and West, R. J. (1983). "Analysis of 500 kV tower failures." Presented at the 1983 meeting of the Pennsylvania Electric Association.

Mulherin, N. D. (1996). "Atmospheric icing and tower collapse in the United States." *Presented at the 7th International Workshop on Atmospheric Icing of Structures*, Chicoutimi, Quebec, Canada, June 3–6.

MEP (1984). "Climatological ice accretion modeling." Canadian Climate Center Report No. 84-10. Prepared by Meteorological and Environmental Planning Limited and Ontario Hydro for the Atmospheric Environmental Service, 195 pp.

National Standard of Canada (1987). "Overhead lines." CAN/CSA-C22.3 No. 1-M87, Canadian Standards Association, Rexdale, Ontario.

NESC (1993). National Electrical Safety Code. National Bureau of Standards, Washington, D.C.

NOAA (1959–Present). *Storm data*. National Oceanic and Atmospheric Administration, Washington, D.C.

NPPD (1976). "The storm of March 29, 1976." Public Relations Department, Nebraska Public Power District.

Peabody, A. B. (1993). "Snow loads on transmission and distribution lines in Alaska." *Proc., 6th Int. Workshop on Atmospheric Icing of Structures*, Budapest, Hungary.

Peterka, J. A. (1992). "Improved extreme wind prediction for the United States." *J. Wind Engineering and Industrial Aerodynamics*, Elsevier Science Publishers B.V., 41-44, 533–541.

Peterka, J. A., Finstad, K., and Pandy, A. K. (1996). *Snow and wind loads for Tyee transmission line*. Cermak Peterka Petersen, Inc. Fort Collins, CO.

Richmond, M. C., Gouze, S. C., and Anderson, R. S. (1977). *Pacific Northwest icing study*. Meterology Research, Inc., Altadena, CA.

Richmond, M. C. (1985). "Meterological evaluation of Bradley Lake hydroelectric project 115kV transmission line route." M. C. Richmond Meteorological Consultant, Torrance, CA.

Richmond, M. C. (1991). "Meteorological evaluation of Tyee Lake hydroelectric project transmission line route, Wrangell to Petersburg." Richmond Meteorological Consulting, Torrance, CA.

Richmond, M. C. (1992). "Meterological evaluation of Tyee Lake hydroelectric project transmission line route, Tyee power plant to Wrangell." Richmond Meterological Consulting, Torrance, CA.

Robbins, C. C., and Cortinas, J. V. Jr. (1996). "A climatology of freezing rain in the contiguous United States: Preliminary results." Preprints, *15th AMS Conference on Weather Analysis and Forecasting*, Norfolk, Virginia, August 19–23.

Sakamoto, Y., Mizushima, K., and Kawanishi, S. (1990). "Dry snow type accretion on overhead wires: Growing mechanism, meteorological conditions under which it occurs and effect on power lines." *Proc., 5th Int. Workshop on Atmospheric Icing of Structures*, Paper 5-9, Tokyo, Japan.

Shan, L., and Marr, L. (1996). *Ice storm data base and ice severity maps*. Electric Power Research Institute, Palo Alto, CA.

Tattelman, P., and Gringorten, I. (1973). "Estimated glaze ice and wind loads at the earth's surface for the contiguous United States." Air Force Cambridge Research Laboratories Report AFCRL-TR-73-0646, 34 pp.

USFS (1994). *Forest Service Handbook FSH6609.14. Telecommunications Handbook, R3 Supplement 6609.14-94-2 Effective 5/2/94*, United States Forest Service, Washington, D.C.

Wang, Q. J. (1991). "The POT model described by the generalized Pareto distribution with Poisson arrival rate." *J. of Hydrology*, 129:263–280.

Yip, T. C. (1993). "Estimating icing amounts caused by freezing precipitation in Canada." *Proc., 6th Int. Workshop on Ice Accretion on Structures*, Budapest, Hungary.

Young, W. R. (1978). "Freezing precipitation in the Southeastern United States." M.S. Thesis, Texas A&M University. 123 pp.

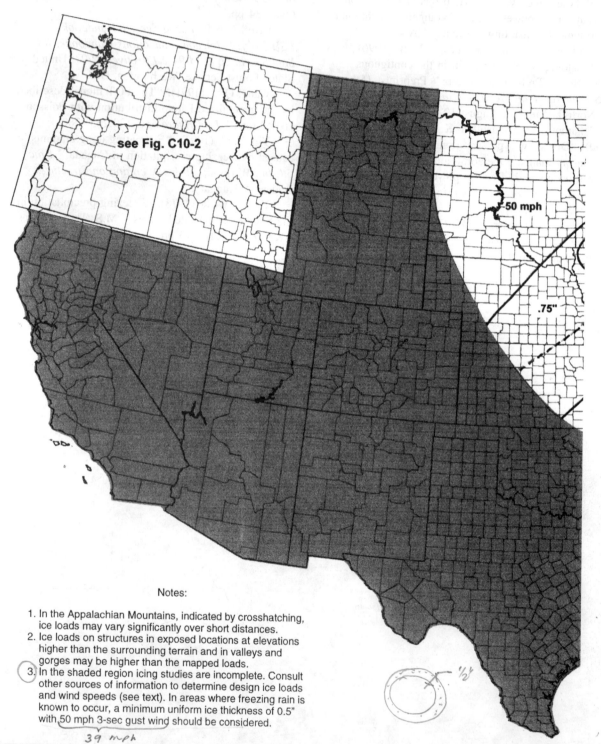

see Fig. C10-2

50 mph

.75"

Notes:

1. In the Appalachian Mountains, indicated by crosshatching, ice loads may vary significantly over short distances.
2. Ice loads on structures in exposed locations at elevations higher than the surrounding terrain and in valleys and gorges may be higher than the mapped loads.
3. In the shaded region icing studies are incomplete. Consult other sources of information to determine design ice loads and wind speeds (see text). In areas where freezing rain is known to occur, a minimum uniform ice thickness of 0.5" with 50 mph 3-sec gust wind should be considered.

39 mph

FIGURE C10-1. 50-year Mean Recurrence Interval Uniform Ice Thicknesses due to Freezing Rain with Concurrent 3-s Gust Wind Speeds: Contiguous 48 States

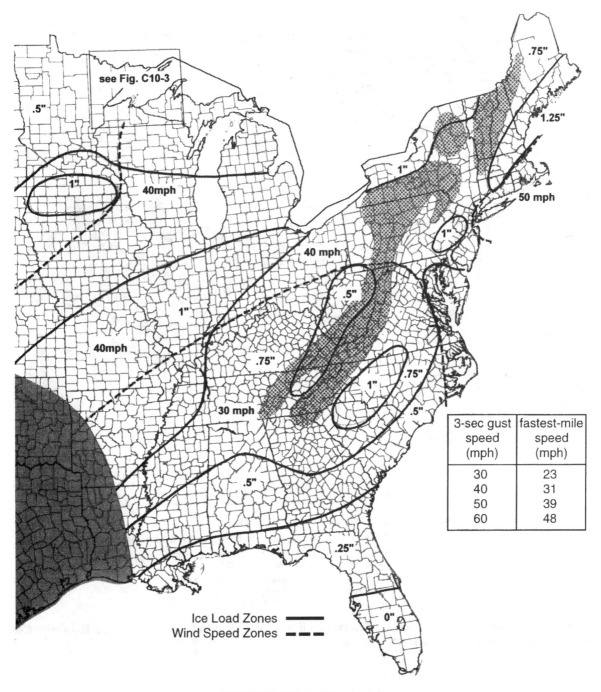

FIGURE C10-1. (*Continued*)

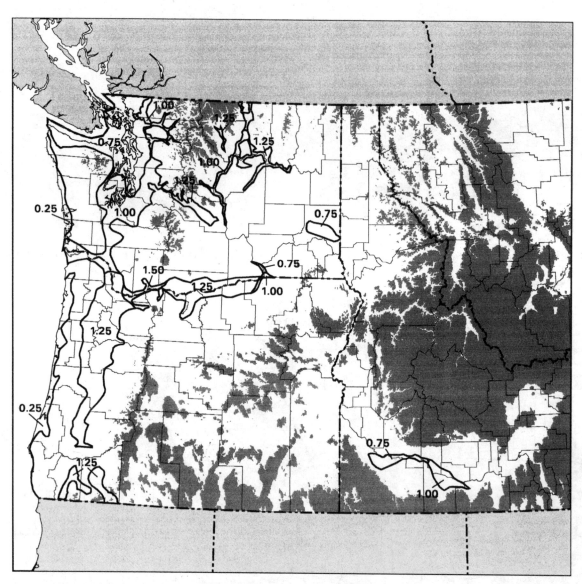

NOTES: 1. Ice thickness is shown in inches.

2. Unless otherwise specified use 0.50 inch ice thicknesses.

3. Freezing rain is unlikely to occur in the shaded mountainous regions above 5,000 feet.

4. Apply a concurrent fastest-mile wind speed of 40 mph, a 3-second gust of 50 mph, to the appropriate ice thicknesses.

FIGURE C10-2. 50-year Mean Recurrence Interval Uniform Ice Thicknesses due to Freezing Rain with Concurrent 3-s Gust Wind Speeds: Pacific Northwest

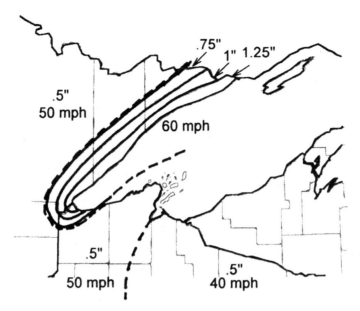

FIGURE C10-3. 50-year Mean Recurrence Interval Uniform Ice Thicknesses due to Freezing Rain with Concurrent 3-s Gust Wind Speeds: Lake Superior

FIGURE C10-4. Weather Stations for Fig. C10-1

TABLE C10-1. Ice and Wind Factors

Mean Recurrence Interval (years)	Factor to Multiply t	Factor to Multiply 3-s Gust Wind Speeds
25	0.8	1.0
50	1.0	1.0
100	1.2	1.0

COMMENTARY APPENDIX B

CB.0 SERVICEABILITY CONSIDERATIONS

Serviceability limit states are conditions in which the functions of a building or other structure are impaired because of local damage, deterioration or deformation of building components or because of occupant discomfort. While safety generally is not an issue with serviceability limit states, they nonetheless may have severe economic consequences. The increasing use of the computer as a design tool, the use of stronger (but not stiffer) construction materials, the use of lighter architectural elements and the uncoupling of the nonstructural elements from the structural frame, may result in building systems that are relatively flexible and lightly damped. Limit states design emphasizes that serviceability criteria are essential to ensure functional performance and economy of design for such building structural systems [CB-1, CB-10, CB-14].

There are three general types of unserviceability that may be experienced:

1. Excessive deflections or rotation that may affect the appearance, functional use or drainage of the structure, or may cause damaging transfer of load to non-load supporting elements and attachments.
2. Excessive vibrations produced by the activities of building occupants, mechanical equipment, or the wind, which may cause occupant discomfort or malfunction of building service equipment.
3. Deterioration, including weathering, corrosion, rotting, and discoloration.

In checking serviceability, the designer is advised to consider appropriate service loads, the response of the structure, and the reaction of the building occupants.

Service loads that may require consideration include static loads from the occupants and their possessions, snow or rain on roofs, temperature fluctuations, and dynamic loads from human activities, wind-induced effects, or the operation of building service equipment. The service loads are those loads that act on the structure at an arbitrary point in time. (In contrast, the nominal loads have a small probability of being exceeded in any year; factored loads have a small probability of being exceeded in 50 years.) Appropriate service loads for checking serviceability limit states may be only a fraction of the nominal loads.

The response of the structure to service loads normally can be analyzed assuming linear elastic behavior. However, members that accumulate residual deformations under service loads may require examination with respect to this long-term behavior. Service loads used in analyzing creep or other long-term effects may not be the same as those used to analyze elastic deflections or other short-term or reversible structural behavior.

Serviceability limits depend on the function of the building and on the perceptions of its occupants. In contrast to the ultimate limit states, it is difficult to specify general serviceability limits that are applicable to all building structures. The serviceability limits presented in Sections CB.1.1, CB.1.2, and CB.1.3 provide general guidance and have usually led to acceptable performance in the past. However, serviceability limits for a specific building should be determined only after a careful analysis by the engineer and architect of all functional and economic requirements and constraints in conjunction with the building owner. It should be recognized that building occupants are able to perceive structural deflections, motion, cracking, or other signs of possible distress at levels that are much lower than those that would indicate that structural failure is impending. Such signs of distress may be taken incorrectly as an indication that the building is unsafe and diminish its commercial value.

CB.1 DEFLECTION, VIBRATION AND DRIFT

CB.1.1 Vertical Deflections and Misalignment

Excessive vertical deflections and misalignment arise primarily from three sources: (1) gravity loads, such as dead, live and snow loads; (2) effects of temperature, creep and differential settlement; and (3) construction tolerances and errors. Such deformations may be visually objectionable, may cause separation, cracking, or leakage of exterior cladding, doors, windows and seals, and may cause damage to interior components and finishes. Appropriate limiting values of deformations depend on the type of structure, detailing, and intended use [CB-16]. Historically, common deflection limits for horizontal members have been 1/360 of the span for floors subjected to full nominal live load and 1/240 of span for roof members. Deflections of about 1/300 of the span (for cantilevers, 1/150 of length) are visible and may lead to general architectural damage or cladding leakage. Deflections greater than 1/200 of the span may impair operation of moveable components such as doors, windows and sliding partitions.

In certain long-span floor systems, it may be necessary to place a limit (independent of span) on the maximum deflection to minimize the possibility of damage of adjacent nonstructural elements [CB-17]. For example, damage to nonload-bearing partitions may occur if vertical deflections exceed more than about 10 mm (3/8 in.) unless special provision is made for differential movement [CB-11]; however, many components can accept larger deformations.

Load combinations for checking static deflections can be developed using first-order reliability analysis [CB-16]. Current static deflection guidelines for floor and roof systems are adequate for limiting surficial damage in most buildings. A combined load with an annual probability of 0.05 of being exceeded would be appropriate in most instances. For serviceability limit states involving visually objectionable deformations, repairable cracking or other damage to interior finishes, and other short-term effects, the suggested load combinations are:

$$D + L \qquad \text{(Eq. CB-1a)}$$

$$D + 0.5S \qquad \text{(Eq. CB-1b)}$$

For serviceability limit states involving creep, settlement or similar long-term or permanent effects, the suggested load combination is:

$$D + 0.5L \qquad \text{(Eq. CB-2)}$$

Live load, L, is defined in Section 4. The dead load effect, D, used in applying Eqs. (CB-1a), (CB-1b) and (CB-2) may be that portion of dead load that occurs following attachment of nonstructural elements. For example, in composite construction, the dead load effects frequently are taken as those imposed after the concrete has cured; in ceilings, the dead load effects may include only those loads placed after the ceiling structure is in place.

CB.1.2 Drift of Walls and Frames

Drifts (lateral deflections) of concern in serviceability checking arise primarily from the effects of wind. Drift limits in common usage for building design are on the order of 1/600 to 1/400 of the building or story height [CB-8]. These limits generally are sufficient to minimize damage to cladding and nonstructural walls and partitions. Smaller drift limits

may be appropriate if the cladding is brittle. An absolute limit on interstory drift may also need to be imposed in light of evidence that damage to nonstructural partitions, cladding and glazing may occur if the interstory drift exceeds about 10 mm (3/8 in.) unless special detailing practices are made to tolerate movement [CB-11, CB-15]. Many components can accept deformations that are significantly larger.

Use of the factored wind load in checking serviceability is excessively conservative. The load combination with an annual probability of 0.05 of being exceeded, which can be used for checking short-term effects, is

$$D + 0.5L + 0.7W \qquad \text{(Eq. CB-3)}$$

obtained using a procedure similar to that used to derive Eqs. (CB-1a) and (CB-1b). Wind load, W, is defined in Section 6. Due to its transient nature, wind load need not be considered in analyzing the effects of creep or other long-term actions.

Deformation limits should apply to the structural assembly as a whole. The stiffening effect of nonstructural walls and partitions may be taken into account in the analysis of drift if substantiating information regarding their effect is available. Where load cycling occurs, consideration should be given to the possibility that increases in residual deformations may lead to incremental structural collapse.

CB.1.3 Vibrations

Structural motions of floors or of the building as a whole can cause the building occupants discomfort. In recent years, the number of complaints about building vibrations has been increasing. This increasing number of complaints is associated in part with the more flexible structures that result from modern construction practice. Traditional static deflection checks are not sufficient to ensure that annoying vibrations of building floor systems or buildings as a whole will not occur [CB-1]. While control of stiffness is one aspect of serviceability, mass distribution and damping are also important in controlling vibrations. The use of new materials and building systems may require that the dynamic response of the system be considered explicitly. Simple dynamic models often are sufficient to determine whether there is a potential problem and to suggest possible remedial measurements [CB-9, CB-13].

Excessive structural motion is mitigated by measures that limit building or floor accelerations to levels that are not disturbing to the occupants or do not damage service equipment. Perception and tolerance of individuals to vibration is dependent on their expectation of building performance (related to building occupancy) and to their level of activity at the time the vibration occurs [CB-7]. Individuals find continuous vibrations more objectionable than transient vibrations. Continuous vibrations (over a period of minutes) with acceleration on the order of 0.005–0.01g are annoying to most people engaged in quiet activities, whereas those engaged in physical activities or spectator events may tolerate steady-state accelerations on the order of 0.02–0.05g. Thresholds of annoyance for transient vibrations (lasting only a few seconds) are considerably higher and depend on the amount of structural damping present [CB-18]. For a finished floor with (typically) 5% damping or more, peak transient accelerations of 0.05–0.1g may be tolerated.

Many common human activities impart dynamic forces to a floor at frequencies (or harmonics) in the range of 2 to 6 Hz [CB-2–CB-5]. If the fundamental frequency of vibration of the floor system is in this range and if the activity is rhythmic in nature (e.g., dancing, aerobic exercise, cheering at spectator events), resonant amplification may occur. To prevent resonance from rhythmic activities, the floor system should be tuned so that its natural frequency is well removed from the harmonics of the excitation frequency. As a general rule, the natural frequency of structural elements and assemblies should be greater than 2.0 times the frequency of any steady-state excitation to which they are exposed unless vibration isolation is provided. Damping is also an effective way of controlling annoying vibration from transient events, as studies have shown that individuals are more tolerant of vibrations that damp out quickly than those that persist [CB-18].

Several recent studies have shown that a simple and relatively effective way to minimize objectionable vibrations to walking and other common human activities is to control the floor stiffness, as measured by the maximum deflection independent of span. Justification for limiting the deflection to an absolute value rather than to some fraction of span can be obtained by considering the dynamic characteristics of a floor system modeled as a uniformly loaded simple span. The fundamental frequency of vibration, f_o, of this system is given by

$$f_o = \frac{\pi}{2l^2} \sqrt{\frac{EI}{\rho}} \qquad \text{(Eq. CB-4)}$$

in which EI = flexural rigidity of the floor, l = span,

and $\rho = w/g$ = mass per unit length; g = acceleration due to gravity (9.81 m/s^2), and w = dead load plus participating live load. The maximum deflection due to w is

$$\delta = (5/384)(wl^4/EI) \qquad \text{(Eq. CB-5)}$$

Substituting EI from this equation into Eq. (CB-4), we obtain

$$f_o \approx 18/\sqrt{\delta} \ (\delta \text{ in mm}) \qquad \text{(Eq. CB-6)}$$

This frequency can be compared to minimum natural frequencies for mitigating walking vibrations in various occupancies [CB-6]. For example, Eq. (CB-6) indicates that the static deflection due to uniform load, w, must be limited to about 5 mm, independent of span, if the fundamental frequency of vibration of the floor system is to be kept above about 8 Hz. Many floors not meeting this guideline are perfectly serviceable; however, this guideline provides a simple means for identifying potentially troublesome situations where additional consideration in design may be warranted.

CB.2 DESIGN FOR LONG-TERM DEFLECTION

Under sustained loading, structural members may exhibit additional time-dependent deformations due to creep, which usually occur at a slow but persistent rate over long periods of time. In certain applications, it may be necessary to limit deflection under long-term loading to specified levels. This can be done by multiplying the immediate deflection by a creep factor, as provided in material standards, that ranges from about 1.5–2.0. This limit state should be checked using load combination (CB-2).

CB.3 CAMBER

Where required, camber should be built into horizontal structural members to give proper appearance and drainage and to counteract anticipated deflection from loading and potential ponding.

CB.4 EXPANSION AND CONTRACTION

Provision should be made in design so that if significant dimensional changes occur, the structure will move as a whole and differential movement of similar parts and members meeting at joints will be a minimum. Design of expansion joints to allow for dimensional changes in portions of a structure separated by such joints should take both reversible and irreversible movements into account. Structural distress in the form of wide cracks has been caused by restraint of thermal, shrinkage and prestressing deformations. Designers are advised to provide for such effects through relief joints or by controlling crack widths.

CB.5 DURABILITY

Buildings and other structures may deteriorate in certain service environments. This deterioration may be visible upon inspection (weathering, corrosion, staining) or may result in undetected changes in the material. The designer should either provide a specific amount of damage tolerance in the design or should specify adequate protection systems and/or planned maintenance to minimize the likelihood that such problems will occur. Water infiltration through poorly constructed or maintained wall or roof cladding is considered beyond the realm of designing for damage tolerance. Waterproofing design is beyond the scope of this standard. For portions of buildings and other structures exposed to weather, the design should eliminate pockets in which moisture can accumulate.

REFERENCES

[CB-1] "Structural serviceability: A critical appraisal and research needs." *J. Struct. Div.*, ASCE, 112(12), 2646–2664, 1986.

[CB-2] Allen, D. E., and Rainer, J. H. "Vibration criteria for long-span floors." *Canadian J. Civ. Engrg.*, 3(2), 165–173, 1976.

[CB-3] Allen, D. E., Rainer, J. H., and Pernica, G. "Vibration criteria for assembly occupancies." *Canadian J. Civ. Engrg.*, 12(3), 617–623, 1985.

[CB-4] Allen, D. E. "Floor vibrations from aerobics." *Canadian J. Civ. Engrg.*, 19(4), 771–779, 1990a.

[CB-5] Allen, D. E. "Building vibrations from human activities." *Concrete International*, 12(6), 66–73, 1990b.

[CB-6] Allen, D. E., and Murray, T. M. "Design criterion for vibrations due to walking." *Engrg. J.*, AISC, 30(4), 117–129, 1993.

[CB-7] American National Standard Guide to the Evaluation of Human Exposure to Vibration in Buildings (ANSI S3.29-1983). Am. Nat. Stds. Inst., New York, NY, 1983.

[CB-8] "Wind drift design of steel-framed buildings: State of the art." *J. Struct. Div.*, ASCE, 114(9), 2085–2108, 1988.

[CB-9] Bachmann, H., and Ammann, W. Vibrations in structures. Structural Engineering, Doc. 3e, International Assoc. for Bridge and Str. Engr., Zurich, Switzerland, 1987.

[CB-10] Commentary A, Serviceability Criteria for deflections and vibrations. National Building Code of Canada-1990, National Research Council, Ottawa, Ontario, 1990.

[CB-11] Cooney, R. C., and King, A. B. Serviceability criteria for buildings. BRANZ Report SR14, Building Research Association of New Zealand, Porirua, New Zealand, 1988.

[CB-12] Ellingwood, B., and Tallin, A. "Structural serviceability: floor vibrations." *J. Struct. Div.*, ASCE, 110(2), 401–418, 1984.

[CB-13] Ellingwood, B. "Serviceability guidelines for steel structures." *AISC Engrg. J.*, 26(1), 1–8, 1989.

[CB-14] Fisher, J. M., and West, M. A. "Serviceability design considerations for low-rise buildings." Steel Design Guide No. 3, American Institute of Steel Construction, Chicago, IL, 1990.

[CB-15] Freeman, S. "Racking tests of high rise building partitions." *J. Struct. Div.*, ASCE, 103(8), 1673–1685, 1977.

[CB-16] Galambos, T. V., and Ellingwood, B. "Serviceability limit states: Deflections." *J. Struct. Div.*, ASCE, 112(1), 67–84, 1986.

[CB-17] "Bases for the design of structures—Deformations of buildings at the serviceability limit states." ISO Standard 4356, 1977.

[CB-18] Murray, T. "Building floor vibrations." *AISC Engrg. J.*, 28(3), 102–109, 1991.

[CB-19] Ohlsson, S. "Ten years of floor vibration research—a review of aspects and some results." *Proc. Symp. on Serviceability of Buildings*, National Research Council of Canada, Ottawa, 435–450, 1988.

[CB-20] Tallin, A. G., and Ellingwood, B. "Serviceability limit states: Wind induced vibrations." *J. Struct. Engrg.*, ASCE, 110(10), 2424–2437, 1984.